HIGH T_c SUPERCONDUCTING THIN FILMS, DEVICES, AND APPLICATIONS

AMERICAN INSTITUTE OF PHYSICS
CONFERENCE PROCEEDINGS NO. **182**
NEW YORK 1989

AMERICAN VACUUM SOCIETY SERIES 6

SERIES EDITOR: GERALD LUCOVSKY
NORTH CAROLINA STATE UNIVERSITY

HIGH T_c SUPERCONDUCTING THIN FILMS, DEVICES, AND APPLICATIONS

ATLANTA, GA 1988

EDITORS:
GIORGIO MARGARITONDO,
ROBERT JOYNT
& MARSHALL ONELLION
UNIVERSITY OF WISCONSIN

Authorization to photocopy items for internal or personal use, beyond the free copying permitted under the 1978 US Copyright Law (see statement below), is granted by the American Institute of Physics for users registered with the Copyright Clearance Center (CCC) Transactional Reporting Service, provided that the base fee of $3.00 per copy is paid directly to CCC, 27 Congress St., Salem, MA 01970. For those organizations that have been granted a photocopy license by CCC, a separate system of payment has been arranged. The fee code for users of the Transactional Reporting Service is: 0094-243X/87 $3.00.

Copyright 1989 American Institute of Physics.

Individual readers of this volume and non-profit libraries, acting for them, are permitted to make fair use of the material in it, such as copying an article for use in teaching or research. Permission is granted to quote from this volume in scientific work with the customary acknowledgment of the source. To reprint a figure, table or other excerpt requires the consent of one of the original authors and notification to AIP. Republication or systematic or multiple reproduction of any material in this volume is permitted only under license from AIP. Address inquiries to Series Editor, AIP Conference Proceedings, AIP, 335 E. 45th St., New York, NY 10017.

L.C. Catalog Card No. 88-83947
ISBN 0-88318-383-8
DOE CONF-881035

Printed in the United States of America.

Contents

Preface ix

I. Thin Film Preparation

Sputtering

Reactive Sputtering of Superconducting Thin Films 2
    Y. Arie and J. R. Matey

Preparation and Characterization of High T_c $YBa_2Cu_3O_{7-x}$ Thin Films on Silicon by DC Magnetron Sputtering from a Stoichiometric Oxide Target 8
    W. Y. Lee, J. Salem, V. Y. Lee, T. C. Huang, R. Karimi, V. Deline, R. Savoy, and D. W. Chung

RF Bias Getter Reactive Sputtering of La–Cu–O 16
    Peter J. Clark and Bill Gardner

Effect of Substrate Temperature and Biasing on the Formation of 110 K Bi–Sr–Ca–Cu–O Superconducting Single Target Sputtered Thin films 26
    N. G. Dhere, R. G. Dhere, and J. Moreland

Deposition of Ceramic Superconductors from Single Spherical Targets 33
    G. K. Wehner, Y. H. Kim, D. H. Kim, and A. M. Goldman

Deposition of 1-2-3 Thin Films over Large Areas 45
    P. N. Arendt, N. E. Elliot, R. E. Muenchausen, M. Nastasi, and T. Archuleta

Sputtered Thin Film $YBa_2Cu_3O_n$ 53
    Kenneth G. Kreider, James P. Cline, Alexander Shapiro, J. L. Pena, A. Rojas, J. A. Azamar, L. Maldonado, and L. Del Castillo

Ion Beam Sputter Deposition $YBa_2Cu_3O_{7-\delta}$: Beam Induced Target Changes and Their Effect on Deposited Film Composition 61
    O. Auciello, M. S. Ameen, T. M. Graettinger, S. H. Rou, C. Soblel, and A. I. Kingon

Investigation of $SrTiO_3$ Barrier Layers for RF Sputter-deposited Y–Ba–Cu–O Films on Si and Sapphire 65
    J. K. Truman, M. Leskela, C. H. Mueller, and P. H. Holloway

Superconducting Tl–Ca–Ba–Cu–O Thin Films by Reactive Magnetron Sputtering 74
    D. H. Chen, R. L. Sabatini, S. L. Qiu, D. Di Marzio, S. M. Heald, and H. Wiesmann

Thin Films of $Y_1Ba_2Cu_3O_x$ Deposited Using Three Target Co-sputtering and Their Applications to Microbridge Junctions and Single-element IR Detectors 82
    J. Y. Josefowicz, D. B. Rensch, A. T. Hunter, H. Kimura, B. M. Clemens, J. Spargo, E. Wiener-Avnir, G. Kerber, J. A. Wilson, W. D. Jack, J. M. Myroszynyk, and R. E. Kvaas

Metallic Alloy Targets for High T_c Superconducting Film Deposition 90
    P. Manini, A. Nigro, P. Romano, and R. Vaglio

RF Magnetron Sputtering of High-T_c Bi–Sr–Ca–Cu–O Thin Film 99
    S. K. Dew, N. R. Osborne, B. T. Sullivan, P. J. Mulhern, and R. R. Parsons
Single-phase High T_c Superconducting $Tl_2Ba_2Ca_2Cu_3O_{10}$ Films 107
    M. Hong, J. Kwo, C. H. Chen, A. R. Kortan, D. D. Bacon, and S. H. Liou
Oriented Growth of $Y_1Ba_2Cu_3O_x$ Thin Films by Dual Ion Beam Sputtering 115
    J. P. Doyle, R. A. Roy, D. S. Yee, and J. J. Cuomo
Superconducting Bi–Sr–Ca–Cu–O Films by Sputtering Using
a Single Oxide Target 122
    M. Hong, J.-J. Yeh, J. Kwo, R. J. Felder, A. Miller, K. Nassau, and
    D. D. Bacon
Effects of Oxygen Partial Pressure of Properties of Y–Ba–Cu–O Films
Prepared by Magnetron Sputtering 130
    T. Miura, Y. Terashima, M. Sagoi, K. Kubo, J. Yoshida, and K. Mizushima

Evaporation

A Flash Evaporation Technique for Oxide Superconductors 140
    Matthew F. Davis, Jaroslaw Wosik, J. C. Wolfe, and
    Christopher L. Lichtenberg
Sequentially Evaporated Thin Y–Ba–Cu–O Superconductor Films:
Composition and Processing Effects 147
    George J. Valco, Norman J. Rohrer, Joseph D. Warner, and Kul B. Bhasin
Epitaxial Growth of Dy–Ba–Cu–O Superconductor on $(100)SrTiO_3$ Using
Reactive and Activated Reactive Evaporation Processes 155
    R. C. Budhani, H. Wiesmann, M. W. Ruckman, and R. L. Sabatini
Rate and Composition Control by Atomic Absorption Spectroscopy for the
Coevaporation of High T_c Superconducting Films 163
    C. Lu, N. Missert, J. E. Mooij, P. Rosenthal, V. Matijasevic, M. R. Beasley,
    and R. H. Hammond

II. Characterization
Transport

High Critical Currents and Flux Creep Effects in e-Gun Deposited Epitaxially
001 Oriented Superconducting $YBa_2Cu_3O_{7-\delta}$ Films 172
    B. Dam, G. M. Stollman, P. Berghuis, S. Q. Guo, C. F. J. Flipse, J. G. Lensink,
    and R. P. Griessen
Factors Influencing Critical Current Densities in High T_c Superconductors 180
    G. Fisanick
Limitations on Critical Currents in High Temperature Superconductors 194
    C. C. Tsuei, J. Mannhart, and D. Dimos

SPECTROSCOPIES

ARXPS-studies of ĉ-Axis Textured and Polycrystalline Superconducting
$YBa_2Cu_3O_x$ 208
    J. Halbritter, P. Walk, H.-J. Mathes, W. Aarnink, and I. Apfelstedt

Novel Bonding Concepts for Superconductive Oxides: an XPS Study 216
    T. L. Barr, C. R. Brundle, A. Klumb, Y. L. Liu, L. M. Chen, and M. P. Yin

An XPS Investigation of the Surface Layer Formed on '123' High T_c
Superconducting Films Annealed in O_2 and CO_2 Atmospheres 232
    R. Caracciolo, M. S. Hedge, J. B. Wachtman, Jr., A. Inam, and T. Venketesan

XPS DOS Studies of Oxygen-plasma Treated $YBa_2Cu_3O_{7-\delta}$ Surfaces as a
Function of Temperature 240
    T. Conard, J. M. Vohs, J. J. Pireaux, and R. Caudano

Temperature Effects in the Near-E_F Electronic Structure of
$Bi_4Ca_3Sr_3Cu_4O_{16+x}$ 248
    Y. Chang, N. G. Stoffel, Ming Tang, R. Zanoni, M. Onellion, Robert Joynt,
    D. L. Huber, G. Margaritondo, P. A. Morris, W. A. Bonner, and
    J. M. Tarascon

Photoemission Resonance Study of Sintered and Single-crystal
$Bi_4Ca_3Sr_3Cu_4O_{16+x}$ 252
    Ming Tang, Y. Chang, R. Zanoni, M. Onellion, Robert Joynt, D. L. Huber,
    G. Margaritondo, P. A. Morris, W. A. Bonner, J. M. Tarascon, and
    N. G. Stoffel

Electron Energy Loss Studies of the New High-T_c
Superconductor $Bi_4Ca_3Sr_3Cu_4O_{16+x}$ 257
    Y. Chang, R. Zanoni, M. Onellion, G. Margaritondo, P. A. Morris,
    W. A. Bonner, J. M. Tarascon, and N. G. Stoffel

Oxygen and Copper Valencies in Oxygen-doped Superconducting $La_2CuO_{4.13}$ ... 262
    N. D. Shinn, J. W. Rogers, Jr., J. E. Schirber, E. L. Venturini, D. S. Ginley,
    and B. Morosin

AES and EELS Analysis of $TlBaCaCuO_x$ Thin Films at 300K and at 100K 269
    A. J. Nelson, A. Swartzlander, L. L. Kazmerski, J. H. Kang, R. T. Kampwirth,
    and K. E. Gray

Resonant Photoemission and Chemisorption Studies of Tl–Ba–Ca–Cu–O 276
    Roger L. Stockbauer, Steven W. Robey, Richard L. Kurtz, D. Mueller,
    A. Shih, A. K. Singh, L. Toth, and M. Osofsky

Photoemission from Single-crystal $EuBa_2Cu_3O_{6+x}$ Cleaved below 20 K;
Metallic-to-insulating Surface Transformation 283
    R. S. List, A. J. Arko, Z. Fisk, S.-W. Cheong, S. D. Conradson,
    J. D. Thompson, C. B. Pierce, D. E. Peterson, R. J. Bartlett, J. A. O'Rourke,
    N. D. Shinn, J. E. Schirber, C. G. Olson, A.-B. Yang, T.-W. Pi, B. W. Veal,
    A. P. Paulikas, and J. C. Campuzano

Characterization of $Bi_2Sr_2Ca_1Cu_2O_8$ 289
    F. J. Himpsel, G. V. Chandrashekhar, A. Taleb-Ibrahimi, A. B. McLean,
    and M. W. Shafer

Surface Analysis of H, C, O, Y, Ba, and Cu on Pressed and
Laser-evaporated YBCO 297
  J. Albert Schultz, Howard K. Schmidt, P. Terence Murray, and Alex Ignatiev

Photon-stimulated Desorption from High-T_c Superconductors 304
  R. A. Rosenberg and C.-R. Wen

Surface and Electronic Structure of Bi–Ca–Sr–Cu–O Superconductors
Studied by LEED, UPS and XPS 312
  Z.-X. Shen, P. A. P. Lindberg, B. O. Wells, I. Lindau, W. E. Spicer,
  D. B. Mitzi, C. B. Eom, A. Kapitulnik, T. H. Geballe, and P. Soukiassian

Optical Properties of High-T_c Superconductors: Who Needed This Anyway 318
  D. E. Aspnes and M. K. Kelly

Photoemission Studies of High Temperature Superconductors 330
  Z.-X. Shen, P. A. P. Lindberg, W. E. Spicer, I. Lindau, and J. W. Allen

SURFACE AND INTERFACE PROCESSES

Characterization of the Ceramic: Substrate Interface of Plasma Sprayed High-
temperature Superconductors 352
  A. S. Byrne, W. F. Stickle, C. Y. Yang, T. Asano, and M. M. Rahman

Reaction between $YBa_2Cu_3O_{7-x}$ and Water 360
  M. W. Ruckman, S. M. Heald, D. DiMarzio, H. Chen, A. R. Moodenbaugh,
  and C. Y. Yang

Interaction of CO, CO_2 and H_2O with Ba and $YBa_2Cu_3O_7$ 368
  S. L. Qiu, C. L. Lin, M. W. Ruckman, J. Chen, D. H. Chen, Youwen Xu,
  A. R. Moodenbaugh, Myron Strongin, D. Nichols, and J. E. Crow

A Nonaqueous Chemical Etch for $YBa_2Cu_3O_{7-x}$ 376
  R. P. Vasquez, M. C. Foote, and B. D. Hunt

Effect of Silver on the Water-degradation of YBaCuO Superconducting Films .. 384
  Chin-An Chang, J. A. Tsai, and C. E. Farrell

Aluminum and Gold Deposition on Cleaved Single Crystals
of $Bi_2CaSr_2Cu_2O_8$ Superconductor 391
  B. O. Wells, P. A. P. Lindberg, Z.-X. Shen, D. S. Dessau, I. Lindau,
  W. E. Spicer, D. B. Mitzi, and A. Kapitulnik

Surface and Interface Properties of High Temperature Superconductors 399
  J. H. Weaver, H. M. Meyer, III, T. J. Wagener, D. M. Hill, and Y. Hu

THEORY

Tight-binding Description of High-temperature Superconductors 410
  Brent A. Richert and Roland E. Allen

Utilization of a Highly-correlated CuO_n Cluster Model to Interpret Electron
Spectoscopic Data for the High-temperature Superconductors 418
  David E. Ramaker

Novel Applications of High Temperature Superconductivity: Neurocomputing . 426
  R. Singh, J. R. Cruz, F. Radpour, and M. J. Semnani

Author Index 433

Preface

This book contains the proceedings of the 1988 Topical Conference on High T_c Superconducting Thin Films, Devices, and Applications of the American Vacuum Society (AVS). This Topical Conference was held in Atlanta, Georgia, immediately preceding the annual symposium of the AVS. This high-T_c meeting is the continuation of the topical conference on Thin Film Processing and Characterization of High-Temperature Superconductors, organized by the AVS in 1987. A third topical conference is planned in conjunction with the 1989 AVS annual symposium in Boston. These meetings reflect the continuing interest of the members of the AVS for high-temperature superconductivity, and the specific, crucial role that their research plays in the exciting field of science and technology.

The subject of high-temperature superconductivity was launched in late 1986 by the paper of Bednorz and Müller announcing the observation of possible superconductivity at 35 K in a compound of La, Ba, Cu and O. The scientific activity since that time has been frenzied, and shows little sign of abatement. It has been driven by the achievement of ever-higher critical temperatures: 95 K in YBaCuO and up to 125 K in TlBaCaCuO, and the establishment of superconductivity in other oxides: BaBiO alloyed with K, and BiSrCaCuO. Attracted both by the intrinsic scientific interest of these materials and by the promise of new technologies, scientists have shifted the focus of materials research and solid state physics to a remarkable extent toward superconductivity.

This field intersects most strongly with the traditional interests of the American Vacuum Society in the area of thin films. The fabrication of high-quality superconducting films is of course an absolute prerequisite for all electronics applications. More surprisingly, the successful passage of very high supercurrent densities in films (much higher than has so far been achieved in bulk samples) has demonstrated the importance of thin films for the fundamental science of high-temperature superconductivity. This was already clear at the 1987 AVS topical conference, where those results were reported. At the 1988 topical conference, the relevance of thin films for the understanding of transport at the microscopic level was demonstrated by studies of current flow across single grain boundaries.

Spectroscopy is the other area in which vacuum science has contributed in an essential way to this new field. No theoretical understanding of the materials is possible until the electronic structure has been clarified. The main features of the bands had already been identified by the time of the 1987 topical conference, and the valency of copper reasonably well established. At the 1988 topical conference observations of a Fermi edge were reported in both YBaCuO and BiSrCaCuO, and the appearance of a superconducting gap in the latter system was also suggested by photoemission studies.

The 1988 topical conference was opened by the keynote speaker, C. W. (Paul) Chu, who reviewed the recent discoveries of new high-T_c materials. The first session was held jointly with the AVS Topical Conference on Probing the Nanometer Scale Properties of Surface and Interfaces, and was chaired by J. Murday. The second invited speaker of the session was A. J. Melmed, who presented the first results obtained with field-ion microscopy on high-T_c samples. The second session was moderated by G. Margaritondo, and was dedicated to photoemission spectroscopy experiments. The invited speakers in this session were W. E. Spicer and J. H. Weaver. The afternoon session was chaired by L. C. Feldman. The first invited speaker was D. E. Aspnes, who reviewed the recent work with optical techniques. The next two invited speakers, C. C. Tsuei and G. J. Fisanick, covered different aspects of the crucial problem of critical current densities. The session was concluded by the invited presentation of G. K. Wehner on a novel sputter deposition technique. The conference had a poster session, which included 59 contributed presentations. The attendance throughout the meeting was one of the highest ever recorded for an AVS topical conference.

As in 1987, all divisions of the AVS sponsored the topical conference. This general interest was reflected by the composition of the Program Committee, chaired by G. Margaritondo, whose members were M. D. Boeckmann, Donald Carmichael, Susan Cohen, H. F. Dylla, L. C. Feldman, Richard J. Gambino, and John T. Grant. We enjoyed many constructive interactions with the AVS Program Committee, and in particular with its Chairman, John Noonan. The organization of the conference was impeccable, thanks to the excellent preparation work done by the AVS Meetings Manager, Marion Churchill, and by Marcia Schlissel. We are also grateful to the Local Arrangements Committee, chaired by John Wendelken with the assistance of Don Santeler. Due to the rapid evolving of high-T_c research, we again adopted a quick publication schedule for these proceedings. The timely processing of the papers would not have been possible without the help of Gerry Lucovsky, Rita Lerner, and of many other people, to whom we are deeply indebted. We would like to mention all members of the Program Committee, who also served as monitors for the reviewing process, and the many anonymous referees who provided accurate reviews in a very short time. We thank Kathleen Strum for supervising the operation of the Editorial Office in Atlanta. The office was run by many volunteers, including several members of the Program Committee, and Doreene Berger, Yeh Chang, Michael Hennelly, Ming Tang, and Tom von Foerster. We are much indebted to Lois Blackbourn for her organization of the final reviewing process, and to Lori Johnson for her assistance.

The work of the authors and of all the people mentioned above has produced a timely review of this exciting area of research. We trust that the book will be as successful as the proceedings of the 1987 topical conference [*Thin Film Processing and Characterization of High-Temperature Superconductors,* edited by J. M. Harper, R. J. Colton and L. C. Feldman, AIP Conference Proceedings no. 165 (American Institute of Physics, New York, 1988)]. Besides conveying information in a timely fashion, it is our hope that this book will provide a permanent record to one of the most exciting periods in the lives of the members of the American Vacuum Society.

Proceedings Editors

Giorgio Margaritondo
Robert Joynt
Marshall Onellion

University of Wisconsin-Madison

November 23, 1988

SPUTTERING

Reactive Sputtering of Superconducting Thin Films 2

Preparation and Characterization of High T_c $YBa_2Cu_3O_{7-x}$ Thin Films on Silicon by DC Magnetron Sputtering from a Stoichiometric Oxide Target 8

RF Bias Getter Reactive Sputtering of La–Cu–O 16

Effect of Substrate Temperature and Biasing on the Formation of 110 K Bi–Sr–Ca–Cu–O Superconducting Single Target Sputtered Thin Films 26

Deposition of Ceramic Superconductors from Single Spherical Targets 33

Deposition of 1–2–3 Thin Films over Large Areas 45

Sputtered Thin Film $YBa_2Cu_3O_n$ 53

Ion Beam Sputter Deposition $YBa_2Cu_3O_{7-\delta}$: Beam Induced Target Changes and Their Effect on Deposited Film Composition 61

Investigation of $SrTiO_3$ Barrier Layers for RF Sputter–deposited Y–Ba–Cu–O Films on Si and Sapphire 65

Superconducting Tl–Ca–Ba–Cu–O Thin Films by Reactive Magnetron Sputtering 74

Thin Films of $Y_1Ba_2Cu_3O_x$ Deposited Using Three Target Co–sputtering and Their Applications to Microbridge Junctions and Single–element IR Detectors ... 82

Metallic Alloy Targets for High T_c Superconducting Film Deposition 90

RF Magnetron Sputtering for High–T_c Bi–Si–Ca–Cu–O Thin Films 99

Single-phase High T_c Superconducting $Tl_2Ba_2Ca_2Cu_3O_{10}$ Films 107

Oriented Growth of $Y_1Ba_2Cu_3O_x$ Thin Films by Dual Ion Beam Sputtering 115

Superconducting Bi–Sr–Ca–Cu–O Films by Sputtering Using a Single Oxide Target 122

Effects of Oxygen Partial Pressure on Properties of Y–Ba–Cu–O Films Prepared by Magnetron Sputtering 130

REACTIVE SPUTTERING OF SUPERCONDUCTING THIN FILMS

Y. Arie and J. R. Matey
David Sarnoff Research Center
Princeton, NJ 08543-5300

ABSTRACT

We report a simple, versatile method for fabrication of superconducting thin films of $Y_1Ba_2Cu_3O_{7-x}$ using reactive RF sputtering from three targets in vacuum and a subsequent anneal in oxygen.

The structure and composition of the films were verified using Rutherford backscattering spectroscopy and x-ray diffraction studies. Measurements of resistivity and susceptibility vs. temperature established the superconducting transition temperatures for the films. Comparison of the susceptibility and resistance transitions demonstrates heterogeneity in the superconducting character of the films.

INTRODUCTION

The new high temperature superconductors present exciting prospects for high performance microwave circuits. As part of a program for the development of high temperature, superconducting, passive microwave devices we have developed techniques for the preparation and characterization of thin films of $Y_1Ba_2Cu_3O_7$.

METHOD OF PREPARATION

We have used sapphire, ceramic alumina, and 9.5 mole % Y-stabilized cubic zirconia substrates in our experiments. The substrate dimensions were 0.300"x0.420". All substrates were ultrasonically cleaned using Microcleaner detergent[1] for 15 minutes. The substrates were then rinsed in deionized water, dipped in Chromerge[2] and rinsed again in deionized water and then in boiling deionized water. The substrates were then boiled in isopropanol and finally baked for 60 minutes at 250 C.

The sapphire and ceramic alumina substrates were then coated with 100 nm of either reactive sputtered ZrO_2 or thermally grown ZrO_2. The thermally grown films were prepared by sputtering Zr and subsequently oxidizing at 500 C in air.

The substrates were transferred into a Perkin Elmer 4400 series sputtering system via a load lock. The $Y_1Ba_2Cu_3O_{7-x}$ was then deposited by reactive sputtering from three targets of yttrium, copper and barium under an atmosphere of 80% argon, 20% oxygen at 12-15 microns pressure. The deposition is carried out at room temperature. The substrate rotates under the three targets in succession, at 15 rpm. The sputtering yield from each target is controlled individually by varying the sputtering power. Hence, the stoichiometry of the film can be varied at will. The total sputtering yield at the nominal $Y_1Ba_2Cu_3O_{7-x}$ composition is 0.03 nm/sec. The films prepared for these experiments were about 2μ thick. The sputtered samples were stored either in vacuum or a nitrogen dry box after preparation.

The annealing furnace used in these experiments is a programmable Lindberg furnace[3]. Our basic annealing procedure is:

- The furnace is ramped up to 900 C over 30 minutes. At this time the samples are in the furnace tube, but outside the heated region.
- After the furnace reaches 900 C, the samples are moved into the heated region at the rate of 2.5 cm/minute.
- The samples remain in the heated region for annealing for a variable length of time.
- At the end of the anneal time the furnace is ramped down to 650 C at the rate of 8 C/minute.
- The sample is then soaked at 650 C for 10 hours.
- The temperature is then ramped down to 400 C at 8 C/minute.
- The sample is soaked at 400 C for 4 hours.
- The furnace is then turned off and allowed to cool to room temperature over about 2 hours.
- The entire procedure is carried out in a flowing stream of dry oxygen.

The annealed samples were stored either in vacuum or a nitrogen dry box after preparation.

MATERIALS CHARACTERIZATION

Films of $Y_1Ba_2Cu_3O_{7-x}$ were deposited on silicon wafers at the same time as the preparation of the sapphire, alumina and zirconia samples. Rutherford backscattering spectra (RBS) of as prepared films on these silicon substrates show a composition of $Y_1Ba_2Cu_3O_8$ with 10% accuracy. Figure 1 shows a typical RBS spectrum. X-ray diffraction studies of as prepared films on strontium titanate substrates indicate that the films are amorphous.

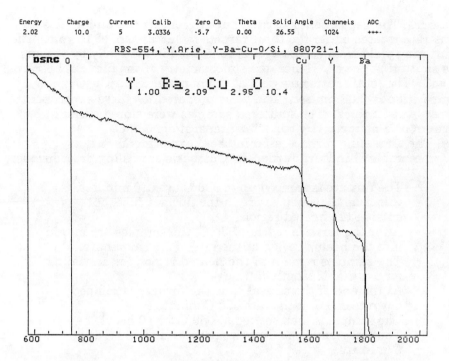

A portion of an annealed sample was scraped from a cubic zirconia substrate and characterized using x-ray powder diffraction. The diffraction pattern indicates that the sample is almost entirely the orthorhombic form of $Y_1Ba_2Cu_3O_{7-x}$. One extra "weak" line may indicate the presence of a small amount (<5%) of $Y_1Ba_2Cu_3O_5$. It was not possible to make a definite identification on the basis of one line.

SUPERCONDUCTING PROPERTIES

We test the superconducting properties of our films using four-terminal, low frequency AC resistance measurements; four-terminal, DC resistance measurements; and low frequency AC susceptibility measurements. The AC resistance measurements are made with a low frequency impedance analyzer[4]. The DC resistance measurements are made with a programmable current source and a nanovoltmeter[5] using standard current reversing procedures to eliminate thermal EMF's. Connections to the samples are made using InHg amalgam.

The susceptibility measurements are made using the thin film susceptometer sketched in figure 2. The mutual inductance and the equivalent series resistance of the coil pair are measured using our low frequency impedance analyzer.

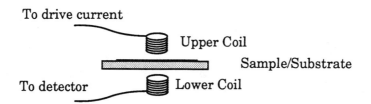

Figure 2

The experiments are carried out in a cryostat of conventional design. The experiments are supported by a thermal resistance at the end of a cold finger within the vacuum chamber of the cryostat. The temperature is measured and controlled using a calibrated silicon diode thermometer and a commercial controller[6]. The cryostat is cooled using liquid nitrogen or liquid helium. For temperatures between 45K and 77K we pump on the nitrogen.

The measurements are made under computer control and the instruments can be switched between samples by the computer. Four thin film susceptibility samples and two resistance samples can be measured during each run. Figure 3 shows the results on a typical sample on cubic zirconia. The four terminal resistance, inductance and equivalent series resistance have been normalized to their values at 300 K for easy comparison. The measurements were carried out at 10 kHz and an applied field of about 10^{-5} tesla.

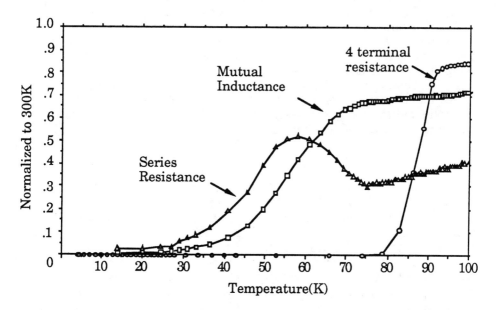

Figure 3

The width of the resistance transition is broad, 10-15 degrees, but comparable with those seen in the literature. We interpret this as a heterogeneous sample, with different parts of the sample undergoing the superconducting transition at different temperatures. Using the Bruggeman effective medium approximation for three dimensions[7] we expect that zero resistance will be reached at the percolation transition, when 33% of the material is superconductive.

Above the percolation transition there are no complete paths of superconductor. Flux lines can move freely across the sample and we expect no change in the mutual inductance. Below the percolation transition flux motion is impeded and the mutual inductance should begin to drop. Percolation paths of non-superconducting material, along which flux may move, will remain until the superconductor fraction exceeds 66%. Naively, we expect that this is the point at which the magnetic shielding should be complete and the mutual inductance should go to zero[8].

In figure 3, we see that the inductance signal varies slowly between 70K and 100K due to temperature dependent shielding of the mutual inductance coils by the copper sample holder. Below the zero resistance transition, we see a strong drop in the mutual inductance signal, going to zero at about 25K.

The series resistance plot shows an interesting peak. Below the percolation transition flux lines may break through the superconducting paths at weak points. This contributes a new source of loss in the film, which is seen in the peak in the series resistance. As the temperature is lowered, more of the sample becomes a superconductor and the strength of the weak points improves.

From these measurements, we infer that the actual widths of the transitions of these films are > 65K -- the onset is seen in four terminal resistance at 90K, the 33% complete point is at 75K, and the 66% complete point is at about 25K.

We have seen similar results on samples on cubic zirconia annealed for times ranging from 12 minutes to 3 hours. We have also seen this behavior on samples prepared using the ceramic alumina and sapphire substrates described above.

DISCUSSION

The resistance and susceptibility data demonstrate that we can prepare thin films of $Y_1Ba_2Cu_3O_{7-x}$ material on a variety of substrates using reactive RF sputtering. The same technique can be easily adapted to more complex materials by the addition of a fourth or fifth target and the associated power control systems.

Our results support the idea that these films are heterogeneous, with a broad range of transition temperatures. Such heterogeneity can have important consequences. Gittleman and Matey[9] have shown that the microwave surface resistance of superconductor/normal composites is a function of the volume fraction of the superconducting phase.

Our results show that characterization of the superconducting properties of HTSC thin film by four terminal resistance measurements alone neglects important aspects of the film.

ACKNOWLEDGEMENTS

Our thanks to R. Paff for the x-ray diffraction study, to L. Hewitt for the RBS measurements, to V. Pendrick for help in preparing the substrates, to J. Gittleman for helpful discussions, and to A. Schujko for technical assistance.

REFERENCES

1. International Products, Trenton, NJ.
2. Fisher Scientific, Fairlawn, NJ.
3. 55000 series with 55114P temperature control console. Lindberg Co., Watertown, WI.
4. Model 4192, Hewlett Packard, Palo Alto, CA.
5. Models 181 and 220, Keithley, Cleveland, OH.
6. DT-470 and Model 805, Lake Shore, Westerville, OH.
7. R. Landauer, " Electrical Conductivity in Inhomogeneous Media" in *Electrical Transport and Optical Properties of Inhomogeneous Media*, J. C. Garland and D. B. Tanner, Ed., AIP Conference Proceedings No. 40 (American Institute of Physics, 1978).
8. There are cases in which the Bruggeman theory needs more attention to the details than we have taken here. For example, if the superconductor was present in the form of thin shells surrounding a normal material, with the shells imbedded in a matrix of normal material, the normal material inside the shells would not participate.
9. J. I. Gittleman and J. R. Matey, "Modeling the Microwave Properties of the $Y_1Ba_2Cu_3O_7$ Superconductors", J. Appl. Phys., to be published.

PREPARATION AND CHARACTERIZATION OF HIGH T_c $YBa_2Cu_3O_{7-x}$ THIN FILMS ON SILICON BY DC MAGNETRON SPUTTERING FROM A STOICHIOMETRIC OXIDE TARGET

W.Y. Lee, J. Salem, V. Y. Lee, T. C. Huang, R. Karimi, V. Deline, and R. Savoy
IBM Almaden Research Center, San Jose, CA 95120.

D. W. Chung
San Jose State University, San Jose, CA 95192.

ABSTRACT

The effects of deposition temperature and O_2 pressure during cool down on the superconducting properties of $YBa_2Cu_3O_{7-x}$ thin films on Si substrates by DC magnetron sputtering from a stoichiometric oxide target are reported. Results form X-ray diffraction analysis indicate that the films deposited at 400°C or lower, at 650°C and cooled down in $<4 \times 10^{-4}$ O_2, and at 650°C and cooled down in >0.16 Torr O_2 have, respectively, amorphous, tetragonal, and orthorhombic structure. Superconducting orthorhombic films with a zero resistance T_c of as high as 76 K have been thus prepared on Si directly, without further heat treatments. The deposition of these films on Si is possible because of the minimal film-substrate interaction at the relatively low deposition temperature used, as indicated by the depth profiles obtained for these films using secondary ion mass spectrometry. Results of cross-sectional transmission electron microscopic studies of some of these films with different T_c and X-ray photoemission spectroscopic studies of the amorphous, tetragonal, and orthorhombic films are also reported.

INTRODUCTION

Thin films of the newly discovered high T_c Y-Ba-Cu oxide superconductors have been successfully deposited on various substrates. Most of these films were deposited on e.g., $SrTiO_3$, and MgO, because of the strong interaction of these films with other substrates (e.g., Al_2O_3 and Si) during the high temperature (~900 °C) post annealing step required to achieve high T_c superconducting state[1]. For microelectronic applications such as interconnects and hybrid semiconductor/superconductor devices, it would be more desirable to deposit these films on Si substrates. It is well recognized that to deposit these films on Si a diffusion barrier[2] or a low temperature process is required to minimize this film-substrate interaction. We reported previously[3] the deposition of these films at 600-700°C on Si wafers with a zero resistance temperature (T_c) as high as 76 K using DC magnetron sputtering from a stoichiometric oxide target. Recently, deposition of these films on Si at the similar temperature range has been reported using activated reactive evaporation[4,5] and laser ablation[6,7]. In this paper, the effect of deposition temperature and oxygen pressure during the cool-down period after sputter deposition is reported. Results of cross-sectional transmission electron microscopic (TEM) and X-ray photoemission spectroscopic (XPS) studies of some of these films are also reported.

EXPERIMENTAL PROCEDURES

© 1989 American Institute of Physics

The preparation of the 2" $YBa_2Cu_3O_{7-x}$ sputtering target, and the substrate-cathode configuration used to achieve stoichiometric film composition were described previously[1]. Briefly, this configuration utilizes an off-centered substrate holder and a positively biased counter electrode (anode) to minimize the preferential resputtering effect due to the presence of negative ions during sputtering[1,8,9]. The sputtering target used typically has a T_c of ~ 90 K and consists dominantly of the $YBa_2Cu_3O_{7-x}$ phase although minor secondary phases such as Y_2BaCuO_x, and $CuBaO_2$ or CuO are detected occasionally, according to X-ray diffraction analysis. Results of electron microprobe analysis show that the target used has mostly a Y:Ba:Cu ratio of (1):(2.1±0.1):(3.0±0.1) with isolated impurity phases.

These films (1-2μm thick) are deposited onto (100) Si substrates at ambient temperature to 700 °C and cooled to room temperature in ~2 hours in the presence of up to 200 Torr O_2. The Si substrates are mechanically clamped to a Cu substrate holder during deposition. The temperature of the Cu holder is controlled to within 1 °C of the desired temperature with an Omega CN9000 or CN2010 temperature controller, and is measured with a mechanically attached chromel-alumel thermocouple. The substrate temperature is taken as the temperature of the Cu holder, and no attempt is made to measure the temperature of the Si substrate directly. The film deposition rate is typically 40 Å/min. at 100W DC input power and 4 mTorr Ar-10% O_2 total pressure.

The technique for measuring the temperature dependence of the resistance of these films was also described previously[1]. In addition, several techniques including electron microprobe, scanning electron microscopy (SEM), TEM, XPS, Secondary Ion Mass Spectrometry (SIMS) and x-ray diffraction techniques are used to characterize these films. Transmission electron microscopic studies are carried out using a JEOL JEM-200CX TEM microscope, operated at 200 KeV. Cross-sectional view samples, prepared with conventional methods, are used to obtain the bright or dark field images and selective area diffraction patterns. X-ray photoemission spectroscopic studies are made with a SSI model 501 small spot ESCA system equipped with a monochromatic Al Kα X-ray source. The reported binding energies were referenced to C 1S at 284.6 eV.

RESULTS AND DISCUSSION

1. The Effect of Deposition Temperature

The superconducting properties of these films are studied as a function of substrate temperature under different cool-down conditions. The films deposited at temperatures between 600 to 700 °C and cooled down in 1-200 Torr of O_2 are superconducting without further heat treatments. As reported previously[3], the films with the highest T_c (76 K) are deposited at 650°C. Results of electron microprobe analysis show that the films deposited at 650 °C have a Y:Ba:Cu ratio of (1):(2.2):(3.05), compared to the ratio of (1):(2.1±0.1):(3.0±0.1) for the sputtering target used. A ratio of (1):(2.4):(3.1) and (1):(2.4):(3.8) is observed for the films deposited at 600 °C and 700 °C, respectively. The films deposited at less than 600°C generally are not superconducting regardless of the cool-down conditions, although the compositions of these films are close to that of the sputtering target. Figure 1 shows the X-ray diffraction patterns for the films deposited at ambient temperature (~ 80°C), 485°C, and 650°C (T_c = 76 K). The film deposited at ambient temperature is amorphous since its diffraction pattern (Fig. 1a) consists of only broad and diffused peaks, except for the Si (002) peak from the substrate. In general, similar amorphous diffraction pattern is observed for the films deposited at up to 400°C. The film deposited at 485 °C is crystalline, judging from the the presence of sharper diffraction peaks shown in Figure 1b. The detailed crystal structure of this film, however, is not yet identified. Similar diffraction pattern is observed for the films deposited at up to 550°C and cooled down in up to 200 Torr O_2. The film deposited at 600°C generally show a diffraction pattern (not shown in Figure 1) with sharp and well defined peaks attributable to a mixture of mostly tetragonal and orthorhombic perovskite structure. The diffraction pattern shown in Figure 1c for the films deposited at 650 °C indicates an orthorhombic perovskite structure with a slight c-axis in-plane preferred orientation. The lattice constants obtained from a least-squares refinement (a = 3.845, b = 3.886, and c = 11.660 Å) are similar to those reported previously for powder $YBa_2Cu_3O_{7-x}$ (Ref. 10).

From these X-ray diffraction results and the electron microprobe data mentioned earlier, it is clear that the films deposited at 650°C have the best crystal structure and stoichiometry required for achieving superconducting state and thus tend to have the highest T_c. As was also reported in our earlier paper[3], substantial x-ray diffraction line broadening and microcracks are observed for some of these films. These are attributed to the strain in the film due to the much higher thermal expansion coefficient for $YBa_2Cu_3O_{7-x}$ (12.9 x 10^{-6}°C^{-1}) (Ref.11) than for Si (3 x 10^{-6}°C^{-1}) (Ref. 12).

To study the interaction between the $YBa_2Cu_3O_x$ films and the Si substrate, SIMS depth profiles are obtained for the films deposited at different temperatures. The data for the films deposited at 600, 650, 700°C, and for the films deposited at 600°C and subjected to a post-anneal at 650°C in 1 atmosphere O_2 are given in Figure 2. No significant diffusion of Si from the substrate into the film is observed for the films deposited at 600 °C (Fig.2 a) and 650 °C (Fig. 2b). For the film deposited at 700°C (Fig. 2c), and the film deposited and post-annealed at 650°C (Fig. 2d), Si signal can be detected about half way through the film. The absence of Si diffusion for the films deposited at 650°C is further confirmed in a more detailed studies where SIMS depth profiles are obtained, under the same conditions, for several films deposited at 650°C and at ambient temperature (~80°C). The Si profile for the films deposited at 650°C typically shows less than 30 Si secondary ion counts throughout the entire thickness of the film, and is practically the same as the one for the film deposited at ambient temperature. The Ba, Cu and Y profiles, however, indicate ~ 3x deeper penetration into the Si substrate for the film deposited at 650°C. Because of this broadening of the film-substrate interface for the film deposited at 650°C, a minimum film thickness of > 5000Å probably is required, under the conditions used in this work, to produce high T_c $YBa_2Cu_3O_x$ films on Si substrate.

2. The Effect of O_2 Pressure during Cool-Down Period

The effect of O_2 pressure during cool-down period is studied for the films deposited at a constant temperature of 650°C. The films cooled down in very low O_2 pressure (< 4 x 10^{-4} Torr) are non-superconducting with room temperature resistivities (ρ) of 1-6 x 10^4 ohm-cm. The films cooled down in 0.16 Torr O_2 have ρ of 10^{-2} to 10^{13} ohm-cm. Here, only the films with ρ of ~ 10^{-2} ohm-cm are superconducting with a T_c of ~ 55 K and a slight negative temperature coefficient of resistivity in the

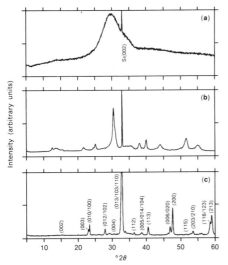

Figure 1. X-ray diffraction patterns for the films deposited on Si at a) ambient temperature (80°C), b) 485°C, and c) 650°C (T_c = 76 K).

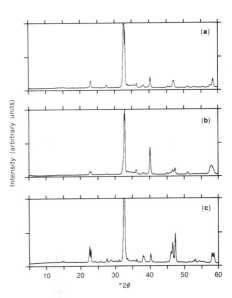

Figure 3. X-ray diffraction patterns for the films deposited on Si at 650°C and cooled down in a) $< 4 \times 10^{-4}$, b) 0.16, and c) 1 Torr (T_c = 74 K) of O_2.

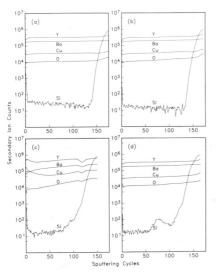

Figure 2. Secondary ion mass spectrometry depth profiles for the films deposited on Si at a) 600°C (T_c = 65 K), b) 650°C (T_c = 71 K), c) 700°C (T_c = 41 K), and d) condition a) plus post-annealed at 650°C (T_c = 59 K).

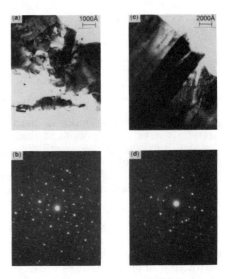

Figure 4. Transmission electron bright field images (a and c), and (001) selective area diffraction patterns (b and d) for the films deposited on Si with a T_c of 41 and 76 K, respectively.

normal state. For the films cooled down in 1-200 Torr O_2, a ρ of $\sim$3-20 x 10^{-3} ohm-cm is observed. These films show a T_c of $\sim$ 68-76 K with generally a metallic (positive) normal state temperature coefficient of resistivity. The films prepared under the three cool-down conditions mentioned above all have the perovskite crystal structure with trace of $BaCuO_2$ or Y_2BaCuO_x secondary phases present occasionally, as can be seen from the X-ray diffraction patterns shown in Figure 3. From detailed examinations of the diffraction profiles at 2θ = 27-29° for these films (Fig. 3), and for the bulk tetragonal and orthorhombic $YBa_2Cu_3O_{7-x}$, it is concluded that the film has tetragonal, a mixture of tetragonal and orthorhombic, and orthorhombic crystal structure when cooled down, respectively, in the increasing O_2 pressure range mentioned above. The lattice constants from a least-square refinement of the diffraction peaks for the tetragonal film are: a= 3.86, and c= 11.65 Å. These values are close to those reported for bulk tetragonal $YBa_2Cu_3O_{7-x}$(Ref. 13).

The results presented above and in Section 1 thus indicate that, with the reactive sputtering technique used here, a critical deposition temperature of 650°C and an O_2 cool-down pressure of > 0.16 Torr are both required to prepare the orthorhombic high T_c $YBa_2Cu_3O_{7-x}$ films. This critical deposition temperature apparently can be further reduced using an oxygen plasma or atomic oxygen assisted laser[6] or e-beam[14] evaporation technique.

3. TEM and XPS Studies

Cross-sectional TEM images and selective area electron diffraction patterns are obtained for the films deposited at 600, 650, and 700°C (cooled down in $\sim$100 Torr O_2) with a zero resistance temperature of 66, 76, and 41 K, respectively. Figure 4 shows the TEM bright field images and the (001) selective area diffraction patterns for the films with a zero resistance T_c of 41 and 76 K. These micrographs reveal a significant difference in the growth morphology of these two films. The 76 K film shows a typical columnar structure of orthorhombic crystals, as indicated by its selective area diffraction pattern shown in Fig. 4d. Both the bright field (Fig. 4 c) and the dark images (not shown) of this film indicate that these orthorhombic crystals are composed of thin lath-shaped crystals and are heavily twined as evidenced by the presence of needle-shaped fringe patterns. These lath-shaped crystals have a fairly uniform thickness extending along the length of the grains which is parallel to the growth direction. On the other hand, the 41 K film shows only a very small amounts of lath-shaped crystals along the growth direction but a high density of stacking faults with very little twins present within the grain. The orthorhombic lattice of the 76 K film seen in the selective area electron diffraction pattern is similar to that observed for the bulk $YBa_2Cu_3O_{7-x}$. The extra spot of the doublet shown in Fig. 4d is from the reflection of a neighboring grain. The 41 K film also has an orthorhombic lattice but with a b-axis twice as large as the a-axis since an extra reflection at ± 0 1/2 0 and $\pm 1/2$ 3/2 0 can be seen in its selective area difraction pattern shown in Fig. 4b. It should be pointed out that an orthorhombic structure similar to that of the 41 K sample was observed for a bulk $YBa_2Cu_3O_x$ superconductor after heat pulsing in a vacuum and was attributed to an oxygen-deficient superstructure present in the superconductor after heat pulsing[15].

Finally, X-ray photoemission spectra are also obtained for some of the films deposited at different temperatures and cooled down under different O_2 pressure. Figures 5 and 6 show, respectively, Ba $3D_{5/2}$ and the valence band spectra for the

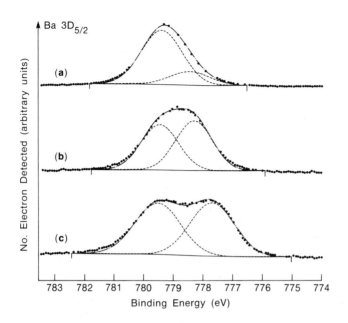

Figure 5. X-ray photoemission Ba $3D_{5/2}$ spectra for the films deposited at a) ambient temperature (~80°C) and cooled down in 200 Torr O_2, b) 650°C and cooled down in < 4 x 10^{-4} O_2, and c) 650°C and cooled down in 200 Torr O_2.

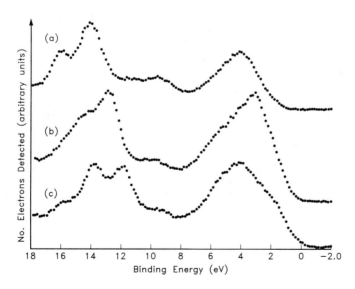

Figure 6. X-ray photoemission valence band spectra for the films deposited at a) ambient temperature (~80°C) and cooled down in 200 Torr O_2, b) 650°C and cooled down in < 4 x 10^{-4} O_2, and c) 650°C and cooled down in 200 Torr O_2.

films deposited at a) ambient temperature and cooled down in 200 Torr of O_2, b) 650°C and cooled down in $<4 \times 10^{-4}$, and c) 650°C and cooled down in 200 Torr of O_2. As was mentioned in Sections 1 and 2, film a), b), and c) has, respectively, amorphous, tetragonal and orthorhombic structure. These films are not superconducting except for film c) which has a T_c of $\sim$ 74 K. From Figure 5, it is seen that the Ba $3D_{5/2}$ transition has a peak at 779.4 eV common to all these three films. This peak was attributed to the contaminants ($BaCO_3$ or $Ba(OH)_2$) present on the surfaces of these films during air exposure[15]. In addition, there is a peak at 778.4 eV for the amorphous and the tetragonal films, and at 777.6 eV for the orthorhombic film. The intensity of the 778.4 eV peak is 2.5x greater for the tetragonal than for the amorphous film. The peak for the tetragonal film thus is shifted by 0.8 eV closer to the contaminant peak. Similar results are observed for the other Ba core levels and for the Ba 5P level (between 10 to 18 eV) in the valence band region shown in Figure 6. There is also a significant difference in the valence band spectra between E_f and 8 eV for these three films. A shoulder at $\sim$ 1.8 eV can be seen for the orthorhombic (Fig. 6c), but not the amorphous (Fig. 6a) or the tetragonal (Fig. 6b) films. The maximum is shifted by 1.2 eV toward E_f for the tetragonal with respect to the orthorhombic and the amorphous films. The extra lower binding energy component (e.g., Ba $3D_{5/2}$ at 777.6 eV, and Ba 5P at 11.9 eV) and the shoulder at $\sim$ 1.8 eV observed for the orthorhombic film thus are characteristic of the film in the superconducting state, similar to those observed previously for superconducting films prepared with a high temperature post annealing step[16]. The spectra shown in Figures 5 and 6 can be obtained only from freshly prepared samples since they are rapidly contaminated during air exposure. For example, the ratio of the intensity of the 778.4 eV to the 779.4 eV peak of the Ba $3D_{5/2}$ spectra is found to decrease rapidly with increasing air exposure, indicating the conversion of Ba in the tetragonal structure to the $BaCO_3$ or $Ba(OH)_2$ species. A concomitant change in the Cu 2P and O 1s spectra of the tetragonal film during air exposure are also observed. Detailed interpretation of these results will be published in a separate paper.

SUMMARY

The effects of deposition temperature and O_2 pressure during cool-down on the superconducting properties of high T_c $YBa_2Cu_3O_{7-x}$ thin films on Si by DC magnetron sputtering from a stoichiometric oxide target are described. Films with amorphous, tetragonal and orthorhombic structure can be prepared by changing the deposition temperature and the O_2 pressure during cool-down period. A critical deposition temperature of $\sim$650°C and O_2 pressure of >0.16 Torr are required to produce the orthorhombic high T_c films. Superconducting films with zero resistance T_c as high as 76 K have thus been prepared on Si directly, without further high temperature post annealing treatments. No significant diffusion of Si from the substrate into the film is detected for the films deposited at 650°C or lower, according to depth profiles obtained using secondary ion mass spectrometry. Results of cross-sectional TEM microanalysis and XPS studies of some of these films are also reported.

ACKNOWLEDGMENTS

The authors are indebted to C. T. Rettner and D. J. Auerbach for making available the T_c measurement apparatus, and to C. R. Brundle and J. E. Baglin for their support of this project and for many useful discussions throughout this work.

REFERENCES

1. W. Y. Lee, J. Salem, V. Lee, C. T. Rettner, G. Lim, R. Savoy, V. Deline, AIP Conf. Proc. No. 165, 95 (1988); and the references cited therein.

2. A. Mogro-Campero and L. G. Turner, Appl. Phys. Lett. 52, 1185 (1988).

3. W. Y. Lee, J. Salem, V. Lee, T. Huang, R. Savoy, V. Deline, and J. Duran, Appl. Phys. Lett. 52, 2263 (1988).

4. R. M. Silver, A. B. Berezin, M. Wendman, and A. L. de Lozanne, Appl. Phys. Lett. 52, 2174 (1988).

5. P. Berberich, J. Tate, W. Dietsche, and H. Kinder, Appl. Phys. Lett. 53, 925 (1988).

6. S. Witanachchi, H. S. Kwok, X. W. Wang, and D. T. Shaw, Appl. Phys. Lett. 53, 234 (1988).

7. T. Venkatesan, E. W. Chase, X. D. Wu, A. Ianm, and C. C. Chang, Appl. Phys. Lett. 53, 243 (1988).

8. S. M. Rossnagel and J. J. Cuomo, AIP Conf. Proc., No. 165, 106 (1988)

9. R. L. Sandstrom, W. J. Gallagher, T. R. Dinger, R. H. Koch, R. B. Laibowitz, A. W. Kleinsasser, R. J. Gambino, B. Bumble, and M. F. Chisholm, Appl. Phy. Lett. 53, 444 (1988).

10. R. Beyers, G. Lim, E. M. Engler, R. J. Savoy, T. Shaw, T. R. Dinger, W. J. Gallagher, and R. L. Sandstrom, Appl. Phys. Lett. 50(26), 1918(1987).

11. H. M. O'Bryan and P. K. Gallagher, Advanced Ceramic Materials 2(3B), 640(1987).

12. R. C. Weast (ed.), CRC Handbook of Chemistry and Physics, CRC Press (Boca Raton, Florida), 66th edn., 1985-1986.

13. W. Wong-Ng, H. F. McMurdie, B. Paretzkin, C. R. Hubbard, and A. L. Dragoo, Powder Diffraction 3, 113 (1988).

14. N. Missert, R. Hammond, J. E. Mooij, V. Matijasevic, P. Rosenthal, T. H. Geballe, A. Kapitulnik, M. R. Beasley, S. S. Laderman, C. Lu, E. Garwin, R. Barton, preprint, August 22, 1988.

15. G. Van Tendeloo and S. Amelinckx, J. Elec. Microscopy Technique 8 (3), 285 (1988).

16 D. C. Miller, D. E. Fowler, C. R. Brundle, and W. Y. Lee, AIP Conf. Proc., No. 165, 336 (1988)

RF BIAS GETTER REACTIVE SPUTTERING OF La-Cu-O

Peter J. Clarke
Sputtered Films, Inc., Santa Barbara, CA 93103

Bill Gardner
Technology of Materials, Santa Barbara, CA 93103

ABSTRACT

The fundamental La-Cu-O system contains all the ingredients needed to construct the new high Tc superconductors and to study the associated process problems. This paper describes the sputtering from a mosaic sputter cathode fabricated from tiles of La and Cu. The composition of the deposited film was checked by primary X-ray emission methods. After the proper atomic ratio was determined it could be tuned by over an 11% range by varying a sputtering gas mixture of Xe and Ze. After the proper atomic ratio was formed O_2 was added to the sputtering gas and the alleged negative ion effect associated with elements with a large electronegativity was studied.[1]

I. INTRODUCTION

The high reactivity of the elements used to form the new high Tc superconductors and the difficulty of obtaining them in their elemental form makes cathode fabrication difficult. Targets formed from the oxides of these elements suffer from the problems associated with oxide targets i.e., sputtering rates depressed by a factor of 10 and poor heat transfer which severely limits the power input to the target when compared to the elemental metal. The reality is that the effective sputtering rate from an oxide can be down by a factor of 100 when compared to the same film deposited by reactive sputtering from the element.[2]

The hope that these films can be deposited in place with superconducting characteristics is going to depend upon its composition and crystalline form.

II. SYSTEM DESCRIPTION AND SETUP PROCEDURES

The sputtering took place in a small and enclosed getter volume onto 100mm Si wafers. No shutter is necessary because the sputter cathode is presputtered into this volume (getter sputtering) before the substrate is introduced into it through a slit. The inert gas was bled in under the anode of the S-Gun and first struck the cathode surface. Oxygen was passed through the center of the anode and first struck the substrate surface. This almost achieves ideal separation of the active and inert gas[3] (Fig. 1).

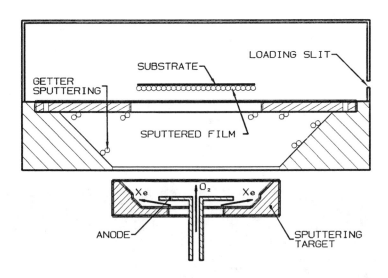

Fig. 1 Getter Sputtering volume showing gas segregation.

The S-Gun cathode was a 7 inch Cu shell in which tiles of Cu and La which were inlaid in a symmetrical pattern. This is shown in Fig. 2. The non-uniformity of the tile target was checked earlier using Al and Cu tiles across a 100mm Si wafer and was found to be +/- 2%.

The coated wafers were kept in argon until analyzed in order to minimize oxidation. Small pieces of the central areas of the wafers were broken out after diamond scribing on the cleavage planes. Several pieces were mounted together on sample stubs and analyzed by primary X-ray emission methods. The resulting primary X-ray emission was analyzed and quantified using an energy analyzer. Quantification was made using a fundamental parameters iteration computer program utilizing a modified Bench-Albee Method. Count intensity references were obtained using pure La and pure Cu metal standards. 10,000 second count collection periods were used. All parameters, i.e. tilt angle, voltage, count ratio, dead time were kept constant for each analysis.

Thickness measurements were made using IBM PC controlled Dektac.

Because of the tenacious oxide on the La tiles it was not possible to clear this oxide by normal inert gas sputtering techniques. This points up the seriousness of oxide contamination of high Tc superconductor targets. The target had to be removed from the vacuum system and the oxide bead blasted away. Fig. 3 shows some of the debris left on the cathode surface after sputtering for several hours. The white spots were analyzed on a La tile and found to contain Mg and Si, the main composition of the beads. This is an important cleaning technique but the debris problem has to be solved.

After sputtering the La and Cu tiles were analyzed. The La tile contained 1% Cu and 2% Fe. The Cu tile contained <1% La and 1% Fe. Fe is a major impurity of La but the supplier of the La cited the tiles were 4 nines pure. The Cu tiles were reported to be 3 nines with <1% Fe.

In all the reported results the Fe has been subtracted from the results.

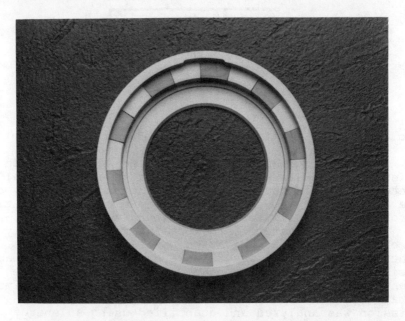

Fig. 2 Mosaic S-Gun sputter cathode.

III. EXPERIMENTAL

Experiment to determine if cross sputtering between the tiles of La or Cu caused any artifacts on the surface of either tile that might modify the sputtering yield.

Much has been reported of seeding a sputtering surface with an impurity that in turn modifies the surface and its sputtering yield.[4] For instance the growth of cones on a Cu surface that has been seeded with Mo. It is reasonable to suspect that the La tiles could effect the Cu tiles or the contrary.

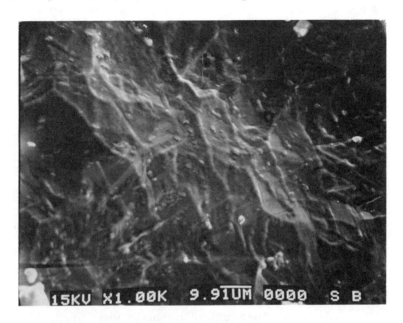

Fig. 3   La sputter cathode surface.

The tiles were sputtered for four hours at 265 volts and 5 amps. The sputtering gases were Ne and Xe. No reactive gas was used in this experiment. Two tiles were removed from the assembled sputter target and examined under a scanning electron microscope.

When removed from the system the La was a bright silver and the Cu a bright copper color. Under the electron microscope the Cu was smooth and showed preferential sputtering according to grain orientation. The La had much smaller crystals and showed a ridge between the grains. We suspect this ridge is caused by oxygen diffusion along the grain boundaries. Neither the Cu or the La showed any evidence of cones or other artifacts on the surface except for the debris that was analyzed as Si and Mg. These surfaces are shown in Fig. 3 and Fig. 4.

From this experiment we conclude the sputtering yield was not modified by any surface artifacts.

Fig. 4 Cu sputter cathode surface.

Experiment to determine the atomic % composition of the material sputtered from the tile target as a function of the inert gas and ratio of tiles.

Table I The effect of tile ratio & sputtering gas on atomic % ratio of La-Cu.

| TILES | | Sputtering Gas | Atomic % | |
# La	# Cu		La	Cu
15	6	Ar	35.6	63.2
15	6	Kr	38.8	60.6
18	3	Ne	57.1	43.0
18	3	Ar	54.5	45.5
18	3	Xe	65.0	35.0

The results tend to show the expected i.e. the La content is proportional to the number of La tiles and the mass of the sputtering gas. The Ne run was reanalyzed with the same result.

Experiment to determine the atomic % composition of the material sputtered from the tile target as a function of a mixture of Ne and Xe as the sputtering gas.

Table II The effect of inert gas composition on the atomic % ratio of La-Cu

Flow Ne SCCM	Flow Xe SCCM	Atomic % La	Cu
0	6.3	65	35.3
5	6.3	65.5	34.5
10	6.3	62.1	37.9
15	6.3	62.8	37.2
20	6.3	60.5	39.5
20	4.3	64.5	35.5
20	2.3	60.4	40.0
20	0	57.1	43.0

This experiment shows explicitly that the composition can be electronically tuned over a range of 11%. This cannot be done with a composition target because its surface reaches an equilibrium composition no matter what inert gas is used. The experiment has implications for those silicide targets that are difficult to obtain in high purity and precise stoichiometry.

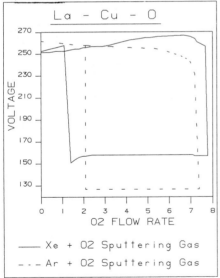

Fig. 5
Voltage vs. Flow Rate

Experiment to determine the variation with oxygen flow of applied voltage for a constant sputtering current of 5 A and a fixed Xe or Ar flow.

The first system to be investigated was the Xe-O_2 gas mixture. The O_2 was introduced slowly and the system first went through a maximum in voltage and then decreased. At a flow of 7.3 SCCM, the S-Gun arced badly and had to be turned off and then ramped up to full current. The voltage immediately stabilized at 157 V. The O_2 flow was decreased and the voltage fell slightly to 151 V and the S-Gun again started to arc. The S-Gun was again switched off and then ramped up to 5 A. The Voltage jumped up to 258 V.

The second system to be investigated was the Ar-O_2 gas mixture because the Xe gas supply became depleted.

The system acted the same except there was a constant decrease in voltage as the O_2 flow was increased.

The important difference will be that the Xe-O_2 can be controlled using the maximum as a control point. There is nothing available like this in the Ar-O_2 system. These curves are shown in Fig. 5.

Experiment to study the negative ion effect that has been reported as a serious impediment to the sputtering of metals like Ba. Ba is just slightly more electronegative than La so the negative ion effect should be similar.

Two films were deposited, one in Xe alone and the other in a mixture of Xe and O_2. The atomic % composition of each film was determined. The Xe-O_2 film was deposited near the maximum on the curve shown in Fig. 5.

Table III The effect of O_2 on the atomic % composition of La-Cu

Vc	Ic	Flow Xe SCCM	Flow O_2 SCCM	Atomic % La	Cu
260	5	6.3	0	65	35.3
267	5	6.3	5.7	63.9	36

Within the limits of experimental error there is no difference in the film composition. With this experiment the negative ion effect was not observed.

The above experiment was repeated using Ar gas and a mixture of Ar-O_2 films gas. The Ar-O_2 films were deposited near the knee of the curve and on the low voltage leg of the hysteris curve as shown in Fig. 4. The atomic % composition and thickness of the first two films was determined.

Table IV The effect of O_2 on the atomic % composition and sputtering rate.

Vc	Ic	Flow Ar SCCM	Flow O_2 SCCM	Atomic % La	Cu	Thcknss KAng	SputConstg Ang/min KW
268	5	20.4	0	54.5	45.5	116	2320
244	5	20.4	6.7	54	45.9	122	2440
142	5	20.4	6.2			10	200

Again within the limits of experimental error there is no difference in the atomic % composition. The third film which was not analyzed for film composition shows the dramatic difference in deposition constant for a metal and a fully oxidized sputter target. For reference a typical constant for an Al film using Ar as the sputtering gas target is 1200 Angstroms/min KW. Again there was no evidence of the negative ion effect.

Experiment to calculate the relative number of tiles needed of each metal to form the composition La_2-Cu by determining the weight loss per tile of La and Cu and assuming both metals have the same sputtering distribution.

Table V   Weight of Tiles

New La - 26.0 gm	Used La - 20.2 gm	difference 5.8 gm
New Cu - 42.1 gm	Used Cu - 36.2 gm	difference 5.9 gm

Within the limits of error both tiles sputtered the same weight. Since the mass of La is 138.9 and of Cu is 63.5 then the Cu tile contributes about 2.2 more atoms than each La tile. To sputter the composition La_2-Cu it would be necessary to have 4.4 more La tiles than Cu tiles. By experiment it is necessary to have a ratio of about 6 to 1 and use Xe gas. From this experiment it can be argued that La has an under cosign distribution when compared to Cu. This result gives an ambiguous answer to the negative ion effect.

It is a puzzle to think the same number of gms were sputtered from the La and Cu tiles. In vacuum a good rule of thumb is - visualize one effect and the opposite happens.

Because the Xe gas supply got depleted only two films were analyzed with a substrate bias.

The first was an $Xe-O_2$ film deposited at the maximum of the curve and to a thickness of about 11 microns. The sample was electrically floating and was measured at 1.5 V. The sample showed large (>1000 Ang.) randomly oriented crystals of Cu_2La and poorly crystalized (<100 Ang.) La_2O_3. There was no CuOx or Cu-La-O compounds.

The second film of unknown atomic % composition was deposited in an $Ar-O_2$ mixture on the bottom leg of the curve which indicates a completely oxidized sputter cathode. The sample was electrically floating and was measured at -.8 V. The structure of the sample was amorphous.

IV. OBSERVATIONS

1. La forms a very tenacious oxide and care must be taken to be sure this oxide is cleared from the sputter cathode.
2. No cross contamination between the tiles that effects the sputtering rate was observed.

3. The La content in the deposited film is a function of the inert sputtering gas.
4. The atomic % ratio of La to Cu can be varied over a range of 11% using a mixture of Ne and Xe.
5. The V vs. $Xe-O_2$ and V vs. $Ar-O_2$ hystersis curves are dramatically different. The $Xe-O_2$ lends itself nicely to feed back control.
6. No support was found for the negative ion effect is observed when La is substituted for Ba in the sputtering target. The difference in the observations may be found in the use of two different types of magnetron sputtering devices used to study this effect i.e., a planar magnetron and the S-Gun. The major difference is the S-Gun has a central anode which has been shown to attract all the electrons of the discharge while the electrons in a planar discharge go to ground through any available grounded surface including the substrate. This effect would also apply to the negative ions.
7. It may be possible to deposit films with the correct crystal orientation. These experiments will be continued.

FOOTNOTES

1. S.M.Rossnagel and J.J.Cuomo: preprint "Negative Ion Effects During Magnetron and Ion Beam Sputtering of $YBa_2Cu_3O_x$"

2. W.D.Sproul: 14th International Conference on Metallurgical Coatings "High Rate Reactive Sputtering Process Control" (March 23-27, 1987)

3. G.Este, W.D.Westwood: J.VacSci.Technol.A, Vol.2, No.3, July-Sept. 1984

4. G.K.Wehner, D.J.Hajicek: J.Appl.Phys.42,1145 (1971)

EFFECT OF SUBSTRATE TEMPERATURE AND BIASING ON THE FORMATION OF 110 K Bi-Sr-Ca-Cu-O SUPERCONDUCTING SINGLE TARGET SPUTTERED THIN FILMS

N. G. Dhere, R. G. Dhere
Solar Energy Research Institute,
1617 Cole Blvd, Golden, CO 80401,
J. Moreland,
National Bureau of Standards, Boulder, CO 80401

This paper presents the study of the effect of the substrate temperature and biasing on the superconducting properties of thin films deposited by RF sputtering from a single Sputter-Gun type ring target prepared by powder pressing of a mixture of the oxides with a nominal metallic composition Bi:Sr:Ca:Cu of 1.47:2:1.8:3.17. Thin films deposited on unheated substrates were smooth, dark, homogeneous, amorphous and insulating. Films became increasingly crystalline and their resistance diminished with increase in substrate temperatures above 300° C and upto 525° C; and increase in substrate biasing upto - 150 V. DC biasing proved its usefulness, even though it caused experimental difficulties. Attempts have been made to replace it by RF biasing. Problems related to simultaneous substrate heating and RF biasing still need to be resolved. As deposited superconducting films were obtained by employing substrate heating during deposition (500-525° C) and negative DC substrate biasing (- 150 V). Partial transformation to 110 K phase was observed when the films were air annealed at 865-870° C (melting) for 20 min, followed by annealing at 850-855° C during 5 h and at 400° C, during 6-8 h. X-ray diffraction analysis showed mostly (001) type reflections from $Bi_2Sr_{3-x}Ca_xCu_2O_{8+y}$ and several reflections from 110 K, $Bi_2Sr_2Ca_2Cu_3O_{10+y}$ phase for which a tetragonal lattice with a = 7.66 A and c = 37.38 A, having ´c´ values consistent with values reported in the literature, has been proposed.

1. INTRODUCTION

Several groups have prepared superconducting thin films of Bi-Sr-Ca-Cu-O with T_{co} of upto 80.5 K, by evaporation from multiple sources, flash evaporation, single and multiple target sputtering, on unheated MgO single crystal substrates followed by heat treatment at 850-865° C[1-3]. Bi based superconducting 110 K phase, discovered by Maeda et al[4] has been found to be more difficult to prepare. Adachi et al[5] obtained Bi based superconducting thin films with T_{co} > 100° C, by single, planar target, RF sputtering on substrates heated to > 800° C, and post-annealing in the range 865-890° C. Proximity to the source was utilized to obtain some plasma exposure during deposition. Kuwahara et al[6] have suggested that 110 K phase is less stable than 75-85 K, $Bi_2Sr_{3-x}Ca_xCu_2O_{8+y}$ phase, at normal annealing temperatures in the range 850-890° C and that final prolonged annealing at 400° C, in oxygen, promotes its growth. Both the groups emphasize the need to obtain prior partial formation of a superconducting phase. Biasing of substrates during sputtering can enhance the substrate bombardment by ionized species. This could reduce the temperatures required for the formation of the superconducting phase which later helps further crystallization of 110 K phase. The composition of 110 K phase is $Bi_2Sr_2Ca_2Cu_3O_{10+y}$. Adachi et al and Kuwahara et al have suggested that its structure is tetragonal and have indicated that the value of ´c´ parameter is approximately 36 or 37.05 A[5-6]. This paper presents the effect of DC and RF substrate biasing on the composition, crystallinity, resistivity and formation of superconducting phases, especially 110 K phase and provides its lattice parameters.

2. EXPERIMENTAL

Bi-Sr-Ca-Cu-O thin films, 2500 A - 1 um thick, were deposited by RF magnetron sputtering in a cryo-pumped sputtering system, on (100) MgO single crystal and fused quartz substrates, heated to upto 650° C. Films on fused quartz substrates were utilized only for analysis. Attempts were made to utilize commercial, flat, thin film Pt resistance, temperature sensors, as substrate heaters by Joule heating. But they became unstable above 350° C. Pyrolytic BN/graphite heaters (Hot film, Union Carbide) were later utilized and proved adequate for substrate heating to over 650° C. Parts of the insulating MgO single crystal substrates were coated with silver or copper thin films, and contacts, for negative DC biasing of MgO substrates were made with copper foils wrapped around part of the metallized region, leaving a part exposed for continuity of electrical contact with the deposited film. Copper films oxidized considerably, in the presence of ionized oxygen, at high temperatures. Moreover, at high temperatures, metal thin films mostly disappeared due to inter-diffusion. An amateur radio trans-receiver was adapted as a RF amplifier, by feeding 12 V peak-to-peak, 13.56 MHz signal from RF sputtering generator. Output from RF amplifier was connected, through a matching network, to ~ 43 mm diameter, backing copper plate, placed on the substrate heater. DC bias developed on the copper backing plate, was measured by a multimeter attached at the RF feed-thru by RF blocking L-C network. The use of an old tube model and lack of feed-back control resulted in drift and some variation of the biasing conditions. Difficulties encountered with simultaneous substrates heating and RF biasing are being resolved, after which DC biasing will be replaced by RF biasing. Oxygen partial pressures, P_{O2}, in the sputtering gas was varied between 1-2 x 10^{-3} Torr (0.13-0.27 Pa), while the total pressure of argon and oxygen was maintained at 3 x 10^{-3} Torr (0.4 Pa). The hot-pressed ring targets, with nominal proportion of Bi:Sr:Ca:Cu:O of 1.47:2:1.8:3.17, made of an unreacted mixture of the oxides (Bi_2O_3, SrO, CaO and CuO), for Sloan S310 type sputter-gun was supplied by Kema. The composition and the crystallographic structure of the films was analyzed respectively by x-ray microanalysis and rotating anode x-ray diffraction system, using Cu K radiation. The sheet resistance was measured with a four point probe. Final resistance versus temperature measurements were carried out at NBS, Boulder Laboratories, to an accuracy of ± 0.1 K. Leads were attached by ultrasonic bonding for this purpose. Spring loaded pressure contacts were utilized for routine resistance versus temperature measurements. The samples were held above the melting temperature, in the range 862-870° C for 20 minutes and then annealed at temperatures in the range 850-855° C for 5-6 hours[1]. In the case of some thin films, extended annealing at 400° C for 6-8 hours was employed, during the cooling cycle, to test the hypothesis of Kuwazara et al[6], regarding the effectiveness of prolonged annealing in oxygen at 400° C.

3. RESULTS AND DISCUSSION

Bi-Sr-Ca-Cu-O thin films, 0.25-1 um, deposited, at rates of 12-20 A.s^{-1}, on unheated substrates were smooth, dark, homogeneous, insulating and amorphous. The sheet resistance of the films deposited at substrate temperatures of upto 300-400° C and with negative DC bias of upto - 30 V, was in the range 0.22-2.6 MΩ/□. Earlier results with annealed films had shown an empirical relationship between the lowering of sheet resistance and the improvement of the degree of crystallinity as well as of the superconducting properties. This was utilized as a quick test. Substrate heating to 450-475° C and negative DC bias upto - 50 V resulted in thin films with sheet resistance of 3-10 kΩ/□. The sheet resistance decreased continuously with an increase in the substrate temperature upto ~ 550° C and DC bias upto - 150° C, whilst, the films became greenish white and insulating at substrate temperatures > 650° C and DC bias of - 200 V. Therefore, substrate temperatures of 500 to 525° C and negative DC bias values of - 100 to - 150 V were chosen. Typical ion currents for this bias were in the range of 30-35 mA. The films became sooty when P_{O2} was increased to 2 x 10^{-3} Torr (0.27 Pa). Hence P_{O2} value of 1 x 10^{-3} Torr (0.13 Pa) were utilized. Modest changes in composition were observed by electron probe x-ray microanalysis due to substrate heating. Compositions of thin films deposited at room temperature on unheated substrates were approximately 10% poorer in Cu and Ca, as compared to the target composition. Increase in the substrate temperature to 500-525° C resulted in further 10% reduction in Cu content, net 5% increase in Ca content and 10-20% increase in Bi content. Negative DC biasing slightly reduced Bi and improved Cu contents. As will be seen below, in DC biasing, actual voltage seen by the plasma are small.

The thicknesses were lower and the compositions altered drastically as a result of RF biasing with effective negative DC bias over - 70 V, net RF power > 10 W. Bismuth contents were reduced to < 1 at.% while those of copper to 6 at.%. This, no doubt, resulted from preferential resputtering. The quantity of oxygen in the films was also reduced due to the reduction of Bi content. Sheet resistance of the films deposited at substrate temperatures of 500-525° C and effective DC bias, as a result of RF biasing, above - 70 V, net RF power of 10 W, was in the range 20-200 kΩ/□. Bismuth and copper depleted less with effective DC bias of - 25 V and net RF power of 3 W. This shows that the actual biasing in the case of DC biasing was small. This may be due to the resistive voltage loss along the plane of the film, as well as the repulsion by the negative charge sheath on the substrates. Further work on RF biased films will be carried out after the modification of the substrate heating and RF biasing set-up.

X-ray diffraction patterns from thin films deposited at substrate temperatures of 350° C and DC bias voltages below - 50 V, consisted of few broad peaks, showing their microcrystalline nature. As the substrate temperatures and bias voltages were increased to 500-525° C and - 150 V, respectively, the peak intensities increased and they became less broad, due no doubt to the polycrystalline nature of the film. The sheet resistance of these films was in the range 700-1000 Ω/□. Resistance versus temperature measurements showed that "as deposited" films were strongly semiconducting. The films deposited at substrate temperatures in the range of 500-525°C and negative DC bias of - 100 to - 150 V, showed characteristic drop in resistivity below 30 K showing that "in situ" superconducting films can be prepared under these conditions (Fig. 1). The films were heat treated in air. It was found that the melting occurred in the range 862-870° C, which is slightly higher than the melting temperature of 862-865° C observed in the case films deposited on unheated substrates. The films were maintained above 870° C for 20 minutes, and annealed at 850-855° C for 5-6 hours. During the cooling cycle, some of the films underwent an extended annealing at 400° C for

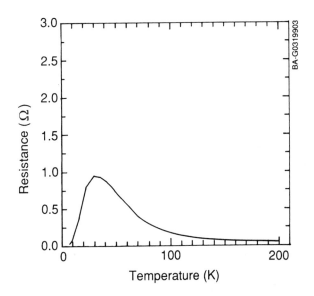

Fig. 1 - Resistance-versus-temperature plot of an unannealed Bi-Sr-Ca-Cu-O thin film, deposited at substrate temperature of 525° C and DC bias voltage of - 150 V.

6-8 hours. Resistance versus temperature measurements of annealed thin films deposited on substrates heated to temperatures below 400° C and with negative DC bias below - 50 V, showed only the presence of 85 K and 6-20 K transition depending on the composition, with a trace of 110 K transition being observed in a few cases. X-ray diffraction patterns of these films showed mostly the presence of 75-85 K, tetragonal, $Bi_2Sr_{2-x}Ca_{1+x}Cu_2O_8$ phase, with a = 3.812 A and c = 30.66 A, proposed by Torrance et al[7]. Partial but significant transition was observed, in the range 113 - 103 K, in post-annealed films deposited at substrate temperatures of 500-525 C and negative DC bias of - 150 V, when the cooling cycle was extended to include annealing at 400° C (Fig. 2). As seen above these films were superconducting with ´T_c on´ below 30 K. X-ray diffraction patterns of the annealed films, corresponding to fig. 2, revealed mostly (001) type reflections from orthorhombic, $Bi_2Sr_{3-x}Ca_xCu_2O_{8+y}$ phase, with a = 5.399 A, b = 5.414 a and c = 30.904 A, proposed by Subramanian et al as the probable 116 K phase[8] (Fig. 3). It was found that several other peaks could be explained on the basis of a tetragonal phase with a = 7.66 A and c = 37.38 A. The composition of 110 K phase is most probably $Bi_2Sr_2Ca_2Cu_3O_{10+y}$, as proposed by Kuwahara et al[6]. Kuwahara et al[6] and Adachi

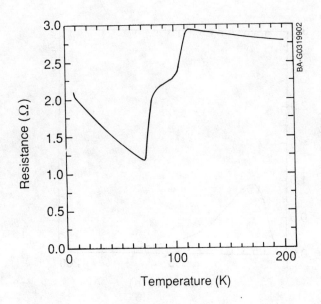

Fig. 2 - Resistance-versus-temperature plot of a Bi-Sr-Ca-Cu-O thin film, deposited at substrate temperature of 500° C and DC bias voltage of - 150 V; post-annealed at 865-870° C for 20 min. (melting), 850-855° C for 6 h, (annealing) and 400° C for 6-8 h.

et al[5] have indicated ´c´ values of 37.05 A and approximately 36 A respectively, for the probable tetragonal structure of this phase. It is proposed that the observed tetragonal structure is due to 110 K, $Bi_2Sr_2Ca_2Cu_3O_{10+y}$ phase. As can be seen the value of ´c´ parameter of the proposed lattice is compatible with the values suggested in the literature[5,6], while the value of ´a´ parameter follows a relation of 2 or 1/2 times with the values of ´a´ and ´a´, ´b´ proposed for the lower T_c tetragonal and orthorhombic phases respectively[7,8].

4. CONCLUSIONS

Substrate heating and effective negative biasing (either DC or RF) of MgO single crystal substrates promotes the formation of low resistivity, crystalline phases, in Bi based, single target sputtered, superconducting thin films. ´In situ´, superconducting thin films, with ´T_c on´ below 30 K, are obtained with substrate temperatures of 500-525° c and negative DC biasing of - 150 V. Post-annealing of these films results in a partial transformation to 110 K phase. A tetragonal lattice with a = 7.66 A and c = 37.38 A has been

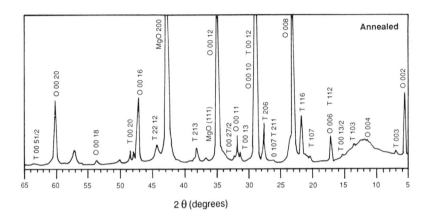

Fig. 3 - X-ray diffraction pattern of Bi-Sr-Ca-Cu-O thin film corresponding to fig. 2.

proposed for 110 K, $Bi_2Sr_2Ca_2Cu_3O_{10+y}$ phase. ´c´ values of the proposed phase are consistent with the values proposed in the literature, while ´a´ value follows 2 or 1/2 times relationship to the values of ´a´ and ´a´, ´b´ parameters of the lower T_c phases. Extended annealing in air, at 400° C for 6-8 h, during the cooling cycle enhances the fraction of 110 K phase.

ACKNOWLEDGEMENTS

This work was supported by the U.S. Department of Energy, under contract DE-AC02-83CH10093. It was also partially supported by the Brazilian Ministry of Education through CAPES. The authors are thankful to Alice R. Mason and John P. Goral for their help in composition analysis. The authors gratefully acknowledge useful discussions, regarding RF biasing of the substrates, with Dr. John W. Coburn, IBM San Jose, CA.

REFERENCES

1. N. G. Dhere, J. P. Goral, A. R. Mason, R. G. Dhere and R. H. Ono, J. Appl. Phys., 64, (1988) to be published in Nov. 15, issue.

2. M. S. Osofsky, P. Lubitz, M. Z. Hartford, A. K. Singh, B. S. Qadri, E. F. Skelton, W. T. Elam, R. J. Soulen Jr., W. L. Letcher and S. A. Wolf, submitted to Nature on May 18, 88.

3. J. H. Kang, R. T. Kampwirth, K. E. Gray, S. Marsh and E. A. Huff, Physics Lett., 128A, 102 (1988).

4. H. Maeda, Y. Tanaka, M. Fukutomi and T. Asano, Jpn. J. Appl. Phys. Lett. 27, 2 (1988).

5. H. Adachi, K. Wasa, Y. Ichikawa, K. Hirochi and K. Setsune, J. Crystal Growth (1988), to be published.

6. K. Kuwahara, S. Yaegashi, K. Kishio, T. Hasegawa and K. Kitazawa, to be published.

7. J. B. Torrance, Y. Tokura, S. J. Laplaca, T. C. Huang, R. J. Savoy and A. I. Nazzal, Solid State Communications, 66, 703 (1988).

8. M. A. Subramanian, C. C. Torardi, J. C. Calabrese, J. Gopalkrishnan, K. J. Morrisey, T. R. Askew, R. B. Flippen, U. Choudhry and A. W. Sleight, Science, 238, 1015 (1988).

DEPOSITION OF CERAMIC SUPERCONDUCTORS FROM SINGLE SPHERICAL TARGETS

G. K. Wehner*, Y. H. Kim*, D. H. Kim**, and A. M. Goldman**
University of Minnesota, Minneapolis, MN 55455

ABSTRACT

In sputtering from a flat multi-component target the film composition is a function of substrate location and is usually different from that of the target. The main reason for this is that different atomic species are sputtered with different angular distributions. An additional problem arises with negative oxygen ions which cause resputtering from substrates located opposite the target. These complications disappear when the sputtering is performed using spherical targets. The deposit is formed from atoms ejected in all directions and this, together with spherical symmetry, results in an automatic integration resulting in the replication of target composition in the deposited film. Furthermore the negative oxygen ions which cause resputtering are ejected radially from the target and distributed over the whole solid angle of the target. Experimental proof of this was obtained by sputtering a sphere of $YBa_2Cu_3O_{7-x}$ in a Hg triode plasma. After the usual post deposition oxygen heat treatment a film with R=0 at 89K was obtained. Recent work at Battelle NW has shown that the Hg plasma had nothing to do with the favorable results. They demonstrated that deposits obtained from a 1-2-3 sphere using a Xe triode plasma were position-independent. A progress report on sputtering using spherical targets including results from the more recently discovered Bi-Sr-Ca-Cu-O superconductors will be presented.

INTRODUCTION

Sputtering has become a widely used technique for the deposition of thin films of the new High-T_c superconducting ceramics. The technique has been important in the fabrication of various superconducting films for a long time. It was used by Gavaler[1] in 1974 to prepare films of Nb_3Ge in the A-15 phase which had transition temperatures of 23K. This high T_c A-15 phase of Nb_3Ge has actually never been produced in bulk form. For the production of Nb_3Ge films it was necessary to keep the sputtering voltage low (<1000V) and the argon pressure high (>0.1 Torr). Sputtering at the present time is the fabrication technique of choice for the production of thin film tunneling junctions with electrodes of either Nb or NbN.[2]

In the synthesis of films with several elemental constituents, it is much simpler in principle to sputter from a single multicomponent target than to use multiple sources. The single-target process is generally not simple as it is necessary to modify the target composition in order to achieve the desired film composition, as the latter is usually different from that of the target. The origin of this complication is that employing a single, flat polycrystalline target, film composition is actually a function of substrate location relative to the normal to the plane of the target. In the case of a target with only two constituents, there must always exist an angle at which a substrate can be positioned which will result in the target composition being retained by the deposit. However this is not necessarily true for targets with three or more constituents, which

* Department of Electrical Engineering
** School of Physics and Astronomy

is unfortunately the case for all of the high-T_c superconductors.

Conservation of matter, and what is known about the sputtering process can be used to guarantee that target and deposition are exactly identical[3]. The main reason for deviations between the atomic compositions of the target and substrate is that different atomic species are ejected from the target with different angular distributions. This had been demonstrated experimentally in studies of sputtering from flat polycrystalline targets immersed in a low pressure plasma like large negative Langmuir probes. The composition analysis of deposits collected under different ejection angles showed that even deposits of the same metal, differing only in isotopic mass, have slightly different angular distributions, indicating that the masses of the atoms play a major role.[4] However, the total flux of ejected material when integrated over all ejection angles adds up to the exact target composition. The actual angular distributions of various constituents sputtered from a multi-component target are a function of bombarding ion energies, as recently confirmed in Ref. 5. In both Refs. 4 and 5 it was found that the general trend at low energies (<300 eV) is that under perpendicular bombardment the lighter atoms are more preferentially ejected in a direction normal to the target surface. This is most likely the result of lighter species being reflected at the head of the cascade by heavier atoms underneath but not vice versa.

The way the composition of the total sputtered flux becomes that of the target is a dynamical process involving the formation of an "altered layer" within the target near its surface. In this layer, from which the sputtered atoms originate, the concentration of the low sputtering rate constituents is sufficiently enriched that the sputtered flux composition is identical to that of the target.. A precondition for this to occur is that the "altered layer" is not replenished from the bulk. If the latter were the case, then there would be no difference between evaporation and sputtering. The argument relating to conservation of matter does not apply to volatile, non-sticking species which are pumped away or are gettered in other parts of the apparatus.

When gas pressures in the sputtering apparatus are increased so that the mean-free paths of the sputtered atoms become shorter than the target-to-substrate distance, the angular distributions of the constituents play less of a role and in effect the angular distribution is integrated over by scattering of the sputtered atoms during travel. However the sputtering rate suffers with higher gas pressure because sputtered atoms are scattered back to the target. The net result of the above considerations is the fact that the location of the substrate relative to the normal to the target is a very important parameter and that optimum locations are a function of ion energy and that they may also change as the target surface topography evolves as the surface is eroded.

EXPERIMENTAL TECHNIQUES

All of the above difficulties disappear if one uses an unconventional approach to sputtering, namely the use of a spherical rather than a flat target. Conservation of matter together with spherical symmetry guarantees that every point at the substrate, independent of distance form the target, receives a deposit with the composition of the target. The reason for this is that the collected material contains atoms which were ejected under all possible ejection angles, resulting, in effect, in an automatic integration. The exact duplication of the target composition at the substrate depends on the fulfillment of several preconditions. These include the following:

1. The metal stem, which supports the spherical sputtering target and provides electrical connection or cooling, disturbs spherical symmetry. Only those substrates which are located out of line-of-site of the stem-to-target connection receive the right composition.

2. The plasma in which the sphere is immersed has to be of uniform density such that the ion bombardment current density is uniform over that part of the sphere which contributes to the desired deposit. If the target is used as the cathode in an abnormal diode glow discharge, this condition is met automatically.

3. The target surface must reach its steady state equilibrium composition, which is different from the bulk. This must occur before a film is collected. This requires pre-sputtering with a shutter.

4. The target must remain below the temperature at which diffusion in the bulk would replenish those constituents which are preferentially sputtered. For high sputtering rates this condition will require cooling of the sphere.

5. The target material should be polycrystalline with no preferential orientation. Alternatively it could be amorphous.

6. The use of magnetic fields for plasma enhancement, although generally desirable, is likely to create unsymmetrical conditions and thus must be avoided.

Diode and magnetron sputtering from flat cuprate targets, either dc or rf, has been the subject of many publications. References 6 -17 list some of the more recent ones and they will provide guidance to many others.

Figure 1 depicts schematically the situation when a flat cuprate target might be

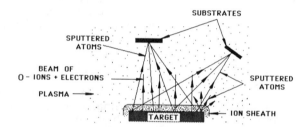

Fig. 1. Schematic of sputtering from a flat target.

sputtered in a low pressure plasma. From the flat surface, neglecting the fringe area near the edges, the mostly neutral sputtered atoms would be ejected generally in a cosine-like distribution. The angular deviations from this for different species such as Bi, Tl, Y, Ba, and Cu which have large mass differences have been widely neglected or overlooked and they are not pictured in the figure. The only work dealing with this subject is that of D. Burbidge et al.[17] who determined using both optical emission spectroscopy and Energy Dispersive X-Ray Diffraction (EDX) analysis of the deposited films that Y is ejected more normally, but Ba more obliquely from a flat surface.

Figure 1 depicts another important feature, namely that in Ba or Sr-containing cuprates some fraction of the oxygen atoms leave as O^- ions which are accelerated in the target ion sheath to the full sputtering voltage to form a beam normal to the target surface. The same of course is true for the electrons released by the ion bombardment. This O^- ion beam is able to re-sputter material from an "on-axis" substrate[7], while the electrons, unless magnetron sputtering is used, contribute to heating the substrate. The necessity of positioning the substrate "off-axis" was recognized early as being essential to achieving acceptable results.[15] The problem with O^- ions would of course not exist if neither the target nor the sputtering gas contained any oxygen or other strongly electronegative elements.

The sputtering studies involving high-T_c superconductors described in the following, evolved from studies of the epitaxial growth of Si films on Si wafers, directed at replacing the usually required high substrate temperatures with low energy ion bombardment.[6] What makes our approach different from others is the use of low pressure Hg triode plasmas for sputtering. These plasmas are created with low voltage, but high current triode discharges in which cathode spots on liquid Hg pools provide an essentially unlimited supply of electrons for plasma formation. These discharges had been the subject of intensive studies in the twenties by Langmuir, Tonks, A. W. Hull, Dushman, Penning etc., and became very popular as rectifiers for handling thousands of amperes, or in the form of ignitrons for controlling dc currents. Their attractive trait is that low discharge voltages are sufficient for sustaining a vacuum arc.

A dc triode discharge combines simplicity with the advantage that important sputtering parameters such as bombarding current density, ion energy, and gas, or vapor pressure can be independently controlled. Furthermore, compared with rf plasma techniques, the measurement of plasma potentials or the substrate biasing voltage is simpler and more straightforward.

The advantages of a Hg plasma over a noble gas plasma are explained in more detail in Ref. 18. Not only does the cathode spot on the liquid Hg pool provide an unlimited electron supply for plasma formation at low discharge voltages, but vapor pressures are very simply controllable with temperature, and Hg atoms, with their high mass and low ionization potential are desirable in sputtering. The often raised concern about Hg contamination in sputter-deposited films is unwarranted because Hg behaves in many respects like a noble gas. As long as substrates are kept at a temperature above about 300^0 C during deposition we were not able to detect with SIMS or SNMS any traces of Hg, not even in gold films which are known to amalgamate readily

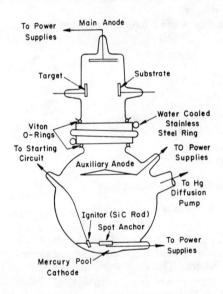

Fig. 2 The Hg triode discharge tube with electrons supplied from a spot on an Hg pool

Figure 2 shows the essential features of the discharge tube. It consists of a demountable Pyrex tube with a lower part which contains a Hg puddle, a SiC cathode spot igniter, a Mo sheet cathode spot anchor, and an umbrella-like auxiliary anode. The latter is used to maintain the cathode spot with a 3A current, and at the same time shields the upper plasma from Hg droplets which might be ejected from the cathode spot. The lower part of the tube is connected to a 12 l/s Hg diffusion pump and sits in a water bath with controlled water temperature for setting the vapor pressure of the plasma. The upper part is pressed to the lower part by the outside atmospheric pressure with the help of a water-cooled stainless steel connecting ring which holds in place upper and lower grooves for Viton O-rings. The upper section contains the main anode and opposing feedthroughs for target and substrate holders. The water temperature was kept at about 15^0 C which sets the Hg vapor pressure at about 5×10^{-4} Torr. After igniting the cathode spot the operating voltage between anode and cathode is about 23V when 4A of main current are drawn. This voltage, however, changes very little if one demands much larger currents because the cathode spot provides practically unlimited current.

APPLICATION TO $YBa_2Cu_3O_{7-x}$

$YBa_2Cu_3O_{7-x}$ and related cuprate compounds provide an ideal case for experimental verification of the advantages of spherical targets over planar targets because the superconducting transition temperature is very composition sensitive and thus is a good diagnostic for the successful transfer of material from the target to the substrate. Furthermore a spherical target has a specific advantage over a planar one in that O^- ions and electrons are ejected radially and thus as was mentioned, any resputtering would be distributed over the full solid angle around the target as illustrated in Fig. 3.

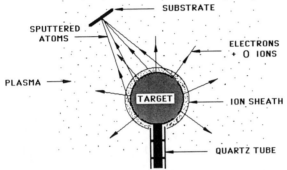

Fig. 3. Schematic of sputtering from a spherical target.

Preliminary studies with small $YBa_2Cu_3O_{7-x}$ flat target discs (1 cm. dia.) and substrates in "off-axis" positions demonstrated the importance of substrate location relative to the target as shown in Fig.4. It was desirable to gain additional information on the actual angular distributions of sputtered Y, Ba, and Cu atoms under actual operating conditions. These were a bombardment by 300eV Hg ions with ion currents in the mA/cm^2 range. With an Hg vapor pressure of only 5×10^{-4} Torr, the mean free path of sputtered atoms is at least five times larger than the distance between target and substrate and therefore interfering gas collisions should become negligible. The target disc was mounted such that its flat side faces a Ti ribbon, which was bent into a semicircle for catching at equal distances from the target the material sputtered at different ejection angles. The result is shown in Fig. 5. In agreement with results obtained under quite different operating conditions (7 cm. dia. magnetron sputtering target and 8 mTorr Ar pressure) in Ref. 17, it is found that Y and Cu are ejected preferentially in the normal direction, while the heavier Ba atoms are ejected more obliquely. If the goal is to achieve a substrate composition with the 1:2:3 ratio of constituents an angle of about 40^0 to the normal would be an ideal substrate location. However one should bear in mind that the observed distributions are influenced by many different parameters, which are not necessarily controlled, such as ionic species, target temperature, ion energy, and surface topography and thus are characteristic of the specific conditions and are not necessarily general. The need for a method to integrate over all possible ejection angles led us to the idea of using a spherical target. This was tested by converting a 1-2-3 target disc using a grinding wheel into a crude hemisphere and a film with the R(T) curve shown in Fig. 6 was

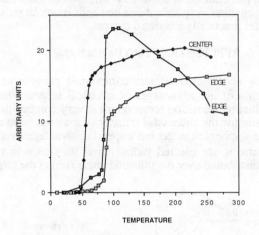

Fig. 4. R(T) curves of films sputtered from flat "1-2-3" targets.

obtained.[19] The target for this work was a 5 mm radius hemisphere which was wedged into a quartz tube and connected by a conducting spring to a vacuum feedthrough. The substrate was a $SrTiO_3$ crystal oriented with the normal to its plane ([100] direction) aligned with the symmetry axis of the hemisphere. The use of a spherical target, which was not yet available, would have been more ideal in achieving all the advantages of this geometry, and in addition would have permitted the positioning of substrates at locations other than this specific symmetry point. The target-to-substrate distance was 1.5 cm. The film was prepared in a Hg vacuum arc with a main anode-cathode current of 4 A and a voltage drop of 23V. The sputtering voltage applied to the target was -300V with respect to the anode potential. The latter is close to the plasma potential in the vicinity of the target or substrate. The target drew a Hg ion current of 3 mA. The floating potential which every insulating substrate automatically acquires was minus 18V with respect to the plasma. The deposition rate from such a small and uncooled target was low, and an overnight run of 17 h. was required to obtain a 9000 Å thick film. During deposition the substrate temperature was about $300^0 C$. The films, as prepared, were black, shiny, and insulating. A conventional post-deposition heat treatment in flowing oxygen resulted in ordered and superconducting films. This treatment involved quick heating to $920^0 C$, a temperature which was maintained for 4 min., followed by slow cooling down to room temperature

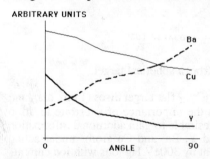

Fig. 5. Angular distribution of "1-2-3" species sputtered from a flat target

at a rate of 4°C/min. The X-ray diffraction pattern of one of the films obtained using a Cu-K$_a$ source and a graphite monochromator is shown in Fig. 7. All of the principal

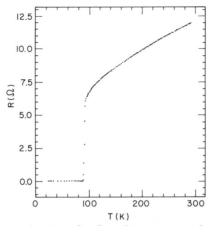

Fig. 6. R(T) curve of a YBa$_2$Cu$_3$O$_{7-x}$ film sputtered from a hemispherical target.

peaks of the YBa$_2$Cu$_3$O$_{7-x}$ (1-2-3) structure as well as the diffraction lines of the SrTiO$_3$ substrate are clearly visible, and no impurity phases can be discerned. From the data one can conclude that the film contains a mixture of domains with both the a and c-axes normal to the plane. The absence of impurity peaks permits one to set an upper limit on the concentration of a given impurity of about 5%.

As often occurs in studies in which many important parameters are relevant, but not controlled, it turned out not to be easy to reproduce the 89K transition temperature in subsequent runs with the same target. This is most likely related to the fact that our apparatus possessed neither a load-lock, nor a shutter, which may be necessary in order to avoid air exposure of the target and for permitting long pre-sputtering (for achieving equilibrium composition in the altered layer) before deposition is begun.

Studies involving Auger analysis were performed which have a bearing on the nature of the previously mentioned altered layer at the target's surface. A bulk sample of the 1-2-3 compound was cleaved in vacuum and various points were Auger analysed as cleaved. As expected, different areas ranged in surface composition being very rich in one of the constituents. Subsequently about 2000Å of material were removed by bombardment with 3.5keV Ar ions. Auger analysis indicated a tendency for the ratios of the various components to equalize after this process. However it was found necessary to remove substantially more material (up to 30mm) before an equilibrium composition was established over the whole surface. As a result of comparison with a standard, it was found that Cu was substantially depleted and Ba was somewhat enriched compared to the ratios in the flux leaving the face. The ratios turned out to be 1 :2.2 :2.1 for Y: Ba: Cu, respectively. One might expect somewhat different ratios for 300eV Hg ions compared with 3.5keV Ar ions, but it is likely that there is a similar alteration of the composition of the surface layer as a result of sputtering.

APPLICATION TO Bi-Sr-Ca-Cu-O COMPOUNDS

Recently Bi-Sr-Ca-Cu-O films were grown on MgO and SrTiO$_3$ substrates. Targets with component ratios close to 2223, 4324, and 4334 were prepared by sintering the correct mixtures of Bi_2O_3, $SrCO_3$, $CaCO_3$ and CuO at 800^0C for 24 h in O_2. After sintering the powder was pressed isostatically at 50000 psi to form slugs

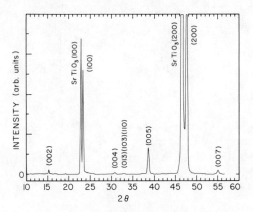

Fig. 7. X-ray diffraction pattern of film of Fig. 6.

which were then fashioned into targets of various shapes.

The first task was to study the angular distributions of the component elements sputtered from a target with a flat surface, as had been done in the case of the 1-2-3 compounds. It was found that the same mass trend appeared. Specifically, Ca, which has the lowest atomic mass of the four elements was more preferentially sputtered in the direction normal to the flat target's surface relative to the heavy atoms of Bi which were sputtered more obliquely.

Then spherical targets with a 1.5 cm dia. were ground. Using them, and a bombardment with 300eV Hg ions and a current density of $1mA/cm^2$, a deposition rate of about 2000Å/h was achieved. The resultant 5000Å thick films had compositions

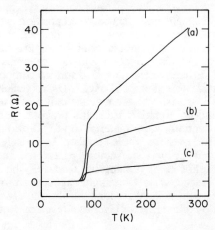

Fig. 8. R(T) curves of several films deposited from spherical BiSrCaCuO targets

which were identical to those of the targets from which they were sputtered. The films were black, shiny and non-superconducting. A post-deposition heat treatment in O_2 was performed by heating for 20 min at 865^0C and subsequently cooling down to room temperature at a rate of $5^0C/min$. This processing resulted in some minor loss of Bi and Sr. Film quality was found to be very sensitive to the maximum annealing temperature as observed by others.[20] The temperature T_c at which zero resistance was observed varied between 60 and 78K, depending on the annealing temperature.

Measurements of R(T) for three representative films grown on [100] MgO crystals are shown in Fig.8. Samples a, b, and c, were sputtered from 4334, 2223, and 4324 targets respectively, Sample a shows a deviation from metallic behavior at 110K in addition to the major drop in resistance around 85K.

A general impression is that the Bi cuprates are easier to produce in the form of films than the 1-2-3 compounds. They also appear to be less sensitive to water vapor exposure. However single-phase films may be harder to produce than in the case of the 1-2-3 compounds.

DISCUSSION

The work discussed above all involves the use of a Hg ion plasma. In principle it should be possible to produce films in a similar manner using more conventional noble gases. The employment of spherical targets should in this instance ensure the formation of stoichiometric films from stoichiometric targets. Recent work in McClanahan's group at Battelle NW[21] has demonstrated this possibility. The apparatus employed is shown schematically in Fig. 9. The cathode for the supply of

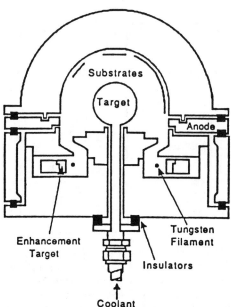

Fig. 9. Schematic of sputtering apparatus using a spherical target in a Xe triode plasma (from Ref. 22)

electrons to the discharge is a tungsten thermionic filament onto which a low electronic work function rare earth metal film is deposited by sputtering during operation. A 5A Xenon discharge was operated with a 25V discharge voltage at a gas pressure of only 1.4 mTorr. The target was a 5 cm. dia. internally-water-cooled Cu sphere which was plasma-spray-coated with a 0.6 mm thick layer of YBaCuO material. The target was sputtered using 300eV Xe ions at a current density of 2.5mA/cm^2. The hemispherically arranged substrates floated electrically at -24V with respect to the anode. A deposition rate of 0.7 mm/h was achieved.

Table I shows that the compositions of the films collected at angles ranging from -10^0 through +90^0 are identical to that of the target. This angular spread would correspond to a planar area of about 200cm^2. The films were not of high quality because the target was actually far from stoichiometric and contained too much Cu. Nevertheless there are a number of important conclusions having a bearing on the problem of fabrication high temperature superconducting films which can be drawn from this work.

First it is clear that the use of an Hg plasma is not a critical feature in the success of the method. It is fairly certain that Hg can be replaced by Xe and probably by any other noble gas. Second, the work at Battelle demonstrates that one can achieve acceptable deposition rates with larger water cooled target spheres. Finally the Battelle work confirms the general idea that in the use of a spherical target film compositions are identical to target compositions. This is true substrate areas which would collect deposits from a fully spherical target surface.

Table I Deposition Composition versus Position (Latitude) in the Coater (from Ref. 22)

Position (deg.)	Composition (at. %)		
	Y	Ba	Cu
90^0 (pole)	12	31	57
70^0	12	30	58
50^0	12	31	57
40^0	13	31	56
30^0	13	30	58
20^0	13	31	56
10^0	12	31	57
0^0 (equator)	12	32	57
-10^0	13	31	56
Target	12	32	56

There is no doubt that a spherical target in a low pressure triode sputtering arrangement eliminates many of the problems which arise using flat stoichiometric targets in the deposition of the cuprate high temperature superconducting compounds. In addition the spherical target approach may turn out to be most useful in other cases where one wants to prepare multicomponent films by sputter deposition and preserve the chemical composition of the target in the resultant films.

SUMMARY

High quality stoichiometric cuprate films of both the YBaCuO and BiSrCaCuO varieties have been grown by sputtering from stoichiometric targets of spherical shape using a Hg vapor dc triode plasma. The low vapor pressure plasma (5×10^{-4} Torr) was sustained with electrons emitted from the cathode spot on a liquid Hg pool. In this way it was possible to create a low voltage (25V) high current vacuum arc. The material sputtered from a spherical instead of a flat target replicates the exact target composition. This is accomplished by automatic integration over all ejection angles which have different angular distributions for the different atomic species in the cuprates.

ACKNOWLEDGEMENTS

The authors would like to thank Drs. M. L. Mecartney and Y-J. Zhang for their helpful discussions relating to the characterization of these materials. They would like to thank Paul Dennis of the Physical Electronics Division of Perkin-Elmer Corporation for his assistance in the Auger studies. This work was supported in part by the Central Administration of the University of Minnesota and by the Air Force Office of Scientific Research under Grant 87-0372.

REFERENCES

1. J. R. Gavaler, J. Vac. Sci Technol. 12, 103 (1975).
2. H. H. Zappe, in "Advances in Superconductivity, Edited by B. Deaver and John Ruvalds, (Plenum Press, New York and London, 1982), p.51.
3. G. Betz and G. K. Wehner, Topics in Applied Physics, Vol. 52, Sputtering by Ion Bombardment II, Chapter 2: Sputtering of Multicomponent Materials, (Springer-Verlag, New York, 1983).
4. R.R. Olsen, M. E. King and G. K. Wehner, J. Appl. Phys. 50, 3677 (1979).
5. W. Wucher and W. Reuter, J. Vac. Sci Technol. A 6, 2316 (1988).
6. K. Wasa, M. Kitabstake, H. Atachi, K. Setsune, and K. Hiroshi, Am. Vac.Soc.Series 3, Thin Film Processing and Characterization of High Temperature Superconductors, in Am. Institute of Physics Conference Proc. #165, New York, (1988), p.38.
7. S. I. Shah and P. F. Carcia, ibid, p 50
8. J. Argana, R. C. Rath, A. M. Kadin, and B. Ballentine, ibid, p.58.
9. Z. Han, L. Burge, H. Li, M. Ulla, W. S. Millman, H. P. Baum, M.R. Xu, P. K. Sarma, M. Levi and B.P. Tonner, ibid, p.66.
10. D. C. Bullock, D. C. Rettner, V. Y. Lee, G. Lim, R. J. Savoy, and D.J. Auerbach, ibid, p. 71.
11. H. Liuo, M. Hong, B. A. Davidson, R. C. Farrow, J. Kwo, T. C. Hsieh, R. M.. Fleming, H. S. Chen, F.C. Feldman, A. R. Kortan, and R. J. Felder, ibid, p79.
12 R. L. Sandstrom, W. J. Gallagher, T. R. Dinger, R. H. Koch, R. B. Laibowitz, A. W. Kleinsasser, R. J. Gambine, B Bumble, and M. R.Chisholm, Appl. Phys. Lett. 53, 444 (1988).
13 H. C. Li, G. Linker, F. Ratzel, R. Smithey, and J. Geerk, Appl. Phys. Lett 52, 1098 (1988).

14. Brian T. Sullivan, N. R. Osborne, W. N. Hardy, J. F. Carolan, B. X. Yang, P. D. Michael, and R. R. Parsons, Appl. Phys. Lett. 52, 1992 (1988).
15. V. Y. Lee, J. Salem, V. Lee, T. Huang, R. Savoy, V. Deline, and J. Duran, Appl. Phys. Lett. 52, 2263 (1988).
16. T. I. Selinder, G. Larsson, U. Helmersson, P. Olsson, and J. E. Sundgren, Appl. Phys. Lett. 52, 2263 (1988).
17. T. Burbidge, P. Mulhern, S. Dew, and R. Parsons, Am. Vac. Soc. Series 3, Thin Film Processing and Characterization of High Temperature Superconductors, in Am. Institute of Physics Conference Proc. #165, New York, 1988, p 87.
18. G. K. Wehner, R. M. WarnerJr., P. D. Wang, and Y. H. Kim, to be published in the 15 December issue of J. Appl. Phys.
19. G. K. Wehner, Y. H. Kim, D. H. Kim, and A. M. Goldman, Appl. Phys. Lett.52, 1187 (1988).
20. J. H. Kang, R. T. Kampwirth and K. E. Gray, Phys. Lett. A 21, 102 (1988).
21. J. P. Prater, E. D. McClanahan, and W. J. Weber, Electronic Materials and Processes, Vol. 2, Society for the Advancement of Material and Process Engineering, Covina Ca 91722 (1988).
22. E. D. McClanahan and R. W. Moss, U. S. Patent 4046666(1977).

Deposition of 1-2-3 Thin Films Over Large Areas

P. N. Arendt, N. E. Elliott, R. E. Muenchausen,
M. Nastasi and T. Archuleta
Los Alamos National Laboratory, Los Alamos, NM. 87545

ABSTRACT

Thin films (0.5-2.0 μms) of $REBa_2Cu_3O_{7-x}$ (where RE=Er,Y) have been deposited by magnetron sputtering and electron beam coevaporation using RE, Cu and BaF_2 starting materials. Post deposition annealing gives films that are fully superconducting between 80-90°K ,on $SrTiO_3$ <100> substrates, as measured by four point probe. Film crystallinity and morphology are examined by x-ray diffraction (XRD) and scanning electron microscopy (SEM) as a function of annealing conditions. The large target-substrate distance of 35 cm employed in the sputtering system allows for stoichiometric deposition to within 5% over a 7.5 cm radius as verified by Rutherford backscattering spectrometry (RBS). The source-substrate distance of 25 cm employed in the electron gun system results in stoichiometric deposition within 5% over a substrate radius of 4 cm.

In separate experiments, YBaCuO films, RF magnetron sputtered from a 1-2-3 oxide target, have been deposited at various pressures and target-substrate distances with the resulting film stoichiometry being measured by RBS. Films of strontium titanate and Pt have also been deposited in an attempt to obtain oriented, large area substrates onto which our 1-2-3 films may be deposited. Results of these experiments as a function of deposition parameters are presented.

INTRODUCTION

Thin films of REBaCuO superconductors have been successfully synthesized by a variety of methods[1,2]. Precise control of the film stoichiometry is one of the parameters necessary to obtain high superconducting transition temperatures[3]. In the case of sputter deposited films by multi-target DC magnetron[4], or by RF magnetron using a composite oxide target[5], this is accomplished by precise control of deposition conditions.

A potential application for HTS films is in RF linear accelerating cavities. For this application, good quality films must be deposited over large areas and curved surfaces. To this end, we have investigated the deposition of films of 1-2-3 materials over large areas using both a three gun magnetron and a single gun magnetron sputter system. The best HTS films are deposited onto single crystal substrates but single crystals are not suitable for our projected application. Thus, we are also investigating thin film substrate materials onto which high quality superconducting films may be deposited.

EXPERIMENTAL SET-UP

The three gun sputter deposition system has a simple load lock for sample introduction onto the rotating substrate platform. This is used to minimize the presputter time needed to condition the rare earth target. When a new target is installed, the power levels to each target are varied and the film stoichiometry from each run is measured by RBS. This process requires several runs to determine the power levels that are needed at each target to obtain 1-2-3 stoichiometric films. Table 1 describes the deposition parameters used to deposit the films as well as the post deposition anneal cycle.

Table 1. Process parameters for three gun sputter system

Targets		10cm-Er,	Cu,	BaF_2
Input power (W)		60 (DC)	31 (DC)	250 (RF)
Target-substrate distance (cm)			35	
Sputter gas/pressure			$Ar/1.5 \times 10^{-3}$ torr	
Film thickness (nm)			200-1000	
Deposition rate (nm/min)			20	
Post anneal:	Heating	Temp.(°C)	Time(hrs)	O_2
	ramp	850	.25	dry
	soak	850	0.5-1.0	wet
	ramp	750	1.0	wet
	ramp	500	3.0	dry
	soak	500	0-5	dry
	ramp	25	2-18	dry

The coevaporation system employs three electron guns in an oil-free, UHV chamber with a source-substrate distance of 25 cm. The substrates are rotated on an 8 cm diameter platform. Typical deposition rates are 120 nm/min at 5×10^{-8} torr. The deposition from each gun is controlled by independent quartz crystal oscillators which are used in a closed loop mode to control the input power to each electron gun. The film stoichiometry is determined by RBS.

The single gun sputter deposition system utilizes composite oxide targets. If desired, the substrates may be rotated and heated. RBS is used to determine film stoichiometry as a function of the deposition parameters. Table 2 describes the deposition parameters used to deposit the films.

Table 2. Deposition parameters for single gun system

Targets	$10cm-SrTiO_3$,	$YBa_2Cu_3O_x$
Input power (W)	200(RF)	200(RF)
Target-substrate distance (cm)	2-5	1-6
Sputter gas/pressure	$Ar/1 \times 10^{-3}$ torr	$Ar/.025-1 \times 10^{-1}$ torr

The structure and morphology of the as deposited and the annealed films, was determined by XRD and SEM.

RESULTS

The stoichiometries of the films deposited by the three gun system are shown for a typical run in Table 3. The stoichiometries are measured by RBS and are normalized to the rare earth element. The power levels for each gun are determined by: fixing the system pressure; fixing the power level of one gun (barium fluoride); varying the power levels of the other two guns; and measuring the sample stoichiometries by RBS. Once an acceptable set of values (within 5% of the desired 1-2-3 composition) for the stoichiometries is obtained, the power levels to each gun are kept constant from run to run. As the targets erode with time, the stoichiometries are rechecked and, if needed, small changes are made to the gun power levels.

Table 3. Stoichiometry vs radial position in three gun system

Radial position (cm)	Stoichiometry Y:Ba:Cu
2.3	1:2.02:3.04
5.0	1:2.03:3.06
7.7	1:1.95:2.98

The stoichiometry of the deposited films from the barium fluoride target is $BaF_{1.6}$. This amount of fluorine indicates 80% molecular sputtering of barium fluoride which is sufficient to routinely get good quality films (superconducting temperatures above 80°K) on <100> strontium titanate using the anneal protocol of Table 1.

An attempt was made to optimize some of the materials and superconducting properties of the films by varying the anneal protocol of Table 1. Several films were coevaporated, in the same run, onto single crystal <100> strontium titanate substrates. RBS analysis of the as deposited films showed an average stoichiometric ratio of $1:1.98\pm.05:2.95\pm.08$. The anneal protocol outlined in Table 1 was then varied for each of these films in that the uppermost anneal temperature for each film was varied in increments of 50°C beginning at 750°C. SEM showed pronounced differences in the film morphology as a function of the peak soak temperature. All films were annealed at this step in moist O_2 flowing at 250 SCCM for 30 min. Below 850°C, only 0.5-1.0 μm, randomly oriented grains were observed on the film. At 850°C, 1 to 3 μm size needles, growing at right angles to each other, formed an open "herringbone" pattern superimposed on the smaller grains. Above 850°C, the herringbone pattern coalesced again showing only submicrometer sized features.

Table 4 lists the change in film orientation with maximum anneal temperature by comparing the ratios of the intensities of the (007) (c axis perpendicular to substrate surface) to the (200) (a axis perpendicular to substrate surface) peaks for each of these films. (This ratio is labeled r_{72}.)

Table 4. Orientation and superconducting temperature
as a function of maximum annealing temperature

Anneal temperature (°C)	r_{72}	$T_c^{R=0}$ (°K)
750	0	insulating
800	0.05	80.9
850	0.19	93.5
900	0.68	91.9
950	0.26	84.6

It is seen that the strongest c axis orientation occurs for the film annealed at 900°C. Also listed in Table 4 is the temperature at which the films become superconducting. The films annealed at 850°C and 900°C became superconductors above 90°K. The film annealed at 750°C was an insulator and did not become a superconductor. When this latter film was reannealed for 30 min. at 850°C, it became a superconductor at 88.1°K.

Sputtering from a composite 1-2-3, 10 cm diameter, target with the system at different pressures, was also done. The target-substrate distance was varied from 1.8 to 5.0 cm. RBS measurements of the film stoichiometries were done for substrates 1 cm apart along a radius whose center was directly below the target center. For a target-substrate separation of 2.5 cm, it is seen in Figure 1 that the Cu/Y ratio goes to a minimum at radii of 2 and 3 cm at the lowest pressure of 2.5×10^{-3} torr. For the higher pressures, the Cu/Y ratio tends to increase with radial distance. In Figure 2 it is seen that the Ba/Y ratio is a minimum at either the 2 or the 3 cm position for all pressures. The results illustrated here for the 2.5 cm target-substrate separation are very similar for measurements made at other target-substrate distances of 1.3, 3.8, and 5.0 cm. The magnetic field for the 10 cm gun used in this study results in the most intense sputtering plasma occurring between 1.7 and 3.4 cm, which encompasses the positions where these minima are observed. Similar sharp variations in stoichiometry in the dense plasma region of sputter guns have been observed[6] and have been attributed to resputtering by reflected negative oxygen ions from the target.

Rotating the substrate holder and offsetting its center from the target center smooths the variations found in the unrotated cases. Figure 3 illustrates the Ba/Y and the Cu/Y ratios as a function of position on a rotating substrate whose center is offset 4 cm from the target center. It is seen that, for this particular case, the film stoichiometry would be closer to a 1-2-3 ratio if the target stoichiometry were 1-2.3-3.1.

In order to obtain large areas of oriented HTS films, large area oriented substrates are needed. Two options for developing such large area substrates have been investigated by us. One was to sputter deposit oriented films of strontium titanate and the other was to deposit oriented <111> Pt films onto non oriented substrates. This latter has been done[7] on single crystal sapphire substrates and <100> MgO. In-situ sputter deposition of 1-2-3 films, at 650°C, onto these oriented Pt films resulted in c-axis oriented superconducting films.

For the sputter deposition of strontium titanate films, we used a compound target of strontium titanate. Various substrate deposition temperatures, from room temperature to 1200°C, were used in an attempt to obtain oriented strontium titanate films. The strontium titanium stoichiometries in the films were verified by RBS as being within 3% of the desired composition. XRD showed the films deposited below 350°C to be amorphous while the films deposited between 350°C and 1000°C were polycrystalline, unoriented $SrTiO_3$. The films deposited above 1000°C showed a titanium deficiency with the XRD pattern indicating an unoriented $Sr_4Ti_3O_{10}$[8] structure. HTS films were deposited and annealed on 1μm thick $SrTiO_3$ films. These were poor HTS films in that they became superconducting at 55°K and

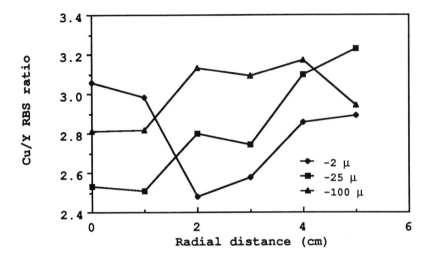

Figure 1. Variation of Cu/Y ratio with position and deposition pressure. Unrotated substrates.

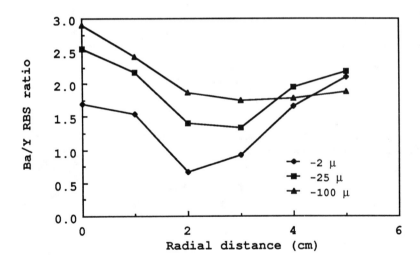

Figure 2. Variation of Ba/Y ratio with position and deposition pressure. Unrotated substrates.

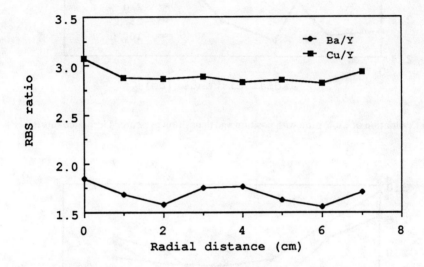

Figure 3. Variation of Ba/Y and Cu/Y ratio with position. System pressure is 25 microns, with rotating substrates. Substrate-target separation is 2.5 cm. Rotator is offset 4.0 cm.

showed a semiconducting behavior before reaching the transition temperature. No HTS films were deposited on the $Sr_4Ti_3O_{10}$ films.

For the Pt film deposition experiments, several substrates were used. These are listed in table 5 and include bulk materials as well as thin films of polycrystalline alumina. The first deposition was a verification of the previously reported results of 100% <111> Pt on single crystal sapphire. An electron beam source was used to evaporate the Pt. In an attempt to remove the requirement of depositing onto a single crystal substrate, Pt was deposited on bulk Ni and on an alumina film. The Pt on the Ni was unoriented. The Pt on the alumina film exhibited roughly 77% <111> orientation. A reaction of the Pt with the Ni is suspected in suppressing orientation. In an attempt to increase the adatom mobility, the substrate temperature was elevated to 1000°C during deposition. No significant change (80%) in orientation was observed, though RBS analysis showed Pt/Al intermixing. Increasing the Pt film thickness also resulted in no significant change (85%) in orientation. In lieu of substrate heating, the adatom energy can also be increased by using ion beam deposition in place of electron beam deposition. The final entry in Table 5 was made by ion beam sputter deposition of the platinum. A quartz glass slide which was near room temperature was used as the substrate. This resulted in 90% <111> Pt film orientation. Thus, films of ion beam deposited platinum have a good fraction of the desired <111> orientation. Subsequent sputter deposition and post annealing of 1-2-3 films on the Pt/sapphire substrates give a poor quality superconducting film with a superconducting temperature of 50°K. The lower temperature in-situ deposition may be the only method of achieving good quality films on the Pt buffer layer. Our experiments of in-situ sputter deposition of 1-2-3 films onto these Pt films are in progress and will be discussed in a future report.

Table 5. Platinum Film Deposition Experiments

Subs./Thickness	Temp. (°C)	Pt Film Thickness	Orientation
Al2O3-SXTAL/1mm	650	200nm	100%
Nickel/1mm	650	200nm	0%
Al_2O_3 Film/100nm	650	200nm	77%
Al_2O_3 Film/100nm	1000	500nm	80%
Al_2O_3 Film/100nm	1000	200nm	85%
SiO_2-Glass/1mm	80	200nm	90%

SUMMARY

A three gun electron beam and a three gun magnetron sputter system have been demonstrated to give stoichiometric depositions, within 5%, of 1-2-3 films over diameters of 8 and 15 cm, respectively. When using a single 10 cm sputter gun and a compound target, the desired stoichiometry may be achieved over diameters of 15 cm if the substrates are rotated and offset and if the sputter target composition is varied from the 1-2-3 composition. Orientation experiments of HTS films, on <100> strontium titanate crystals, as a function of maximum anneal temperature indicate that the greatest c-axis orientation occurs when a 900°C anneal temperature is used. Strontium titanate films deposited at various temperatures are unoriented with titanium loss occurring at substrate temperatures above 1000°C. A second film substrate candidate for 1-2-3 HTS films, <111> platinum, is most easily deposited in a <111> orientation, onto non oriented substrates, using ion gun deposition.

ACKNOWLEDGEMENTS

We wish to acknowledge the Los Alamos Ion Beam Materials Laboratory for supporting the RBS analysis, J. Gray and P. Mombourquette for technical assistance and helpful discussions with R. Springer, F. Garzon, J. Beery and I. Raistrick.

REFERENCES

1. R. H. Hammond, *Phys. Today* **41**(1), S69 (1988).

2. P. M. Mankiewich, et al., *Appl. Phys. Lett.* **51**, 1753, (1987)

3. R. H. Hammond et al., "Superconducting Thin Films of the Perovskite Superconductors by Electron Beam Deposition", Extended Abstract, MRS Spring Meeting, Anaheim, California, April 23-24, 1987.

4. R. E. Somekh, et al., *Nature* **326**, 857 (1987).

5. M. Hong, et al., *Appl. Phys. Lett.* **51**, 694 (1987).

6. L. R. Gilbert, R. Messier, and R. Roy, *Thin Solid Films*, **54**, 129 (1978).

7. S. Hatta, et al., *Appl. Phys. Lett.*, **53**(2), 148 (1988).

8. G. J. McCarthy, W. B. White and R. Roy, *Jour. Amer. Cer Soc.*, **52**, 463, (1969).

SPUTTERED THIN FILM YBa$_2$Cu$_3$O$_n$

Kenneth G. Kreider
Center for Chemical Technology

James P. Cline, Alexander Shapiro
Institute for Materials Science and Engineering
National Institute of Standards and Technology
Gaithersburg, MD 20899

J. L. Pena, A. Rojas, J. A. Azamar, L. Maldonado
and L. Del Castillo, CINVESTAV-IPN, U. Merida
A.P. 73 Cordemex, Yucatan 97310, Mexico

ABSTRACT

This study was carried out to determine the effect of substrate temperature, target to substrate angle, and the partial pressure of oxygen in the sputtering atmosphere on the stoichiometry of the deposition. Films were deposited by a planar magnetron sputtering of stoichiometric 1:2:3 pressed and sintered targets on MgO, ZrO$_2$(Y) and ZrO$_2$ coated Al$_2$O$_3$ substrates. EDX and WDS were used to determine the chemical composition of the thin films. X-ray diffraction was used to identify the structure after film crystallization at 1175K and oxidation at 800K. The partial pressure of oxygen appears to have the most profound effect on the stoichiometry by lowering the barium content. Negative ion resputtering of the growing film apparently also slows film growth rate. Films deposited on alumina circuitboard with a ZrO$_2$ barrier layer have demonstrated sharp transitions to superconducting behavior at 95K when the stoichiometric ratio is preserved.

INTRODUCTION

One of the most useful forms for the new copper oxide-based superconductors is a thin film. This form would be appropriate for computer interconnects[1], magnetic sensors, infrared sensors, and other electronic devices. Thin films of the system Y-Ba-Cu-O have been produced by evaporation[2,3], laser ablation[4] and by ion sputtering[5-13] from both single and multiple targets. Perhaps the simplest of these approaches and the method used in our research is single source sputtering onto a suitable substrate for electronic applications. This approach involves target fabrication from the oxide (or carbonate) powders, planar magnetron sputtering, and post deposition heat treatments to obtain the high temperature superconducting orthorhombic perovskitic phase.

Several critical steps are involved in this processing and

none is more critical than obtaining the correct stoichiometric ratio of yttrium, barium, and copper during the deposition of the film. Problems in uneven transfer of the metallic atoms in the yttrium, barium and copper system have been documented by numerous papers. Michikami[10] et al. and Argana[13] et al. relate losses of copper and barium to substrate temperature and Scheurman[6] et al. relate losses in barium to negative ion resputtering which in turn is sensitive to the angle of deposition. In fact, Cuomo and Rossnagel[14,15] have studied the negative ion effects during magnetron and ion beam sputtering and have related the resputtering of barium not only to the geometric relationship between target and substrate but also the oxygen partial pressure.

Our intent in this paper was to explore the planar magnetron sputtering of YBaCu 1:2:3 thin films and determine the effect of oxygen partial pressure, substrate temperature, and target/substrate geometry on the stoichiometry of the films. In order to confirm the validity of the approach we also studied the crystal structure of films which had been heat treated to obtain superconducting properties and we measured resistance versus temperature to determine Tc.

EXPERIMENTAL

Sputtering targets were prepared from Y_2O_3 (99.9%), CuO (99.999%), and $BaCO_3$ (Tech.) powders by milling, calcining at 1175K, grinding, and firing in air at 1225K to form a strong, flat, 5 cm diameter disc. The targets were sputtered using 115-170 volts and 0.6 amperes direct current in 0.6-1.0 Pa of argon or argon plus oxygen. Substrates included alumina circuitboard (0.5 mm) and yttria stabilized cubic zirconia ceramic. Glass and silicon wafers were also used to insure flat smooth surfaces for compositional analysis in the as sputtered condition. The substrate holders included an aluminum block heater and a water cooled copper block. The post deposition heat treatments were conducted in an oxygen atmosphere and typically included a four hour ramp to 1175K, a one hour hold, and a slow cool to 775K with a one hour arrest to enrich the oxygen content followed by a slow cool to room temperature. Alumina substrates were coated with approximately 0.5 μm of reactively sputtered ZrO_2. Typically 90 min. at 360 V, 0.6 A, and 0.6 Pa Ar plus 20% O_2 was used with a Zr target to coat the Al_2O_3 substrates.

Energy dispersive X-ray analysis was used to measure the composition using a commercial standardless quantitative method. The 20KeV beam size was 0.5 mm except for the focused spot which was 1 μm diameter. We used copper, yttria, and a barium glass as standards to check the accuracy ($\pm$1%) of the method. No corrections for thickness were made because all films were 2.5-6.0 μm thick.

X-ray diffraction analysis was carried out on samples to

determine the phases present and to characterize the nature of their crystallinity. Diffraction geometry included both a standard, $\theta - 2\theta$, scan and an asymmetric, 2θ scan. Both types of analysis were carried out on automated Philips goniometers utilizing Cu Kα radiation and graphite diffracted beam monochrometers. The 2θ scans were conducted on a modified machine equipped with a low speed sample spinner. The modification was such that the θ shaft could be fixed at any desired angle to the incident beam with the detector assembly being allowed to operate normally. The small glancing angle allows for examination of the thin film only, as the penetration of the incident beam is restricted to the surface layer. Samples analyzed in the 2θ geometry were embedded in acrylic sealing compound within the disc shaped sample holder. In order to prevent scattering from the portion of the incident beam not intercepted by the sample, the compound was mixed with either tungsten powder or a finely ground lead glass.

Resistance versus temperature tests were performed in the usual four probe geometry in a liquid nitrogen cryostat with a copper block to which the sample was mechanically attached. The temperature was measured with a type K calibrated thermocouple in mechanical contact with the sample. The electrical contacts were made using silver paint. For the resistivity measurements a 0.7 micro-amp DC current was applied producing a current density of 10^{-3} A/cm^2. The ρ versus temperature tests were performed several times in periods over two months without appreciable change in the results.

RESULTS

The effect of adding oxygen to the 1 Pa sputtering gas is reported in Table I. For sputtering runs with 10% and 6% oxygen added the EDX results clearly show that very little barium remains on the film. The most likely explanation is that the oxygen negative ion is resputtering the barium from the substrate when the oxygen reaches a critical partial pressure under these conditions. The substrate temperatures between 310K and 435K, however, appears to have little effect at the power levels, deposition rate, and substrate distances used in this study. These substrate temperatures were chosen in comparison to the Michikami et al.[2] work in which the barium loss appeared to be a factor of six at 265K.

The effect of geometry between target and substrate is reported for a separation of 10 cm in Fig. 1 and 5 cm in Fig. 2. Although no pronounced effect was measured on the specimen at 1.0 Pa and 10 cm a definite barium depletion was measured on the one which was sputtered at a lower pressure at 5 cm (Table I). The geometric effect is related to the magnetic and electric fields of the magnetron which focuses charged particles along the target-substrate axis leading to resputtering. The depositions for the two specimens at 0.6 Pa included a 76 mm Si wafer substrate on which the central 3 cm spot was clearly visible and a scanning electron micrograph of that area of the sample deposited at 405K is presented in Fig. 3. Low barium in this region would be expected, however, EDX compositional analysis indicated 24.1%, 25.5%, and 25.6% barium in the dimples, on the base, and on the hill regions of Fig. 3. The circular surface features appear to relate to a growth (or erosion) habit rather than a second phase of contrasting composition.

Thin films sputtered on zirconia substrates or alumina substrates with sputtered zirconia buffer layers can be converted to the orthorhombic perovskitic structure. Evidence of this phase formation is given in Fig. 4 and Fig. 5. The pattern displayed in Fig. 4 is of the sputtering run at 415K on an Al_2O_3 substrate collected in 2θ geometry. The stick pattern superimposed on this data is a reference pattern of the orthorhombic BYC superconductor. Tungsten was admixed with the sealing compound giving rise to the broad, squared off, peak at 40.5 degrees two theta. The peak at 59 degrees is also affected by the tungsten.

Due to the low glancing angle of the incident beam, only the thin film is illuminated, thus any pattern originating from the substrate is not observed. A consequence of this geometry is a loss in resolution resulting in the broadening of the peaks. However, a clear correspondence is noted between the reference BYC pattern and observed pattern. This specimen probably consists of a mixture of both the tetragonal and orthorhombic phases, though clear differentiation is prevented by the breadth of the peaks. An impurity peak is observed at 29 degrees two theta, this is associated with a phase that is low in barium relative to the desired BYC phase.

Data shown in Fig. 5 consist of a $\theta - 2\theta$ scan of the 405K sputtering run on an Al_2O_3 substrate. A 2θ scan of the same sample is displaced upwards on the Y axis. The comparison of the patterns collected with the two geometries allows for a clear differentiation between phases associated with the surface film and those associated with the substrate. Examination of the standard scan indicates clearly that the major phase is the orthorhombic BYC superconductor. Of particular interest is the system of peaks located about the 47 degree 2θ region. The tetragonal phase would display one peak, the [200], located at 47.05 degrees. With the formation of the orthorhombic phase this peak splits into the [020], 46.73 degrees, and the [200], 47.58 degrees. The presence of the latter two peaks, and the absence of the tetragonal peak, is clear in the data.

The pattern of the Al_2O_3 is present in the standard scan, but absent in the two theta. The diffuse peak at 29 degrees 2θ is that of cubic yttrium oxide, the appearance of this peak in the 2θ scan indicates this phase is contained in the surface layer. This may be the yttrium oxide boundary phase or, because the film is barium deficient Y_2O_3 may be in the film itself. The breadth of the peaks especially that in the standard scan, indicates this phase is fine grained or poorly crystallized. The peaks located at 36 and 38.5 degrees 2θ indicate copper oxide, CuO, this phase is located in the surface film.

The results of the resistance versus temperature testing of sample B29-A1 are presented in Fig. 6. This sample included 0.8μm of ZrO_2 as a buffer layer over the Al_2O_3 circuitboard substrate and demonstrated metallic behavior above the superconducting transition behavior with a 273K resistivity of 3.5×10^{-4} Ω cm. The results in Fig. 6 indicate that if the conditions as described in Table I are used for a premixed 1:2:3 target sharp transition to superconducting behavior at 95K can be obtained by this process.

CONCLUSIONS

Stoichiometric $YBa_2Cu_3O_{7-x}$ can be sputtered from a 1:2:3 target at 1 Pa argon pressure and 300-435K substrate temperatures. These films can be converted to orthorhombic perovskite on $ZrO_2(Y)$ or zirconia coated alumina substrates by annealing and oxidation. At pressures below 1 Pa some Ba depletion was measured within 2-3 cm of the target axis. The most important disturbances from stoichiometry were found with oxygen addition to the sputtering gas. With 0.1 Pa of oxygen added most of the barium was removed from the film. Stoichiometric films converted to orthorhombic perovskite were measured to have a Tc of 95K.

REFERENCES

1. F. Bedard, The Physical Basis and Applications of High Temperature Superconductivity, Proceedings of Colloqium; High Temperature Superconductivity. Prospects and Challenges, Inst. for Technology and Strategic Research, Washington, D. C., Oct. 9, 1987.
2. R. B. Laibowitz, R. H. Koch, P. Chaudhari, and R. J. Gambino, Phys. Rev. B35, 8821 (1987).
3. A. Mogro-Campero and L. G. Turner, Appl. Phys. Lett. 52(14), p. 1187 (1988).
4. D. Dijkkamp, T. Veukatesan, X. D. Wu, S. A. Shaheen, N. Jisrawi, Y. H. Min Lee, W. L. McLeen, and M. Croft, Appl. Phys. Lett. 51(8), 619 (1987).
5. K. G. Kreider, Sputtered Thin Film Ba_2 Y Cu_3 O_n, Proceedings of Colloquium High Temperature Superconductivity Prospects and Challenges, Inst. for Technology and Strategic Research, Wash., D. C., Oct. 9, 1987.
6. M. Scheuerman, C. C. Chi, C. C. Tsuei, D. S. Yee, J. J. Cuomo, P. B. Laibowitz, R. H. Koch, B. Baren, R. Sninvasan, and M. M. Plechaty, Appl. Phys. Lett. 51(23), p. 1951

TABLE I Effect of film parameters on %Ba

Power W	Temp K	Pres PA	O_2 %	T_s CM	Ba %A
80	345	1.0	0	10	31
90	345	1.0	10	10	1
70	345	1.0	6	10	2
90	415	1.0	0	5	34
90	310	1.0	0	5	32
90	425	1.0	0	5	34
100	435	0.6	0	5	32
60	405	0.6	0	5	33

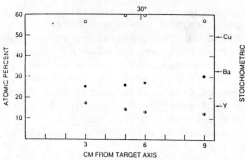

Fig. 1. EDX composition as sputtered at 345K with 1.0 Pa of Argon 10 cm from target. Open circles are Cu, closed Ba.

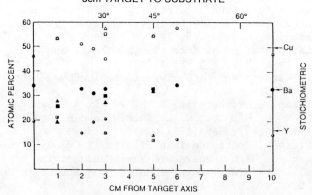

Fig. 2. EDX composition as sputtered 5 cm from target. Circles refer to the specimen at 310K, triangles 435K, and squares 405K (TAble I).

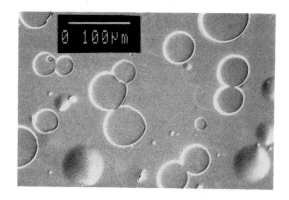

Fig. 3. Scanning electron micrograph of Ba deficient zone in specimen made at 405K, 0.6 Pa Ar.

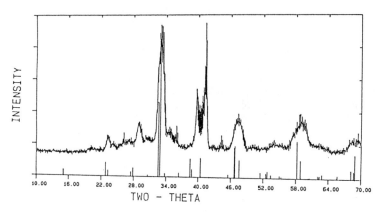

Fig. 4. X-ray diffraction pattern of specimen made at 415K, 1.0 Pa Ar on Al_2O_3 in 2θ geometry.

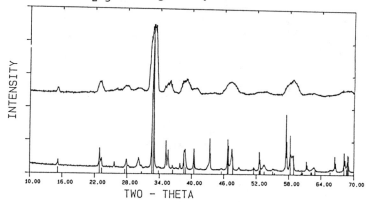

Fig. 5. X-ray diffraction pattern of specimen made at 405K, 0.6 Pa Ar on Al2O3 in both geometries.

(1987).
7. P. H. Kobriu, J. F. DeNatale, R. M. Honsley, J. F. Flintoff andA. B. Horker, Advanced Ceramic Materials, Vol. 2, No. 3B, p. 430 (1987).
8. S. J. Lee, E. D. Rippert, B. Y. Jiu, S. N. Soug, S. J. Hwu, K. Poeppelmeier, and J. B. Ketterson, Appl. Phys. Lett., 51(15), p. 1194 (1987).
9. G. K. Wehuer, Y. H. Kim,. D. H. Kim, and A. M. Goldman, Appl. Phys. Lett., 52(14), p. 1187 (1988).
10. O. Michikama, H. Asauo, Y. Katoh, S. Kubo, and K. Tanabe, Jap. Jul. of Appl. Phys., 26, No. 7, p. L1199 (1987).
11. H. C. Li, G. Linker, F. Ratzel, R. Smithey, and J. Geerk, Appl. Phys. Lett. 52(13), p. 1098 (1988).
12. R. E. Somekh, M. G. Blamire, Z. H. Barber, K. Butler, J. H. James, G. W. Morris, E. J. Tomlinson, A. P. Scharzenberger, W. M. Stobbs, and J. E. Evetts, Nature, Vol. 326, April 1987, p. 857.
13. J. Argana, R. C. Rath, A. M. Kadin, and P. H. Ballentine, AVS Topical Conference on High Temperature Superconductors, Anaheim, CA, Nov. 1987.
14. J. J. Cuomo, R. J. Gambino, J. M. E. Harper, J. D. Kuptsis, and J. C. Weber, J. Vac. Sci. and Tech, 15, 281 (1978).
15. S. M. Rossnagel and J. J. Cuomo, in AIP Conference Preceedings, Thin Film Deposition and Characteristics of High Temperature Superconductors, editetd by J. Harper, L. Feldman, and R. Colton (in Press).
16. Y. Shintarir, K. Nakanishi, T. Takawaki, and O. Tada, Jpn. J. Appl. Phys., 14, 1875 (1975).

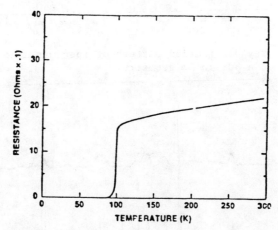

Fig. 6. Resistance of specimen made at 415K, 1.0 Pa Ar.

ION BEAM SPUTTER DEPOSITION $YBa_2Cu_3O_{7-\delta}$: BEAM INDUCED TARGET CHANGES AND THEIR EFFECT ON DEPOSITED FILM COMPOSITION

O. Auciello, M. S. Ameen, T. M. Graettinger, S. H. Rou, C. Soble, and A. I. Kingon, North Carolina State University, Raleigh, NC 27695-7907.

ABSTRACT

Ion beam sputtering is presently used to deposit films from single phase $YBa_2Cu_3O_{7-\delta}$ targets. Generally, Ar+ ion beams (~1500 eV) produced by Kaufman-type ion sources are used for this purpose. It has been observed that these ion beams induce compositional and morphological changes on the polycrystalline ceramic target surface, which results in the composition of sputtered flux displaying a time-dependent behavior. This in turn may lead to undesirably long times for reaching steady state conditions in the sputtering process.

From the literature, it appears that only incomplete studies of these effects have been performed during experiments directed mainly at producing and characterizing high Tc films.

Therefore, the studies reported in this paper have been directed at examining in some detail the effects mentioned above as a function of two important parameters, i.e., ion beam energy and dose deposited in the targets during the sputtering process.

The analysis techniques used to characterize the target changes include electron microscopy, scanning Auger microprobe, and X-ray diffraction techniques. Correlations between target initial conditions and ion-induced changes and film compositions are discussed.

The results indicate copper depletion in the sputtered targets and copper enrichment in the deposited films. The sputter conditions play a critical role in determining the surface topography evolution. No correlations between surface microstructure and film compositions were observed.

INTRODUCTION

In literature on the deposition of Y-Ba-Cu-O films, the difficulty in achieving stoichiometric films has been highlighted.[1-3] This is important as, in contrast to bulk superconductors processed by ceramic processing routes, it has been shown that small deviations from the required overall composition produce dramatically broadened transitions (i.e. large ΔT_c)[4].
This observation also applies to sputter deposition, where there have been numerous reports of nonstoichiometric (i.e. mixed phase) films formed from the sputtering of a single phase, stoichiometric target.[5-8] A number of reasons have been given, and pertinent observations made, to rationalize this phenomenon.

Considering the target first, barium has been reported to sputter preferentially.[9,10] Selinder et al.[3] have shown that as much at .5 μm of target material must be sputtered before a steady state regime is reached ("steady state" refers to the equilibrium condition reached where the sputtered flux is of constant composition, corresponding to the composition of the target). A solution to this problem is to pre-sputter the target, but this is undesirable for commercial application. Reasons for preferential sputtering before achieving steady state have not been explored in these superconductive systems.

Turning to the deposition process, it has been suggested that compositional control of the films is complicated by a low sticking coefficient for copper, particularly at elevated substrate temperatures.

In the case of plasma sputtering techniques (not including ion beam sputter deposition), negative ion effects [11] and electron bombardment [9] have also been shown to result in compositional variations of the film.

The common solution to the above problems which result in compositional deviations from the desired $YBa_2 Cu_3O_{7-\delta}$, has been to (a) operate essentially in the non - steady state regime; and (b) to make an empirical compensation for the deviations. The latter has been achieved by altering the composition of the targets or by sputtering simultaneously from a stoichiometric $YBa_2 Cu_3O_{7-\delta}$ target and an elemental target (for example, copper).

The problem with the above solution is that it is necessarily specific to a given system and set of sputtering conditions. Therefore, the present work is part of a complete study of the factors affecting to compositional changes of the deposited films. The initial objective addressed in this paper is to detail the ion induced surface compositional and topographic changes taking place on the target, and to correlate these with the compositions of the deposited films. The relative sticking coefficients of the sputtered species have not been addressed in this work.

EXPERIMENTAL TECHNIQUE

Figure 1 shows the experimental system used in this study. The experiments were performed in a cryopumped system with a base pressure less than 10^{-7} torr. The ion source was capable of producing beams up to 100 mA and 1500 eV, though typical operating conditions were below these values. Argon was used as an in source gas, such that the target chamber pressure was 8×10^5 torr and 2×10^{-4} torr without and with the ion beam on, respectively. Films were deposited on single crystal MgO substrates or sintered disks of 3% yittria stabilized zirconia (YSZ). The distance between the target and substrate was approximately 10 cm. No substrate heating was used; thermocouples at the substrate position indicated the temperature during deposition did not exceed 40°C. Films were annealed in O_2 at 900°C and cooled slowly in the furnace. Typical film thicknesses were < 2000Å; annealed films possessed a black mirror-like finish.

The targets and deposited films were analyzed by SEM and AES. The Y level detected by Auger was very small compared to the Ba and Cu peaks due to the small Auger sensitivity factor of the Y. Values of Y surface concentration were therefore difficult to ascertain in this study. The relative surface concentrations of Ba and Cu were evaluated on a comparative basis only with the unsputtered target surface as a reference.

RESULTS AND DISCUSSION

Figure 2 a-d shows the evolution of microstructure of the target as a function of beam energy. The beam current was held constant at 4 mA/cm^2 for an exposure time of 30 minutes, resulting in a calculated dose of about 3×10^{20} ions/cm^2. The ion beam was incident on the target at 45°. During the experiments we observed that the initial surface microstructure that may result from different target preparation procedures played a crucial role in determining the evolution of surface morphology during ion bombardment. The factors influencing the final topography include original grain size (dependent on preparation techniques), surface morphology (i.e. polished vs. as sintered surfaces), and the presence of surface contamination. Selinder et al.[3] have reported a slowly changing oxygen content on a sputtered target surface due to both preferential sputtering and bulk diffusion processes. The reactivity of (123) with O_2 and H_2O is well known [12]; the extent of these reactions can influence the response of the surface to ion bombardment. We note that in any study of this nature the past history of the target must be taken into account and controlled to obtain meaningful data.

Our study indicates that sputtering of the (123) surface, as prepared for this particular set of experiments, leads to a striated alignment of surface features dependent not on beam energy but on the direction of incidence of the ion beam (Fig. 2). Both the 800 eV and 1350 eV photos show similar structural features indicating that ion energy is not a dominant factor in topography formation in this voltage or dose regime. These features have been discussed previously by several authors[13,14], and are typical sputtered surfaces. The micrograph of the target exposed to the 1350 eV beam (Fig. 2d) shows some cone formation, a phenomena that has been examined in more detail in the (123) system by Auciello and Krauss[15]. Indeed, we have identified several cone-like formations in our irradiated targets, although the density and morphology of these cones are somewhat different than those observed in reference 15. A possible reason for these differences could be the large variation in ion beam current and energy between Auciello and Krauss' conditions (Ar$^+$,10 keV, ~7μA) and ours (Ar$^+$, .1-1.3 kcV, 10-25 mA). Further studies are necessary to clarify this point.

Table I.

Auger Results of Target and Film Analysis

		Ba/Cu	Ba/O	Cu/O
Unsputtered target surface		1.56	.35	.23
TARGET	1350 eV	2.38	.25	.11
	1000 eV	3.93	.49	.12
	800 eV	1.05	.34	.32
	600 eV	3.46	.37	.11
FILM	1350 eV	.86	.26	.30
	1000 eV	1.22	.27	.22
	800 eV	1.31	.31	.24
	600 eV	1.29	.2	.25

Table I shows the values obtained from a scanning Auger analysis of the virgin and sputtered targets and the corresponding film values. Figure 3 shows a representative Auger spectrum. From this data we can see that the copper level is consistently lower on the sputtered target surface than on a virgin (unbombarded) surface. A general enrichment of copper appears in the deposited films. In contrast, other published results have indicated copper depletion in the deposited films. We suggest that the depletion in these cases are due to altered sticking coefficients of Cu at elevated substrate temperatures, and not preferential sputtering effects. No substrate heating was used in the present study.

The Ba/Cu ratio on the deposited film using an ion dose of ~ 4×10^{20} ions/cm^2 has been plotted as a function of beam energy in Figure 4, indicating a strong dependence of film composition upon beam energy. The target microstructure (Figure 2a-d) does not exhibit this dependence. Presumably the beam is changing the surface composition on the target, although the scatter in the Auger data is too large to observe this effect. Alloy systems and other materials show a large dependency of the depth of the altered layer as well as the magnitude of the composition changes on beam energy.

During the course of the study, several different target morphologies were observed. Observations of the morphology of the targets done with a 5 mA experiment at an exposure of 1 hour showed a different surface topography as seen in Figure 5. The micrograph does not exhibit the surface "erosion" effects observed in the 25 mA experiments. In fact, a form of preferential etching along the grain boundary is observed in this case, suggesting that there may be current thresholds in the sputtering of this material, or alternatively, that these effects may only be observed in the lower current regimes. Auciello and Krauss[15] have observed large amounts of cone formation for bombardment with very low current densities. Figure 5 suggests that if further sputtering were to be done in this regime, severe cone formation would occur. The Auger data of films formed in this regime again indicates a low Ba/Cu ratio.

CONCLUSIONS

We have shown that sputtering (123) films results in copper depletion of the target and copper enrichment of the films in all regimes examined. The beam energy plays a critical role in determining the magnitude of this effect while current seems to play a more deciding role in the surface microstructure that is developed during sputtering. No correlations between target surface microstructure and thin film stoichiometry were observed. Surface erosion, smoothing, cone formation, and preferential etching along the grain boundaries are artifacts observed depending on the processing conditions used.

ACKNOWLEDGEMENT

We would like to acknowledge support from DARPA through contract # N00014-88-K-0525.

REFERENCES

1. J. Fujita, T. Yoshitake, A. Kamijo, T. Satoh, and H. Igarashi, J. Appl. Phys. **64**, 1292(1988).
2. H. Asano, K. Tawabe, Y. Katoh, S. Kubo, and O. Michikami, Jpn. J. Appl. Phys. Lett. **26**, L1221 (1987).
3. T. I. Selinder, G. Larsson, U. Helmersson, P. Olsson, and J.-E. Sundgren, Appl. Phys. Lett. **52**, 1907(1988).
4. J. H. Kang, R. T. Kampwirth, K. E. Gray, A. Wagner, and E. A. Huff, Extended Abstracts, MRS Symposium on High Temperature Superconductors II, p. 209 (1988).
5. W. Y. Lee, J. Salem, V. Lee, T. Huang, R. Savoy, V. Deline, and J. Duran, Appl. Phys. Lett. **52**, 2263(1988).
6. O. Michikami, H. Asano, Y. Katoh, S. Kubo, and K. Tanabe, Jpn. J. Appl. Phys. Lett. **26**, L1199 (1987).
7. Y. Saito, K. Sakabe, K. Ishihara, K. Manaka, and S. Suganomata, Jpn. J. Appl. Phys. **27**, 1103 (1988).
8. S. I. Shah and P. F. Carcia, Appl. Phys. Lett. **51**, 2146 (1987).
9. N. Terada, H. Ihara, M. Jo, M. Hirabayashi, Y. Kimura, K. Matsutani, K. Hirata, I. Ohno, R. Sugise, and F. Kawashima, Jpn. J. Appl. Phys. **27**, 639 (1988).
10. S. I. Shah and P. F. Carcia, Amer. Inst. Phys. Conf. Proc. **165**, 50(1988).
11. S. M. Rossnagel and J. J. Cuomo, Amer. Inst. Phys. Conf. Proc. **165**, 106(1988).
12. G. F. Holland, et al. ACS Symposium Series **351**, 102 (1987).
13. R. M. Bradley and J. M. E. Harper, J. Vac. Sci. Technol. **A6**, 2390(1988).
14. G. Carter, M. J. Nobes, F. Paton, J. S. W. Williams, and J. L. Whitton, Radiat. Eff. **33**, 65(1977).
15. O. Auciello and A. R. Krauss, Amer. Inst. Phys. Conf. Proc. **165**, 114(1988).

INVESTIGATION OF $SrTiO_3$ BARRIER LAYERS FOR RF SPUTTER-DEPOSITED Y- Ba-Cu-O FILMS ON Si AND SAPPHIRE

J.K. Truman, M. Leskela[*], C.H. Mueller, and P.H. Holloway,
Dept. of Materials Science and Engineering, University of
Florida, Gainesville, FL 32611

[*]On sabbatical leave from Department of Chemistry,
University of Turku, SF-20500 Turku, Finland

ABSTRACT

Films of Y-Ba-Cu-O were deposited by RF planar magnetron sputtering from single sintered oxide targets onto silicon and sapphire substrates with and without RF sputter-deposited $SrTiO_3$ barrier layers. The composition of YBCO films was determined by electron microprobe analysis. The effectiveness of the barrrier layers was determined from AES, XRD, room temperature resistivity, and optical and electron microscopy data. Films of $SrTiO_3$ were found to exhibit {111} preferential orientation parallel to {100} Si, but no preferential orientation was observed on sapphire. Silicon was prevented by $SrTiO_3$ from reaching the Y-Ba-Cu-O, but was not a suitable barrier layer since Ba migrated to the Si interface. For sapphire, $SrTiO_3$ prevented interaction between Y-Ba-Cu-O and the substrate, although Al did migrate a limited distance into the $SrTiO_3$.

INTRODUCTION

Most studies which have reported successful growth of superconducting $YBa_2Cu_3O_7$ thin films have utilized single crystal oxide ceramic substrates such as $SrTiO_3$, ZrO_2, and MgO [1,2]. As a result of lattice matching between Y-Ba-Cu-O (hereafter referred to as YBCO) and the substrates, strongly oriented films have been obtained, sometimes with complete epitaxy, particulary on <100> and <110> $SrTiO_3$. However, from a cost and applications point of view, it is more desirable to grow superconducting YBCO films on more affordable and practical substrates such as Al_2O_3 (including sapphire) and particularly Si.

As is now well-documented [3,4], for a YBCO film growth process which employs the typical post-deposition heat treatment in oxygen of ≥800 °C for 1 hour or more, superconducting films are only obtained on sapphire for relatively thick (≥ 1 μm) YBCO films, while films on Si are insulating. The obvious method of preventing deleterious interaction bewteen the YBCO film and the substrate has been the insertion of a barrier layer. On Si substrates the best results have been obtained with a ZrO_2 barrier layer [5], whereas for sapphire improvements in the superconducting properties of YBCO have been obtained for various barrier layers including Ag [6] and ZrO_2 [7]. Curiously, some of these barrier layers, e.g. Ag, are in fact not stable against interaction with the YBCO films. However, their incorporation into YBCO does not seem to cause degradation of the superconducting properties for polycrystalline YBCO. In fact, in the case of Ag, it actually improves properties apparently by lowering the grain boundry resistance.

On many devices however, a conductive barrier layer will be unacceptable. The logical coupling of the benefits of $SrTiO_3$ substrates and the incorporation of a layer between the YBCO thin film lead us to investigate the suitability of $SrTiO_3$ as a barrier between YBCO thin films and (100) Si or (1012) sapphire substrates. The effectiveness of the barrier layer was determined in terms of AES depth profile, X-ray diffraction (XRD), microscopy, and room-temperature resistivity data. To date, no one has reported $SrTiO_3$ barrier layers on Si, while only one group has reported the use of $SrTiO_3$ barrier layers on sapphire [8]. The latter report only made a short mention of the ineffectiveness of $SrTiO_3$ barrier layers, and few details were given. We found that $SrTiO_3$ will limit the reactions bewteen YBCO and Si substrates and stop the reaction between YBCO and sapphire. We also report an epitaxial relationship between $SrTiO_3$ and Si, but this rellationship has limited utility in growing oriented sputter deposited YBCO.

EXPERIMENTAL

The YBCO sputter targets, two inch in diameter and about 1/8 inch thick, were fabricated in-house [9]. The target process used Y_2O_3, CuO, and $BaCO_3$ or BaO_2 powders mixed with acetone and ground to $\leq 5\mu M$ diameter using ZrO_2 grinding media in a nalgene jar. A water/polyvinyl alcohol mixture was then added to the powders as a binding agent. The powders were pressed in stainless steel die using pressures of typically 14,000 psi. Green targets were then sintered in static air in a small box furnace with MgO and Y_2BaCuO_5 supports. The sintering time was typically 12 hours and the temperature ranged from 500 to 950 °C, depending on the desired phases in the target. Targets were then affixed to a Cu backing plate with silver epoxy to ensure adequate cooling.

Both the YBCO and the $SrTiO_3$ films were deposited by RF planar magnetron sputter depostion in the same multisource deposition system, designed and built at the University of Florida, containing two each US RF Guns and US DC Guns [10]. For all depositions the substrate-target distance was 6 cm and the substrates were electrically floating. The ultimate pressure was brought to $\leq 1 \times 10^{-6}$ Torr. Substrates were nominally unheated, although due to plasma interaction the temperature may have reached 150 °C. Silicon and sapphire substrates were prepared by degreasing, chemically etching in 20:1 H_2O:HF and 10:1:1 H_2O:HF:HNO_3, respectively, rinsing in deionized water, and drying with N_2. For $SrTiO_3$ deposition, 100 W RF power was used in conjunction with 10 mTorr total gas pressure, the gas being 10% O_2 and the balance Ar. Early attempts at deposition with 100% Ar sputter gas lead to highly stressed and exfoliated films. The resulting deposition rate was about 7 nm/min and the $SrTiO_3$ thickness was 0.4 - 0.6 μm.

The YBCO films were deposited immediately after the $SrTiO_3$ layer without breaking vacuum. For YBCO deposition, a variety of target compositions and deposition parameters were investigated. However, for the YBCO films used with the $SrTiO_3$ barrier layers in this paper, a Y:Ba:Cu oxide 1:4:4 target composition was used since it gave consistent film composition values (see below). As a result of the 780 °C sintering temperature and the Ba source being BaO, the phases present in the target were $BaCuO_2$ and Y_2O_3 [9]. The sputter gas was Ar, at a pressure of 10 mTorr. The incident RF power was 50 W. The deposition rate was typically 4 - 5 nm/min and YBCO film thicknesses were 0.6 - 1.0 μm.

Samples were removed from the chamber, loaded onto a quartz boat, and placed into a tube furnace within 5 minutes. The typical heat treatment in 80 sccm flowing O_2 was 1 hour heating time from room temperature to 850 °C, 1 hour at 850 °C, and followed by slow cool via natural furnace cooling to below 150 °C in 12 hours (average cooling rate of about 1 °C/minute).

The YBCO film compositions were determined by electron microprobe analysis (EMPA). The morphology of films was determined by SEM and optical microscopy. The crystallinity and preferred orientation of $SrTiO_3$ and YBCO films were determined by X-ray diffraction (XRD) analysis. Interdiffusion of components of YBCO, $SrTiO_3$ barrier layers, and Si or sapphire substrates was studied with AES depth profiles. Room temperature resistivity data was determined by a standard DC four point probe method.

RESULTS AND DISCUSSION

From EMPA data, YBCO films from the 1:4:4 target had a composition of approximately Y:Ba:Cu 1.0:2.0:4.5 with a typical sample to sample variation of Ba/Y and Cu/Y of ±5%. For targets with a Y:Ba:Cu 1:2:3 composition and composed of the 123 phase, our films were not Cu deficient, unlike most literature reports, but were rather quite Ba deficient, in agreement with Rossnagel and Cuomo [11]. Efforts to compensate for the Ba deficiency of the films led to the 1:4:4 target composition. In an effort to reduce the excess Cu in the films, this component was reduced in the target, but so far the film compositions from these targets have been rather inconsistent. Our data does not clearly indicate what mechanism(s) is (are) causing the composition deviations, although oxygen resputtering [11] seems the most likely explanation. Finally, it is clear that the 123 orhtorhombic superconducting phase will react with Si or sapphire substrates since superconductivity is lost when Si or Al is found in the films. While the deposited films were not exclusively the 123 phase, if the $SrTiO_3$ barrier layer restricts reaction when the film consists of approximately 75% 123 phase plus 25% CuO, then our conclusions are still valid. The CuO could not prevent a 123/substrate reaction, since if anything, CuO should be reduced by Si. However, not even a reaction with CuO was observed. The degradation occured by Ba moving from the YBCO, further substantiating that CuO did not interfere with our conclusions.

SrTiO₃/Si and SrTiO₃/sapphire

In the as-deposited condition, SrTiO$_3$ exhibited a very strong preferred orientation of the {111} plane parallel to {100} Si, as shown by XRD data in Figure 1a. This preferred orientation is sensible because the lattice mismatch between the {100} Si and {111} SrTiO$_3$ is only 1.5%. After 850 °C heat treatment, most films became even more strongly oriented, whereas some films lost the preferential {111} orientation, perhaps due to stress relief. Even though approximately 1 nm of amorphous native oxide is expected on Si before deposition of SrTiO$_3$ which could possibly destroy the atom matching required for the epitaxial relationship, Ti and/or Sr would be expected to reduce the native SiO$_2$, especially in conjunction with the impingement of energetic particles on the substrate during sputtering. Thus the establishment of epitaxy is reasonable.

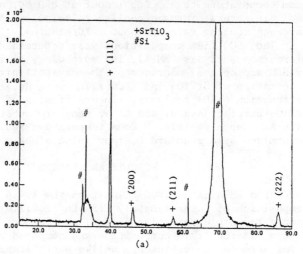
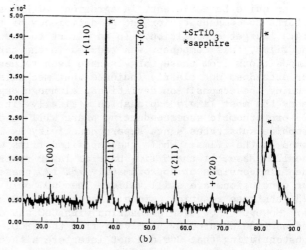

Figure 1. XRD data for (a) as-deposited SrTiO$_3$/Si indicating {111} orientation and for (b) heat treated SrTiO$_3$/sapphire

Furthermore, AES depth profile data of $SrTiO_3$ on Si after heat treatment indicated no intermixing of components at the interface, and particularly no Si could be found in the $SrTiO_3$ film. This was supported by the XRD data in Figure 1a, in which no reaction products and only $SrTiO_3$ and Si are seen. This is in contrast to YBCO directly on Si, where heat treatment caused a strong interaction, with Si migration into YBCO and Ba (and possibly Cu) pile-up at the YBCO interface [3,5]. Hence it appeared that $SrTiO_3$ could prevent interaction of Si and YBCO.

On sapphire substrates the highly transparent $SrTiO_3$ films did not exhibit any preferred orientation in the as-deposited or heat-treated conditions, the latter of which is shown in Figure 1b. The XRD data gave no indication of reaction between $SrTiO_3$ and sapphire. From AES data, a small amount of Al diffused into $SrTiO_3$, but only slightly beyond the interface (roughly 0.05 μm) into the 0.4 μm thick $SrTiO_3$. The extent of this interaction is far less than we and others [1,2] have seen for YBCO on sapphire.

Films of $SrTiO_3$ thicker than 0.5 μm sometimes exhibited microcracks, blisters, and even exfoliation, particularly after heating on sapphire. Even well-adhering 0.4μm thick $SrTiO_3$ films on Si exhibited features which to optical microscopy appeared to be microcracks and blisters, as shown in Figure 2a. The "microcracks" observed with optical microscopy are oriented randomly and criss-cross to form "cells" roughly 10 to 50 μm wide. However, under SEM examination, as shown in Figure 2b, these features were not observed as cracks emerging at the surface. This suggests that they are stress ridges or possibly delaminations at the Si interface which are visible through the transparent $SrTiO_3$ with optical microscopy. The SEM data indicates the $SrTiO_3$ film is continuous and supports the AES data showing it is a barrier layer which resists YBCO/substrate reactions. Stress in the $SrTiO_3$ was reduced by adding O_2 to the Ar, as noted above. Heating the substrate during $SrTiO_3$ deposition may also reduce stress as is known for other thin films [12], but we haven't attempted this yet in our study.

YBCO on Si or sapphire without a barrier layer

Depth profiles of as-deposited samples show all components to be uniformly distributed throughout the layers. Resistivity measurements showed the as-deposited YBCO to be insulating.

After heat treatment of YBCO/Si at 850 °C, in agreement with reported results [3], Ba piles up at the Si interface due to the formation of $BaSiO_3$, as confirmed by AES and XRD data. The Cu-Si interfacial phases reported by Lee et al [5] were not observed. Four point probe resistivity data showed YBCO to be insulating.

For heat treated YBCO/sapphire, XRD and AES data indicate an interfacial $BaAl_2O_4$ layer, in agreement with the literature [4]. The AES data clearly show that Ba piles-up at the substrate interface and the Ba peak shifts as the compound changes from YBCO 123 to $BaAlO_2$. The XRD data also confirmed the presence of the 123 phase and CuO, but no preferred orientation of the 123 was found. Also, Al by AES is found to diffuse into YBCO, but only as far as the

Figure 2.
Optical micrograph (a) of
SrTiO$_3$ on Si showing apparent
microcracks and SEM micrograph (b) of same region
indicating no microcracks
emerging at surface

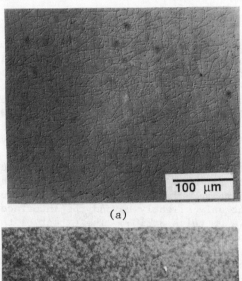

(a)

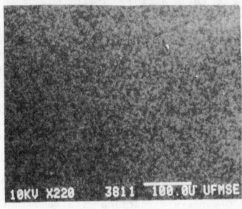

(b)

BaAl$_2$O$_4$ layer, whereas Venkatesan et al. [2] report Al throughout the 123 film. Venkatesan et al. obtained their data from SIMS, whereas our AES signal would be consistent with ≤ 1% Al in YBCO. It is important to note that since the interfacial reaction is limited in depth [4], with a thick enough YBCO film, e.g. ≥ 1 μm, superconducting YBCO can be obtained in the top section of the film.

YBCO/SrTiO$_3$/Si:

In the as-deposited condition, the YBCO films were insulating and AES and XRD data indicated no interaction bewteen YBCO and Si had occured. After heat treatment, as shown in the AES depth profile in Figure 3a, SrTiO$_3$ prevented the interaction of Y, Cu, and Si, but allowed Ba to concentrate at the Si interface. As confirmed by XRD data, Ba and Si reacted to form BaSiO$_3$. The oxygen peak drops suddenly due to sample charging at the SrTiO$_3$ interface. The depth profile also shows that Ba is partially depleted from the YBCO layer, but concentrated at and into the SrTiO$_3$ at both the SrTiO$_3$

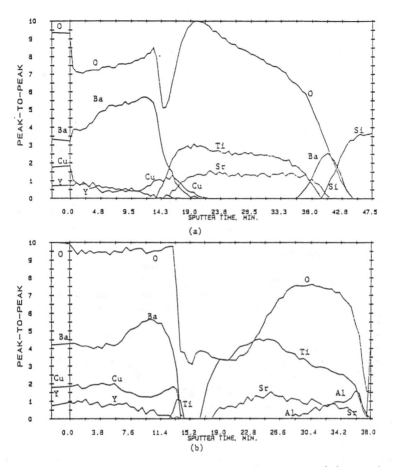

Figure 3. AES depth profiles for heat-treated (a) YBCO/SrTiO$_3$/Si and (b) YBCO/SrTiO$_3$/sapphire

and Si interfaces. This Ba profile is consistent with the fact that BaTiO$_3$ forms a complete solid solution with SrTiO$_3$ [13]. Hence Ba diffused through the SrTiO$_3$, and has not simply passed through the apparent microcracks discussed above. This was confirmed by an AES line scan across 200 μm, which only showed a uniform concentration of Ba or Si, i.e. no enhanced concentration at microcracks. Detection of a Y$_2$Cu$_2$O$_5$ phase by XRD supports Ba depletion from YBCO. The prescence of CuO was also confirmed. Also, both Sr and Ti have migrated into YBCO, in agreement with reported YBCO/SrTiO$_3$ data [4]. Finally, in contrast to the data from YBCO on Si, room temperature resistivity data showed that YBCO with a SrTiO$_3$ layer was much more conducting, with resistivity values ranging from 0.03 to 100 ohm-cm. The XRD data suggest that the 123 phase is present, but overlapping by Si and SrTiO$_3$ peaks confuse this analysis.

Thus, a layer of $SrTiO_3$ does not stop but does limit the reaction between YBCO and Si. This study suggests that successful barrier layer choices for Si should, among other properties, have low solubility for the component elements of YBCO. An example is the ZrO_2 barrier first reported by Lee et al [5].

YBCO/$SrTiO_3$/sapphire:

In the as-deposited condition, the YBCO films were insulating, and AES and XRD data gave no indication of interaction. After heat treatment, as illustrated in the AES depth profile in Figure 3b, $SrTiO_3$ prevented reaction bewteen YBCO and sapphire. No Al reached the YBCO, within the AES detectibility limit of $\approx$ 1%, and no Ba reached the sapphire. This was supported by XRD data in that no $BaAl_2O_4$ was detected and only the prescence of the 123 phase, $SrTiO_3$, CuO, and sapphire were confirmed. Also, no preferred orientation of the YBCO was found by XRD. The AES data show that Al migrated into $SrTiO_3$, but only about 0.20 μm into the 0.4 μm thick layer. The large drop in all the peaks at the YBCO/$SrTiO_3$ interface is due to substrate charging at the interface, and the apparent Ti buildup is likely a resultant artifact. Finally, the YBCO film was found to be conducting with a room temperature resistivity of about 0.02 ohm-cm, supporting the presence of the 123 phase detected by XRD.

Therefore, $SrTiO_3$ is effective in stopping any reactions between YBCO and sapphire. Since YBCO reacts with sapphire over a limited depth, giving the false impression that a reaction hasn't occured, a sufficiently thick YBCO will still be superconducting. The important fact is that $SrTiO_3$ prevents any reaction between YBCO and sapphire. Thus, rather than having part of the YBCO become a sacrificial buffer, the $SrTiO_3$ permits the use of a thinner YBCO film, e.g. 0.5 μm or less rather than the typically reported 1μm or more. Further, since YBCO does not grow in a preferred orientation on sapphire, this potential benefit of film growth on a single crystal substrate is not lost due to the polycrystalline $SrTiO_3$ barrier layer.

SUMMARY

The suitability of RF sputter deposited $SrTiO_3$ barrier layers for use between RF sputter deposited Y-Ba-Cu-O films and Si and sapphire substrates has been investigated. The effectiveness of the $SrTiO_3$ barriers was determined by AES, XRD, optical and electron microscopy, and room temperature resistivity data. The $SrTiO_3$ was found to exhibit {111} preferred orientation parallel to {100} Si. For Si, $SrTiO_3$ was not a suitable barrier layer since although it did prevent Si from reaching the YBCO film, Ba still reacted with the Si after migrating through the layer. For sapphire, $SrTiO_3$ was a suitable barrier layer in that it prevented any interaction between the YBCO and sapphire.

ACKNOWLEDGEMENTS

The financial support of DARPA contract number MDA9788J1006 is gratefully acknowledged.

REFERENCES

[1] M. Naito, R.H. Hammond, B. Oh, M.R. Hahn, J.W.P. Hsu, P. Rosenthal, A.F. Marshall, T.H. Geballe, and Kapitalunk, J. Mater. Res., 2, 713, (1987).

[2] T. Venkatesan, C.C. Chang, D. Dijkkamp, S.B. Ogale, E.W. Chase, L.A. Farrow, D.M. Hwang, P.F. Micelli, S.A. Schwarz, J.M. Tarascon, X.D. Wu, and A. Inam, J. Appl. Phys., 63, 4591, (1988).

[3] A. Mogro-Campero, B.D. Hunt, L.G. Turner, M.S. Burrell, and W.E. Balz, Appl. Phys. Lett., 52, 584, (1988).

[4] J.J. Cuomo, M.F. Chisolm, D.S. Lee, D.J. Mikalsen, P.B. Madakson, R.A. Roy, E. Geiss, and G. Scilla, Thin Film Processing and Characterization of High Temperature Superconductors, ed. J.M.E. Harper, R.J. Colton, and L.C. Feldman, AIP Conf. Proc. No. 165, (AIP, N .Y., 1988), p. 141.

[5] S.Y. Lee, B. Murdock, D. Chin, and T. VanDuzer, Thin Film Processing and Characterization of High Temperature Superconductors, ed. J.M.E. Harper, R.J. Colton, and L.C. Feldman, AIP Conf. Proc. No. 165, (AIP, N .Y., 1988), p. 427.

[6] M. Gurvitch and A.T. Fiory, MRS Proc. Vol. 99, "High Temperature Superconductors," ed. M.B. Brodsky, R.C. Dynes, K. Kitazawa, and H.L. Tuller, (MRS, Pittsburgh, 1988), p. 297.

[7] A. Stamper, D.W. Greve, D. Wang, and T.E. Schlesinger, Appl. Phys. Lett., 52, 1746, (1988).

[8] H.C. Li, G. Linker, F. Ratzel, R. Smithey, and J. Geerk, Appl. Phys. Lett., 52, 1098, (1988).

[9] M.L. Leskela, C.H. Mueller, J.K. Truman, and P.H. Holloway, Mat. Res. Bull., 23, 1469, (1988).

[10] US Inc., Campbell, CA

[11] S.N. Rossnagel and J.J. Cuomo, Thin Film Processing and Characterization of High Temperature Superconductors, ed. J.M.E. Harper, R.J. Colton, and L.C. Feldman, AIP Conf. Proc. No. 165, (AIP, N .Y., 1988), p. 141.

[12] D.S. Campbell, in Handbook of Thin Film Technology, ed. I. Maissel and R. Glang, (McGraw-Hill, N. Y., 1970), ch. 12.

[13] E.M. Levin, C.R. Robbins, and H.W. McMurdie, Phase Diagrams for Ceramists, (American Ceramic Society, Columbus, OH, 1974), p. 195.

SUPERCONDUCTING Tl-Ca-Ba-Cu-O THIN FILMS BY REACTIVE MAGNETRON SPUTTERING

D. H. Chen, R. L. Sabatini, S. L. Qiu, D. Di Marzio,
S. M. Heald, and H. Wiesmann
Brookhaven National Laboratory, Upton, NY 11973

ABSTRACT

Superconducting Tl-Ca-Ba-Cu-O thin films with T_c onsets of 115 K and T_c (R=0) of 95 K have been prepared by reactive magnetron sputtering using Tl, Cu and Ca/Ba metal targets. It was found that proper thallium content is crucial for obtaining a high transition temperature. Wet oxygen and a sealed gold tube with additional thallium compounds were used to reduce the loss of thallium during annealing. X-ray diffraction spectra show that films with the sharpest transition at 95 K are predominantly c-axis oriented. XANES also shows a preferred c-axis orientation for the superconducting film, while for a nonsuperconducting film the near edge structure suggests greater disorder. X-ray microprobe fluorescence measurements indicate that these films are close to the 2122 stoichiometry. Scanning electron microscopy on these films is also presented.

INTRODUCTION

The discovery[1] of superconductivity in the Tl-Ca-Ba-Cu-O systems has resulted in the highest superconducting transition temperatures reported to date.[2] This class of superconductors contains no rare earth elements making applications of the superconductor more practical and cost effective. In contrast to the yttrium based superconductors there have been only a few reports in the literature describing the fabrication of the thallium based thin films. This is partly due to the toxicity of thallium and its compounds. Epitaxial and polycrystalline films have potential applications in integrated circuits, SQUIDS and IR detectors. Some of the deposition techniques which have been employed for fabrication of thallium based thin films are RF magnetron sputtering of a single composite bulk target,[3] sequential electron beam evaporation,[4] off axis RF diode sputtering of bulk targets[5] and simultaneous reactive metal magnetron sputtering.[6] In this paper we discuss the preparation of Tl-Ca-Ba-Cu-O thin films in a manner similar to that employed in Ref. 6. X-ray diffraction data are presented showing that the films with the sharpest superconducting transitions have the greatest degree of preferred orientation. In addition to resistance versus temperature measurements, SEM photomicrographs, x-ray fluourescence microprobe results and x-ray absorption near edge structure (XANES) are also included.

EXPERIMENTAL

The Tl-Ca-Ba-Cu-O films were fabricated in a conventional

commercial sputter deposition system which has been described in our previous report.[7] All the depositions were performed under identical conditions. The argon gas flow was fixed at 13.0 sccm and oxygen flow rate at 0.20 sccm with the gas flow rates controlled via electronic mass flow controllers. The total gas pressure during sputtering was 5 microns and the base pressure of the vacuum system prior to deposition was in the range $1-5 \times 10^{-6}$. The substrate temperature during deposition was 300°C.

All the targets were presputtered for approximately 1 hour prior to deposition. A quartz crystal rate monitor was fixed next to the substrate holder and used to calibrate the deposition rates of the individual targets prior to deposition. The thallium and copper targets were sputtered with dc power supplies while the Ba/Ca (1:1) target was sputtered using a 13.56 MHz power supply.

It was found that the proper thallium content is crucial for obtaining films with high superconducting transition temperatures. In order to reduce the loss of thallium during annealing the as-deposited films were placed in a sealed gold tube with additional Tl_2O_3. Water vapor was introduced into the furnace during the annealing cycle in combination with oxygen. The presence of water vapor resulted in films with higher transition temperatures than films annealed in dry oxygen. Two annealing steps were employed. The gold tube (containing the films) was inserted into the furnace for 2-3 minutes at 850°C, removed quickly and allowed to cool to room temperature. The film was reinserted into the furnace for 2-5 minutes at 800-820°C and furnace cooled. After heat treatment the films were 0.5-1.0 micrometers thick.

Two different substrates were used, single crystal sapphire and single crystal yttrium stabilized cubic zirconia with (100) orientation. The zirconia substrates gave superior results and all of the data shown here were for films grown on this substrate. The superconducting transition was measured using the recommended four probe resistance technique. Four silver strips were painted onto the surface of the film and copper wires were embedded in the silver strips prior to drying and hardening. The area encompassed by the voltage sensing strips was approximately 3 mm x 4 mm for all samples. The current density during measurement was approximately 1 Amp/cm^2.

RESULTS AND DISCUSSION

Figure 1 shows resistance versus temperature for three samples which were annealed under different conditions (see Fig. 1 caption). For sample A the onset occurs at 115°K and the transition is complete at 95°K. The remaining films, B and C, show deterioration in the slope of the resistance versus temperature, T_c onset, and the temperature at which the transition is complete. X-ray diffraction measurements were performed on a Philips powder diffractometer using CuKα radiation and the results are shown in Fig. 2. The curves are labelled A, B, and C and correspond to the sample labels in Fig. 1. Indexing of the 2θ scans identifies all the films as belonging to the 2122 phase. All three 2θ scans are dominated by (00ℓ) reflections consistent with a preferred

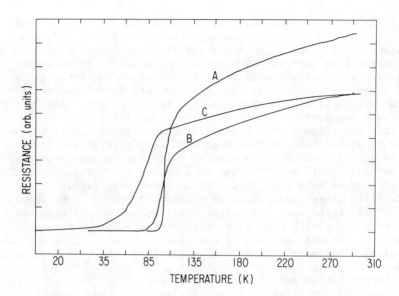

Fig. 1. Resistance vs temperature for Tl-Ca-Ba-Cu-O films on yttrium stabilized ZrO_2 (100). The films were annealed (A) at 850°C for 2 minutes followed by rapid cooling and subsequently annealed at 820°C for 2 minutes then furnace cooled to room temperature, (B) at 850°C for 3 minutes and rapid cooled to room temperature, (C) same as (A) but the subsequent annealing was followed by rapid cooling instead of furnace cooling.

orientation wherein the c-axis of the films is perpendicular to the surface of the substrate. Located at the bottom of Fig. 2 is a computer generated 2θ scan showing the location and intensity of the (00ℓ) reflections for a film with a c-axis orientation perpendicular to the surface of the sampe. There is excellent agreement between the computer generated scan and the experimental data. There is also a correspondence between the degree of orientation and the sharpness of the superconducting transition. Referring to Fig. 1 we observe that the film with the sharpest superconducting transition also exhibits the greatest degree of preferred orientation. As the quality of the superconducting transition deteriorates so does the degree of preferred orientation as evidenced by samples B and C. Examination of the 2θ diffraction scans reveals only a few impurity reflections of small intensity. The dominant reflection for polycrystalline 2122 phase material is located at 31.5 degrees. A decrease in the degree of preferred orientation is juxtaposed by an increase in the intensity of this reflection. This is consistent with the small quantity of impurity phases present. X-ray fluorescence microprobe measurements were used to determine the elemental composition of each of the films. Wavelength dispersive spectroscopy was employed and separate standards for each of the individual elements were used to

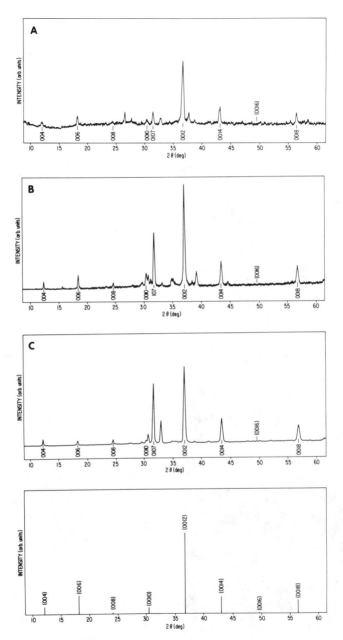

Fig. 2. X-ray diffraction pattern from three samples labeled A, B, and C which correspond to the sample labels in Fig. 1. For comparison, a computer generated 2θ scan is shown at the bottom of this figure which shows the location and intensity of the (00ℓ) reflections.

Table I. The Atomic Composition for Tl-Ca-Ba-Cu-O Thin Films.

Sample	Tl	Ca	Ba	Cu
A	2.33	1.89	2.89	3.17
B	2.29	1.45	2.27	2.89
C	4.03	1.92	3.37	4.08

calibrate the spectrometer crystals. The results are shown in Table I. There is substantial deviation from the atomic composition expected of a film which consists of the stoichiometric 2122 phase. We are investigating this further.

SEM photomicrographs of samples A, B, and C are shown in Fig. 3. The magnification is 5000x. The films appear to be extremely porous. The reason for this porosity is not understood but may be related to the addition of water vapor to the flowing oxygen used in the annealing of the films.

X-ray absorption near edge structure (XANES) displays large modulations of the atomic absorption coefficient and therefore is sensitive to local atomic structure. An energy range from -20 eV to 40 eV (with the edge defined as 0 eV) is typical and it encompasses both pre-edge and post-edge features as well as the edge and main peak. For the case of the Tl-Ca-Ba-Cu-O thin films considered here, the XANES from the Cu k-edge (1s→4p transition) was measured. This was done at beam line X-11A at the National Synchrotron Light Source (NSLS) at Brookhaven National Laboratory. A Si(111) double-crystal monochromator with a nominal energy resolution of ~2.0 eV was used. An advantage of synchrotron radiation is its polarization (in the horizontal plane), which can be used to determine the orientation of anisotropic materials and to probe

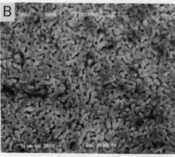

Fig. 3. Scanning electron micrograph of the same films as in Figs. 1 and 2.

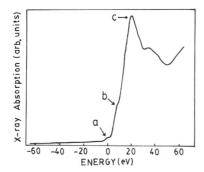
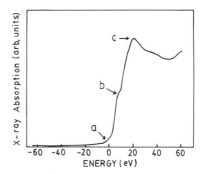

Fig. 4. The absorption edge of superconducting $Tl_2Ca_1Ba_2Cu_2O_x$ thin film with the x-ray electric field vector parallel to the substrate surface.

Fig. 5. The absorption edge of a Tl deficient nonsuperconducting film with the x-ray electric field vector parallel to the substrate surface.

electronic structures along a particular direction. The Tl thin film was positioned so that the substrate normal was 30° from horizontal and the x-ray electric field vector was parallel to the film surface. X-ray fluorescence was measured with a detector placed above the film.

Figure 4 shows the absorption edge for sample B. The pre-edge feature marked a, is characteristic of the $1s \rightarrow 3d_{x^2-y^2}$ (antibonding) transition.[8] This transition, which is dipole forbidden but quadrupole allowed, is weak. The shoulder marked b, represents the dipole allowed $1s \rightarrow 4p$ transition with a shakedown of charge from the occupied $O_{3p\sigma}$ state to the empty Cu $3d_{x^2-y^2}$ state.[9] This charge transfer screens the 1s hole and lowers the $1s \rightarrow 4p$ transition energy. The main peak marked c is the unscreened $1s \rightarrow 4p$ transition. If the x-ray electric field vector is parallel (ê⊥c-axis) to the CuO_2 plane, then the main transition is $1s \rightarrow 4p\sigma$, while for the electric vector perpendicular (ê||c-axis) to the CuO_2 plane, the transition is $1s \rightarrow 4p\pi$.[8] It has been observed for oriented $Tl_2Ca_1Ba_2Cu_2O_x$ powder that a weak shakedown shoulder appears when ê⊥c-axis, while for ê||c-axis a strong shoulder appears.[10] This is consistent with near edge data on $CuCl_4^{-2}$ complexes.[8] For the oriented powder the shoulder height for ê⊥c-axis is ~29% of the total c peak height, while for ê||c-axis it is ~53% of the c peak height. In Fig. 4, the shoulder marked b is ~32% of the total c peak height, which is close to the value of 29% for ê⊥c-axis for the oriented powder. This suggests that the c-axis is perpendicular to the substrate. For comparison, the absorption edge of a Tl deficient nonsuperconducting film is shown in Fig. 5. Here the height of shoulder b is ~51% of the total peak height and the weaker structure above the edge suggests greater disorder. In addition, the ê⊥c-axis polarization for the oriented powders show a strong and relatively narrow main peak c, as is the case in Fig. 4. The pre-edge $1s \rightarrow 3d_{x^2-y^2}$ feature a in Fig. 4 is also stronger in the

$\hat{e} \perp c$-axis oriented powder than in $\hat{e} || c$-axis.[8]

CONCLUSIONS

We have fabricated thin films of $Tl_2Ca_1Ba_2Cu_2O_x$ on yttrium stabilized cubic zirconia by the technique of simultaneous reactive metal magnetron sputtering using three metal targets. Superconducting onsets of 115°K with $T_{c(R=0)}$ of 95°K have been achieved. A correlation has been observed between the quality of the superconducting transition and the degree of preferred orientation in the films. The sharpest transition is exhibited by films having the greatest degree of preferred orientation. XANES show structure consistent with a preferred orientation of the c-axis perpendicular to the substrate plane. X-ray microprobe fluorescence measurements of the film compositions show that the films are close to the 2122 stoichiometry except for the presence of excess copper. The films exhibit a rather porous microstructure at 5000x magnification. This microstructure is believed to be an artifact of the annealing procedure. Future work will be concetrated on achieving films with improved superconducting transition and a more homogenous microstructure.

ACKNOWLEDGEMENTS

We wish to thank the staff of the National Synchrotron Light Source at Brookhaven National Laboratory, where the XANES measurements were performed. This work was performed under the auspices of the U.S. Department of Energy, Division of Materials Science, Office of Basic Energy Sciences under Contract No. DE-AC02-76CH00016.

REFERENCES

1. Z. Z. Sheng, A. M. Hermann, A. El Ali, C. Almasan, J. Estrada, T. Datta, and R. J. Matson, Phys. Rev. Lett. 60, 937 (1988).
2. S. S. P. Parkin, V. Y. Lee, E. M. Engler, A. I. Nazzal, T. C. Huang, G. Gorman, R. Savoy, and R. Beyers, Phys. Rev. Lett. 60, 2539 (1988).
3. M. Nakao, R. Yuasa, M. Nemoto, H. Kuwahara, H. Mukaida, and A. Mizukami, Jpn. J. Appl. Phys. 27, L849 (1988).
4. D. S. Ginley, J. F. Kwak, R. P. Hellmer, R. J. Baughman, E. L. Venturini, and B. Morosin, Appl. Phys. Lett. 53, 406 (1988).
5. W. Y. Lee, V. Y. Lee, J. Salem, T. C. Huang, R. Savoy, D. C. Bullock, and S. S. P. Parkin, Appl. Phys. Lett. 53, 329 (1988).
6. J. H. Kang, R. T. Kampwirth, and K. E. Gray, Phys. Lett. A 131, 208 (1988).
7. H. Wiesmann, De Huai Chen, R. L. Sabatini, J. Hurst, J. Ochab, and M. W. Ruckman, J. Appl. Phys., to be published.
8. N. Kosugi, T. Yokoyama, K. Asakura, and H. Kuroda, Chem. Phys. 91, 249 (1984).

9. R. A. Bair and W. A. Goddard III, Phys. Rev. B <u>22</u>, 2767 (1980).
10. S.M. Heald, J. M. Tranquada, C. Y. Yang, Y. Xu, A. R. Moodenbaugh, M. A. Subramanian, and A. W. Sleight, <u>Proc. Intern. Conf. EXAFS, Seattle, WA, Aug. 1988.</u>

THIN FILMS OF $Y_1Ba_2Cu_3O_x$ DEPOSITED USING THREE TARGET CO-SPUTTERING AND THEIR APPLICATIONS TO MICROBRIDGE JUNCTIONS AND SINGLE-ELEMENT IR DETECTORS

J.Y. Josefowicz, D.B. Rensch, A.T. Hunter, H. Kimura, and B.M. Clemens
Hughes Research Laboratories
Malibu, CA 90265

J. Spargo, E. Wiener-Avnir, and G. Kerber
Hughes Microelectronics Center
Carlsbad, CA 92008-4835

J.A. Wilson, W.D. Jack, J.M. Myroszynyk, and R.E. Kvaas
Santa Barbara Research Center
Goleta, CA 93117-3090

ABSTRACT

We have produced superconducting thin films of $Y_1Ba_2Cu_3O_x$ (YBCO) on $SrTiO_3$, $ZrO_2(Y_2O_3)$, and Al_2O_3 having T_c's of 90K and sharp ΔT_c's (0.3K for 90% to 10% on $SrTiO_3$ and 3.3×10^5 A/cm^2 at 78K). Samples were fabricated in a three target magnetron co-sputtering, computer controlled system using separate BaF_2, Y, and Cu targets. Rutherford backscattering (RBS) and resistivity measurements showed that the thickness and composition uniformity was 99% across a 2-in.-diameter substrate. Films were deposited at room temperature and were subsequently annealed in wet O_2 at 850°C for 1.0 hr. Analysis of the films by x-ray diffraction indicates highly oriented crystalline structure with a and c axes perpendicular to the substrate surface for (100) $SrTiO_3$, and randomly oriented $Y_1Ba_2Cu_3O_x$ polycrystalline structure for ZrO_2 and Al_2O_3.

Both laser ablation and photolithographic/ion milling processes were developed to pattern the high T_c superconducting films. Bridge structures were fabricated in thin films deposited on sapphire and ZrO_2 substrates. Weak link Josephson behavior was observed from 4.2K up to T_c.

We have also fabricated single-element detectors using thin films of superconducting YBaCuO and demonstrated IR sensitivity with responsivities $\cong 3 \times 10^3$ V/W.

INTRODUCTION

Recently there has been considerable activity focused on $Y_1Ba_2Cu_3O_x$ superconducting films formed by deposition from fluorine containing sources[1-4]. These films appear to have greater stability against environmental degradation, particularly in the unannealed state, compared with those formed from nonfluorine containing sources. The superconducting phase is formed by annealing at high temperatures (800 to 900°C) in the presence of wet oxygen, where the water aids in converting BaF_2 to the oxide phase. In this report we present results of simultaneous co-sputtering from Y, BaF_2, Cu targets which produced superconducting thin films reproducibly. These films appear stable even after exposure to moisture and air. We report on the utilization of laser direct writing to fabricate bridge structures in high T_c

superconducting (HTS) thin films of $Y_1Ba_2Cu_3O_x$ with $T_c > 90K$, that demonstrated weak link Josephson behavior up to temperatures close to T_c.

There has been considerable interest in the use of the new class of HTS materials for long-wavelength IR detectors with expected cutoff wavelengths of ~50 μm.[5,6] We present results for IR detectors made in granular YBCO films on $SrTiO_3$ and patterned by ion milling. These results indicate two detection mechanisms: a Josephson coupling mode and a bolometric response.

EXPERIMENTAL RESULTS AND DISCUSSION

A Hughes designed co-sputtering system[7] with three separate targets composed of Y, BaF_2, Cu was used to deposit YBCO thin films. This system is designed for controlled simultaneous sputtering from each of the three targets which are configured to direct the sputtered material onto the substrate. The HTS thin films of YBCO were deposited on 2-in. substrates of Al_2O_3, ZrO_2(Y_2O_3 stabilized) and smaller substrates of $SrTiO_3$ (100). Thickness and composition uniformity was 99% across a 2-in.-diameter substrate as determined by RBS and resistivity measurements. The films ranged from 300 to 1000 nm in thickness. The films were post annealed in wet O_2 at 850°C for 1 hr and further cooled slowly in pure O_2 at a rate of 1.5°C/min. All the films demonstrated metallic behavior above the transition temperature and had superconducting transitions, measured resistively, starting at temperatures above 90K. The transition width ranged from 0.3K for films on $SrTiO_3$ to 21K for films on Al_2O_3 substrates, as shown in Figure 1.

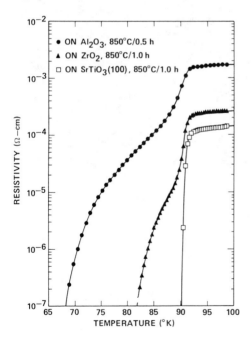

Fig. 1. Resistivity versus temperature for $Y_{1.4}Ba_2Cu_{3.3}O_x$ films deposited on Al_2O_3, ZrO_2 and $SrTiO_3$ (100) substrates.

Analysis of films on $SrTiO_3$ substrates, using x-ray diffraction and cross-section transmission electron microscopy,[8] showed highly oriented structures with a and c axes perpendicular to the film plane as shown in Figure 2. The films on Al_2O_3 and $ZrO_2(Y_2O_3$ stabilized) substrates were determined to be primarily polycrystalline in nature with some evidence of a small degree of c axis orientation.

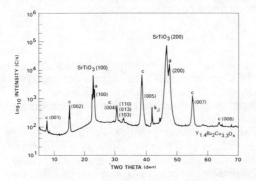

Fig. 2. X-ray diffraction result for $Y_{1.4}Ba_2Cu_{3.3}O_x$ annealed film on a $SrTiO_3$ (100) substrate.

The surface morphology of $Y_{1.3}Ba_2Cu_{3.3}O_x$ films deposited on Al_2O_3, ZrO_2 and $SrTiO_3$ (100) annealed at 850°C, is shown in Figure 3.

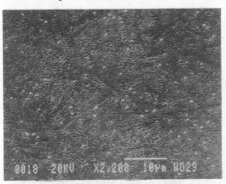

SAPPHIRE SUBSTRATE

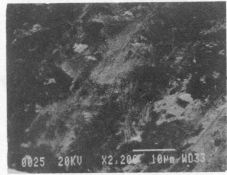

ZrO_2 SUBSTRATE

(100) SrTiO_3 SUBSTRATE

Fig. 3. SEM Photographs showing the surface morphology of $Y_{1.4}Ba_2Cu_{3.3}O_x$ films annealed on Al_2O_3, ZrO_2, and $SrTiO_3$ (100) substrates. For $SrTiO_3$, the basket weave material is a-axis oriented out of the plane with c axis oriented out of the plane beneath it; as determined by cross-section TEM.[8]

The YBCO films were patterned successfully by two methods: laser ablation and photolithography combined with ion milling. A continuous wave YAG laser operating in Q-switched mode at a wavelength of 530 nm was used to pattern the YBCO films. Clean TEM_{00} pulses of 6 μs duration were focused to diffraction limited spots on the sample surface. A high precision x-y table (accuracy and repeatability in position of 1 μm) was used to scan the laser.[9] The other patterning approach was to use standard photolithography followed by Ar ion milling to remove the YBCO film. Examples of patterned films using both approaches are shown in Figure 4.

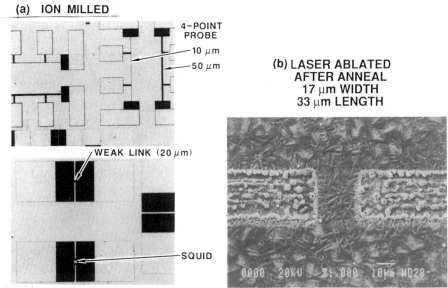

Fig. 4. (a) Ion milled 4-point probe and microbridge patterns for YBCO film on a $SrTiO_3$ (100) substrate. (b) Laser ablated microbridge of YBCO film on a sapphire substrate.

Critical current densities of 3.3×10^5 A/cm^2 at 78K for HTS films on $SrTiO_3$ were determined using the 4-point probe patterns shown in Figure 4(a). These line widths were measured to be 6 and 50 μm having aspect ratios of 200:1 and 25:1, respectively, as shown in Figure 5. Figure 6 shows a 4-point probe I-V characteristic curve of a 35 μm wide, 30 μm long laser ablated YBCO bridge on sapphire at 4.2K [Figure 4(b)]. A symmetrical I-V curve typical of Josephson weak link devices[10,11] can be seen. Supercurrents of about 120 μA with $R_n \cong 0.5$ Ω, and current densities above 400 A/cm^2 were typical of these bridges at 4.2K. The nonlinearity associated with Josephson weak link behavior was visible in the I-V curves up to 76 K, which was above the film's zero resistance transition. In order to assure that the Josephson tunneling mechanism was responsible for the electrical behavior of the laser patterned bridges we looked for the AC Josephson effect. When microwave radiation at 9.375 GHz was coupled to the same bridge, Shapiro steps were observed (Figure 7).[9] Up to 13 steps could be distinguished, indicating a high degree

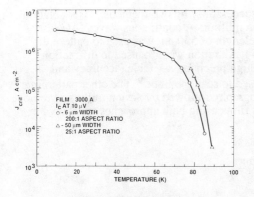

Fig. 5. Critical current versus temperature for $Y_1Ba_2Cu_{3.4}O_x$ films on a $SrTiO_3$ (100) substrate.

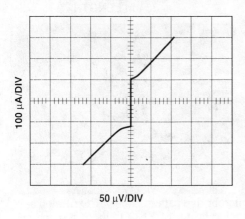

Fig. 6. I-V characteristic at 4.2K for a YBCO microbridge on Al_2O_3.

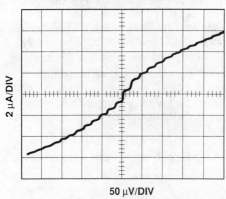

Fig. 7. AC Josephson effect in a YBCO microbridge on sapphire induced by 9.375 GHz microwave radiation at 4.2K.

of phase locking of the fundamental and harmonics when the condition $2eV = n\hbar w_D$ is fulfilled. No subharmonic steps were observed, even when high microwave fields were applied to the junction. The Shapiro steps could be distinguished at temperatures up to 60K. At this temperature three distinct steps (Figure 8) could be observed for the bridge shown in Figure 4(b).

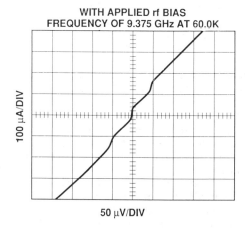

Fig. 8. AC Josephson effect in YBCO bridge induced 9.5 GHz radiation at 60K.

Similar Josephson I-V curves and microwave induced Shapiro steps were also observed for laser patterned YBCO thin films deposited on $ZrO_2(Y_2O_3$ stabilized). In these bridges, the Josephson weak link nonlinearity was observed up to 84K, which is above the zero resistance transition point. No Josephson behavior was detected in the laser patterned bridges in the highly oriented YBCO thin films sputtered on $SrTiO_3$ (100). Further analysis of the YBCO thin films by SEM showed a high degree of granularity for films on Al_2O_3 and ZrO_2 substrates; whereas, a high degree of alignment and little evidence of grain boundaries was observed for YBCO films on $SrTiO_3$ (100) substrates. This would support the conclusion that the grain boundaries in the polycrystalline sputtered YBCO thin films function as barriers for Josephson tunneling between grains in the superconductive state.

Test arrays of single-element detectors, patterned into 50 μm × 200 μm strips and 50 μm × 2800 μm meander strips were fabricated using an ion milling process. Detector element resistances ranged from a few ohms to 20,000 Ω and contact resistance was negligible by comparison. Contact was made with Au using a 4-point probe pattern. The detectors were fabricated in $Y_{0.8}Ba_2Cu_3O_x$ films deposited on $SrTiO_3$ (100) substrates. Films with this intentional non-stoichiometry were highly granular having a T_c onset of 75K and $T_c(0)$ of 34K. Photoresponse was measured by viewing a 500, 800, or 1000K blackbody through a KRS5 window. I-V curves were measured directly by recording the voltage drop across the inner two contacts while sweeping the bias current applied across the outer two contacts. A lock-in amplifier was used to record response as a function of blackbody chopping frequency, temperature, and applied bias.

Measurements of the I-V curve for devices made at temperatures below $T_{c,0}$, exhibited nonhysteretic superconducting weak-link type behavior with zero voltage currents ranging from 1 μA to 1 mA. A typical characteristic for a meander sensor, shown in Figure 9, when exposed to the blackbody source shows a characteristic shift in the I-V curve. At a current bias of 300 μA, the voltage shift is nearly 0.1 V. Voltage shifts scale with detector length as expected in a weakly coupled grain model.

Photoresponse is strongly bias- and temperature-dependent. The bias-dependent response and I-V curves are shown in Figure 10 for nonmeander, 50 μm × 200 μm element. At low biases in the region of zero voltage current flow, the photoresponse is zero. It rises sharply as the I-V curve begins to show dynamic resistance and peaks at a bias current of about 750 μA. It continues to show photoresponse as bias is increased until the I-V curve transitions to the normal state curve estimated to occur at bias voltages in excess of 1 V.

Temperature dependence of the photoresponse is shown in Figure 11 for the meander geometry detector biased at 1 mA. This is near the bias that gives the maximum photoresponse at 9.8K. In Figure 11, the response is maximum at the lowest temperature, and decreases as temperature is raised until the region of $T_{c,onset}$ is reached, where there is a pronounce increase in response. At the low chopping frequency (11 Hz) of this curve, a bolometric response is expected in the region near T_c. A measurement of the temperature deriative of resistance, which increased to a maximum near T_c, was found to give a temperature dependence of the bolometric component of the response expected for the element in Figure 11. For temperatures well below T_c, the response may be caused by modulation of the inter-grain coupling through optically induced changes in the quasi-particle density. The increasing response with decreasing temperature may reflect an increase in the quasi-particle lifetime, as observed by Parker and Williams for Sn and Pb.[12]

Photoresponse at temperatures well below the transition temperature, measured by recording the shift in I-V curve, indicates responsivities in the range of 2 to 3x10³ V/W for nonmeander and meander geometry detectors. The detectors show photoresponse out to bias voltages on the order of volts, indicating there are on the order of hundreds of optically active grain boundaries along current path through the detector. For the submicrometer grain sizes of these films, this indicates that a majority of the grain interfaces are optically active.

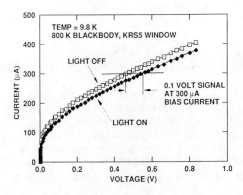

Fig. 9. I-V curves of meander geometry element showing shift due to 800K blackbody source.

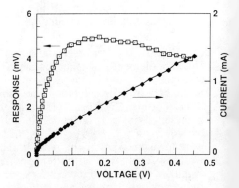

Fig. 10. I-V curve and response versus bias voltage for element 50 μm × 200 μm. T = 9.8K.

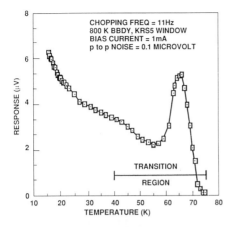

Fig. 11. Response versus temperature for element 50 μm × 200 μm.

REFERENCES

1. P.M. Mankiewich, J.H. Scofield, W.L. Skocpol, R.E. Howard, A.H. Dayem, and E. Good, Appl. Phys. Lett. **51**(21), 1753 (1987).

2. A.F.J. Levi, J.M. Vandenberg, C.E. Rice, A.P. Ramirez, K.W. Baldwin, M. Anzlowar, A.E. White, and K. Short, J. Cryst. Growth (Submitted March 1988).

3. A.F. Marshall, R.W. Barton, K. Char, A. Kapitulnik, B. Oh, R.H. Hammond, and S.S. Laderman, Phys. Rev. B (Submitted February 1988).

4. A. Gupta, R. Jagannathan, E.I. Cooper, E.A. Giess, J.I. Landman, and B. W. Hussey, Appl. Phys. Lett. (Submitted April 1988).

5. Y. Enomoto and T. Murakami, J. Appl. Phys., **59**, 3807 (1986).

6. M. Leung, P.R. Broussard, J.H. Claasen, M. Osofsky, S.A. Wolf, and U. Strom, Appl. Phys. Lett. **51,** 2046 (1987).

7. J.Y. Josefowicz, D.B. Rensch, A.T. Hunter, B.M. Clemens, and H. Kimura, "Co-sputtering of $Y_1Ba_2Cu_3O_x$ films," (to be published).

8. B.M. Clemens, C.W. Nieh, J.A. Kittl, W.L. Johnson, J.Y. Josefowicz, and A.T. Hunter, "Nucleation and growth of YBCO on $SrTiO_3$,"Appl. Phys. Lett. (accepted).

9. E. Wiener-Avnir, J.E. Cooper, G.L. Kerber, J.W. Spargo, A.G. Toth, J.Y. Josefowicz, D.B. Rensch, B.M. Clemens, and A.T. Hunter, "Laser patterning of YBaCuO weak link bridges," IEEE Transactions on Magnetics (accepted).

10. K.K. Likharev, "Superconducting weak links," Rev. Mod Phys. **51**, pp. 101-159, January 1959.

11. K.K. Likharev, "Dynamics of Josephson junctions and circuits," New York: Gordon and Beach, 1986, Ch. 4, pp. 89-128, Ch. 11 pp. 332-375.

12. W.H. Parker and W.D. Williams, Phys. Rev. Lett., **29**, 924 (1972).

METALLIC ALLOY TARGETS FOR HIGH T_c SUPERCONDUCTING FILM DEPOSITION

P. Manini
SAES GETTERS S.p.A., Via Gallarate 215, 20151 Milan, Italy

A. Nigro, P. Romano and R. Vaglio
Dip. Fisica, Univ. Salerno, 84081 Baronissi, Salerno, Italy

ABSTRACT

Many experiments are nowadays conducted worldwide on superconducting films based on the recently developed high T_c superconductor materials (YBCO, BISCO, etc). There are different ways to produce these films, among which sputtering and evaporation are most popular. Normally, use is made of oxides, pure metals or compounds as material sources. In the present paper we describe the fabrication process and the physico-chemical characteristics of various metallic alloy components for both sputtering and evaporation processes which show various advantages in terms of stability, easiness of use, purity, flexibility in composition and shape and allow good process control. Deposition techniques and experimental results obtained on thin films of the new superconductors realized starting from these alloys are also reported.

INTRODUCTION

The discovery of high T_c superconductive oxides (YBCO, BISCO etc.) has sparked off a vigorous scientific and technological research effort worldwide. The excitement generated by these new materials relies on their intriguing chemico-physical and structural properties as well as on their very promising potential applications in many areas such as electronics, computer science, transport and energy storage. Other major practical applications are foreseen in the field of microelectronics (interconnections, SQUID's etc.) so that a great deal of work is nowadays devoted to thin film deposition and processing.

High T_c perovskite thin films can be deposited using various techniques including thermal e-beam evaporation[1,2]. Molecular Beam Epitaxy[3] (MBE), laser ablation[4] and sputtering[5,6,7].

Sputtered films are commonly fabricated starting from single[5,6], or multi-target oxides[7] or pure metals[8] as material sources, also alloys of the metals involved have been considered for this purpose[9].

In the present paper an investigation has been made concerning preparation and characterization of a variety of binary and ternary alloys for both YBCO and BISCO film deposition. The advantages of these alloys over more conventional material sources and some of their chemico-physical characteristics are illustrated. Experimental results concerning the use of metallic alloys for sputtering/evaporation co-deposition processes are also reported and discussed.

ADVANTAGES OF ALLOYS FOR YBCO/BISCO THIN FILM DEPOSITION

Sputtering from single oxide targets is one of the most straight forward methods to deposit thin films of the new high T_c materials. However, some drawbacks may be outlined, the major of which being represented by re-sputtering phenomena which take place during film growth as a consequence of the great abundance of negative oxygen ions released by the target[10,11]. These ions are generated at the cathode surface of the target during the sputtering process and subsequently accelerated towards the substrate. Depending on the experimental set up, substrate bombardment can result in severe etching of as-deposited materials. This in turn prolongs deposition time and decreases target sputtering yield in that only a reduced number of films can be grown with the same source. In addition, due to its preferential action, re-sputtering phenomena may also lead to appreciable deviations in film composition. This effect must be carefully taken into account and properly counter-balanced. Compensated oxide targets or specific geometrical arrangements of the sputtering chamber are usually adopted[11,12,13].

Some difficulties in achieving a good level of reproducibility of sputtered films have also been noticed when using oxide targets. Films obtained from the same target are quite reproducible while films deposited from different targets of same nominal stoichiometry may show unpredictable deviations in final composition. This has been ascribed by some authors to differences in targets microstructure induced by densification processes[12].

Also, deposition from pure metals presents some drawbacks. It allows a high film composition flexibility and process yield but it is intrinsically more complex to control due to the higher number of material sources. In addition some metals like the alkaline-earths are not stable and difficult to sputter.

Metallic alloys offer a very good compromise between the two above mentioned possibilities. In fact, alloys are not inherently affected by re-sputtering phenomena. High deposition rates as well as prolonged targets life can therefore be achieved. The use of two alloy targes (e.g. Ba_1Cu_1, Y_1Cu_1) in place of one single oxide target ($Y_1Ba_2Cu_3O_7$) assures a certain degree of flexibility in film composition while maintaining a superior easiness of sputtering with respect to a 3 source deposition process. This aspect is still more relevant in the case of BISCO system (2 vs. 4 sources). Stability of the alloys is usually good for most practical applications. In addition, their metallic nature also assures good thermal and electrical conductivity, thus making possible their successful use both in D.C. and R.F. sputtering machines.

ALLOY PROPERTIES AND EXAMPLES OF APPLICATION

To meet the demand for simple and reliable high quality superconductive thin film sputter deposition a variety of high purity alloys for YBCO and BISCO have been developed and prepared, based upon the application of specific vacuum metallurgy expertize. Some of them are shown in Table I and Table II respectively. Chemical composition, melting point, metallographic structure and density are therein reported.

Table I Alloys for YBCO deposition

Alloy	Composition (% by wt)	melting point (°C)	Metallographic structure	density* (g/cm³)
Y_1Cu_1	58.32-41.68	935	Y_1Cu_1	5.80
Y_1Cu_3	31.80-68.20	925	$Y_1Cu_2+Y_2Cu_7$	5.90
Ba_1Cu_1	68.40-31.60	570	Ba_1Cu_1	4.60
Ba_2Cu_3	59.03-40.97	600	$Ba_1Cu_1+Ba_1Cu_{13}$	4.65

(*) Experimental error affecting density measurements is estimated to be less than 10%

Table II Alloys for BISCO deposition

Alloy	Composition (% by wt)	melting point (°C)	Metallographic structure	density (g/cm^3)
Bi_2Ca	91.25-8.75	735	$Bi_2Ca_3+Bi_3Ca$	7.00
Ca_1Cu_1	38.68-61.32	567	Ca_1Cu_1	2.90
Sr_1Cu_1	57.96-42.04	586	Sr_1Cu_1	3.85
$Sr_2Cu_2Ca_1$	51.18-37.12-11.70	(+)	(+)	(+)
Bi-Cu	every	-	solid solution	-

(+) Not yet determined

All alloys are prepared in melting raw materials under vacuum or an inert atmosphere by means of an R.F. induction furnace. The appropriate choice of crucibles assures negligible contamination of starting materials during melting. Purity of the resulting alloys typically ranges from 99.0% to better than 99.95% depending on the raw material grade. After melting, the alloys are machined into disks (or other shapes) of different dimension (1" to 4" or more) and sealed under inert atmosphere.

Stability of the Bi-Cu and Y-Cu alloys is very good. Also calcium-based materials such as Bi-Ca and Ca-Cu alloys show a relatively good stability when exposed to the atmosphere. For these alloys it typically takes some hours before noticing a change in the colour of the surface, which indicates that an oxidation process is taking place. Barium and, to a lesser degree, strontium-copper alloys are more reactive and oxidize more rapidly in the presence of humidity. Typical times required to change fresh material surface colour from yellow-gold (Ba-Cu alloys) or silver (Sr-Cu alloys) to a darker colour is about 30 seconds and some minutes respectively. However, also in this case, alloys are much more stable than pure metals and the thin oxide film eventually covering the surface can be removed with a pre-sputtering cleaning process.

Some examples of applications of YBCO and BISCO sputter film deposition are reported in Table III.

Table III   Examples of alloys use for sputter deposition

Sputtering source	deposited film
Ba_2Cu_3 + Y ⁻	YBa_2Cu_3
(2) Ba_1Cu_1 + (1) Y_1Cu_1	$Y_1Ba_2Cu_3$
(n) Ba_1Cu_1 + (m) Y_1Cu_1	$Y_mBa_nCu_{n+m}$
(1) Bi_2Ca + (2) Sr_1Cu_1	$Bi_2CaSr_2Cu_2$
(n) Bi_2Ca + (m) Sr_1Cu_1	$Bi_{2n}Ca_nSr_mCu_m$

( ) In this configuration Y can also be replaced with other rare earths.

Alloys can also be used for sputter/evaporation co-deposition processes as shown in Table IV.

Table IV   Alloys for sputtering/evaporation co-deposition processes

Metal / alloy sources	deposited film
Ba (evaporation) + Y_1Cu_3 (sputtering)	$Y_1Ba_2Cu_3$
Bi (evaporation) + $CaSr_2Cu_2$ (sputtering)	$Bi_2CaSr_2Cu_2$

Also in this case they give a significant contribution to simplification of the whole process as illustrated in the following.

METALLIC ALLOYS FOR SPUTTERING/EVAPORATION CODEPOSITION PROCESSES: SOME EXPERIMENTAL RESULTS

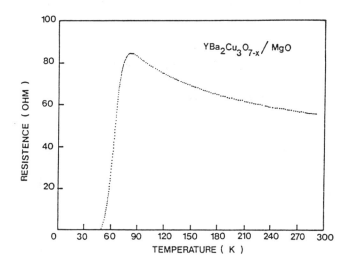

Fig. 1. Resistivity curve for a $Y_1Ba_2Cu_3O_7$ film on MgO. T_c (onset) $\sim$ 80 K, T_c (R=0) $\sim$ 50 K.

In Fig. 1, the superconducting resistive transition for a $Y_1Ba_2Cu_3O_{7-x}$ polycristalline film deposited onto an MgO substrate in our Laboratories by the Y_1Cu_3 (sputtering) and Ba (evaporation) codeposition process is reported (Table IV).

The one-inch Y_1Cu_3 target was sputtered in pure argon at 4×10^{-3} mbar by a D.C. magnetron assembled in a vacuum station based on a cryogenic pump (base pressure 10^{-7} mbar).

Metallic barium was thermally released by an effusion Knudsen cell controlled by an oscillating quartz thickness monitor. The overall deposition rate was 10 Å/sec.

During deposition the substrate was held at room temperature. An oxygen post-treatment annealing at 900°C for 10 minutes was performed ex-situ just after the deposition.

Even though the Y-Cu and Ba relative rates and the oxygen treatment were not yet optimized in these preliminary depositions, these results are encouraging especially in terms of high film compactness, as checked by electron microscopy and surface profilometry, and long term film stability, also upon long exposure to humid air.

Good results were also obtained using different substrates such as Al_2O_3 and $Gd_3Ga_5O_{12}$.

However, the main advantage of the process described lies in its great simplicity since it involves only two metallic sources and the possibility of obtaining very high rates (up to 50 Å/sec. in our system) essential for some applications[14].

This last feature is related to the optimum D.C. sputtering performance of the Y_1Cu_3 target which shows good electrical and thermal conductivity and relatively high melting point.

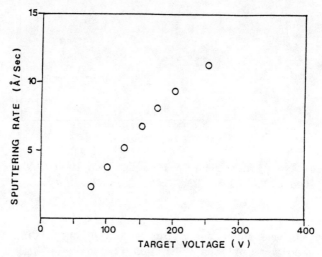

Fig. 2. Y_1Cu_3 deposition rate as a function of target voltage.

In Fig. 2, the deposition rate of the Y_1Cu_3 alloy versus target voltage is reported for a 1 inch target at a distance of 9 cm from the substrate. Argon pressure inside the chamber and target current were 4×10^{-3} mbar and 1A respectively.

Sputtering rates even much higher than those of Fig. 2 can be achieved by simply changing deposition parameters.

It is believed that further work aimed to properly optimize the sputtering/evaporation deposition parameters as well as the subsequent oxygen annealing cycle could be beneficial for the electrical and structural characteristics of deposited films. A suitable choice of substrate material (e.g. monocristalline $SrTiO_3$ in place of MgO) should also contribute to improve film quality.

Based upon these positive results and the above mentioned considerations, experimental work is now in progress to produce

YBCO thin films of better quality. Some preliminary tests are also being carried out to grow $Bi_2CaSr_2Cu_2$ thin films by the same co-deposition method (Table IV).

CONCLUSIONS

In the present paper the use and properties of a variety of metallic alloys for both YBCO and BISCO thin film deposition have been illustrated.

These alloys are particularly suited to meet the increasing demand for simple and reliable high quality thin film deposition techniques.

Some of their advantages over more conventional material sources (oxides/pure metals) have been reviewed. Preliminary experimental results based upon the use of alloys in sputtering/evaporation codeposition processes have been also presented. These results are encouraging since they show high deposition rates and inherent process simplicity.

It is the Authors' opinion that further work to optimize both deposition parameters and oxygen post-treatments should result in improved film quality.

ACKNOWLEDGEMENTS

One of the authors (P. Manini) wishes to thank Dr. C. Boffito of SAES GETTERS for useful discussions and Mr. E. Cogliati for technical support. A. Nigro greatfully acknowledges the financial support of a scholarship received from Ansaldo Ricerche, Genoa, Italy.

REFERENCES

1. R.B. Laibowitz, R.H. Koch, P. Chaudhari and R.J. Gambino, Phys. Rev., B 35, 8821 (1987)
2. Siu-Wai Chan, L.M. Greene, W.L. Feldmann, P.F. Miceli and B.G. Bagley, American Institute of Physics, Conference Proceedings n° 165 New York, 1988, p. 28
3. E.S. Hellman, D.G. Schlom, N. Missert, K. Char, J.S. Harris Jr., M.R. Beasley, A. Kapitulnik, T.M. Geballe, J.N. Eckstein, S-L. Weng and C. Webb, J. Vac. Sci. Technol. B 6(2) 799, Mar/Apr. 1988
4. D. Dijkkanp, T. Venkatesan, X.D. Wu, S.A. Shaheen, N. Jisrawi, Y.H. Min-Lee, W.L. McLean and M. Croft, Appl. Phys. Lett. 51, 619 (1987)
5. M. Hong, S.H. Liou, B.A. Davison, Appl. Phys. Lett. 51(9), 694 (1987)
6. S.J. Lee, E.D. Rippert, B.Y. Jin, B.N. Song, S.J. HWU, K. Poeppelmeier and J.B. Ketterson, Appl. Phys. Lett. 51(15), 1194 (1987)
7. H. Asano, K. Tanabe, T. Katom, S. Kubo and O. Michikami, Jpn. J. Appl. Phys. 26(7), L1221 (1987)
8. R.E. Somekh, M.G. Blamire, Z.M. Barber, K. Butler, J.H. James, G.W. Morris, E.J. Tomlinson, A.P. Schwarzenberger, W.M. Stobbs and J.E. Evetts, Nature 326, 857 (1987)
9. M. Scheuermann, Cheng-Chung Chi, Chang C. Tsuei and Jochen D. Mannhart, Proceedings of Conference on High T_c Superconductivity: Thin Films and Devices, ed. Chi, van Dover, SPIE Proceedings Vol. 948, paper 948-13 (1988)
10. S.M. Rossnagel and J.J. Cuomo, American Institute of Physics, Conference Proceedings n° 165, New York 1988, p. 106
11. S.I. Shah and P.F. Carcia, Ibid., p. 50
12. R.L. Sandstrom, W.J. Gallagher, T.R. Dinger, R.TH. Koch, R.B. Laibowitz, A.W. Kleinsassek, R.J. Gambino, B. Bumble, and M.F. Chisholm, to be published.
13. W.Y. Lee, J. Salem, V. Lee, C.T. Rettner, G. Lim, R. Savoy, V. Deline, American Institute of Physics, Conference Proceedings n° 165, New York 1988, p. 95
14. P. Romano A. Nigro, R. Vaglio, E. Signorelli, K.E. Gray, Appl. Supercond. Conference, S. Francisco, Ca. (1988) to be published.

RF MAGNETRON SPUTTERING OF HIGH-Tc Bi-Sr-Ca-Cu-O THIN FILMS

S.K. Dew, N.R. Osborne, B.T. Sullivan,
P.J. Mulhern and R.R. Parsons
Dept. of Physics, University of British Columbia
Vancouver, B.C., Canada, V6T 2A6

ABSTRACT

Thin films of Bi-Sr-Ca-Cu-O were produced by RF magnetron sputtering onto single crystal (100) MgO substrates. Substrate temperature and deposition atmosphere were found to affect the thickness and composition of as-deposited films. Subsequent annealing of the films resulted in highly oriented superconductors, which were usually multiphase with 2223, 2212, and 2201 phases. Film properties were highly dependent on annealing conditions.

INTRODUCTION

Several families of high critical temperature oxide superconductors have recently been discovered by Bednorz and Muller[1], Chu[2], Hermann[3], and Maeda[4]. One of the most promising of these for technological applications is the Bi-Sr-Ca-Cu-O system discovered by Maeda[4] and investigated by others.[5-7] Since many device applications will require superconducting films, work[8-12] has been done by several groups to develop techniques for producing thin films of superconducting Bi-Sr-Ca-Cu-O. The approach used in this work involves RF magnetron sputtering of a superconductor oxide target.

Several superconducting phases in the Bi-Sr-Ca-Cu-O system have been observed, each having a different critical temperature. These have been identified by Tarascon[7] to belong to a structural family described by the formula $Bi_2Sr_2Ca_{n-1}Cu_nO_x$ — where n=1 phase (2201) is believed to become superconducting around 10K, n=2 (2212) around 85K, and n=3 (2223) around 110K. Clearly, on the basis of T_c, the 2223 phase is the most desirable. Unfortunately, it is very difficult[7,8] to obtain single phase 2223 as the 2212 phase tends to form more readily.

The purpose of this paper is to discuss factors important to increasing the amount of the 2223 phase present as well as to characterize the RF sputter deposition of films in the Bi-Sr-Ca-Cu-O system in general.

APPARATUS

Targets:

Sputtering targets were prepared from stoichiometric amounts of high purity (5N's) Bi_2O_3, $SrCO_3$, $CaCO_3$, and CuO powders in various cation ratios. The powders were typically mixed, fired in air at 780°C for 12 hours and then mechanically reground in an agate mortar. This process was repeated at 790°C and 795°C. Then the mix was pressed at 1.4 kbar into a 50 mm diameter target about 3-4 mm thick. The disk was then sintered at 825°C for 12 hours. Finally, the target was bonded to a copper backing plate using silver epoxy and mounted in a water cooled Corona Model C-2" sputtering source. Except where noted, all results below are from a target with a 2223 (Bi:Sr:Ca:Cu) composition.

Substrates:

From previous work[8], it was seen that MgO substrates did not strongly react with Bi-Sr-Ca-Cu-O films during the annealing process, so (100) orientation single crystal MgO substrates (typically 5x10x1mm) were chosen for most of this work. However, these crystals tended to have relatively rough surfaces, and smoother substrates such as Corning #7059 glass and fused silica were needed for the profilometry and microprobe characterization of the unannealed films.

Deposition & Equipment:

The sputtering chamber was a modified Corona Model C-50 system which was typically evacuated to a base pressure on the order of 10^{-6} Torr. Prior to deposition, the diffusion pump was throttled and high purity gas introduced. Typical sputtering gas pressures of 0.50±0.02 Pa were monitored using a Baratron capacitance manometer. Our previous work with the Bi-Sr-Ca-Cu-O films employed DC sputtering, but problems with discharge stability prompted us to switch to RF. This greatly improved the plasma stability and resulted in the targets lasting much longer. Furthermore, the erosion ring was much more uniform and no target discoloration or pitting was observed. The RF system operated at 13.56MHz and the reflected power was reduced to essentially zero using a load-matching circuit. The forward power was 80W except for the films deposited on heated substrates which were deposited using 40W of RF power. Deposition times were 2 hours (not including a 15 minute presputter), and rates were typically 7-8 nm/min. The bias voltage developed on the target varied somewhat with target composition, but was typically 145-230V for 80W of RF power.

Target-to-substrate distance was 4 cm and the substrates could be either mounted on a shuttered rotating table or on an 85W substrate heater. In either case, the substrate holder was grounded.

The substrate heater consisted of a stainless steel block with an imbedded 85W cartridge heater. The edge of the substrate was clamped to the block using a crystal of MgO. Due to the RF discharge in the chamber, temperature could not be monitored during the film deposition. Instead, the heater was calibrated in the absence of a discharge with a Chromel-Alumel thermocouple attached to the exposed substrate surface with silver paint. We had already found that heating due to substrate bombardment was neglible.

Annealing system:

The Bi-Sr-Ca-Cu-O films were annealed in a 50 mm diameter quartz tube furnace. High purity argon and oxygen could be introduced at a slight overpressure and controlled rates. Except when an air atmosphere was being used, the tube was flushed with argon for 15 minutes before the furnace was turned on. The controller zone temperature was calibrated against the temperature at the sample location using a Chromel-Alumel thermocouple.

RESULTS

Effect of oxygen in sputtering gas mixture:

Most sputtering was carried out in a pure argon atmosphere, but the effects of varying oxygen-argon mixtures on deposition rate onto unheated substrates were briefly investigated. The results indicated that partial pressures of oxygen up to about 0.1 Pa (20% of the total) do not seem to adversely affect the deposition rate. Above this, however, the rate was seen to fall off rapidly. No effects other than the reduction of rate were observed in the films due to the presence of oxygen.

Radial variations:

It was observed[13] in the 1-2-3 films that composition varied with distance from the center of the discharge. With this in mind, the as-deposited Bi-Sr-Ca-Cu-O films (on glass) were studied using an electron microprobe for radial variation in the cation ratios. Figure 1 shows that the copper concentration is higher than optimum at the center of the discharge and lower than optimum farther out. The relative fractions of the other elements remain quite constant.

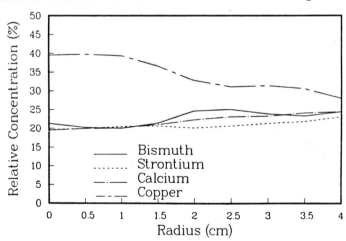

Fig. 1. Variation in cation concentration away from the center of the discharge. Nominal values (2223) are 22% for Bi, Sr, and Ca, 33% for Cu.

Another radial effect is seen in the deposited film thickness profile (Figure 2), which shows that there is a dip in the center of the film. This result strongly suggests that there is some film resputtering similar to that observed in Y-Ba-Cu-O films[14]. This effect was not as prominent with all target compositions and may be correlated with the higher target bias voltages. This decrease in thickness in the center is in contrast with the optical density obtained by a scanning densitometer which shows a monotonic decrease in film opacity with increasing radius.

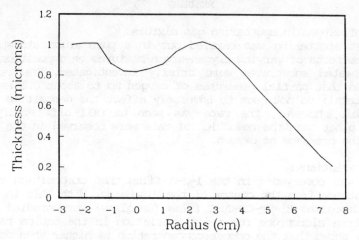

Fig. 2. Thickness profile of a 2223 film (as deposited). The center of the discharge is at R=0.

Heated substrates:

The production of as-deposited superconducting thin films is an important goal of thin film research since it simplifies the process and tends to reduce overall temperatures. Experience with $YBa_2Cu_3O_x$ films shows that this can be accomplished simply by heating the substrates. For this reason, the consequences of substrate heating during film deposition were investigated. The effects of temperature on deposition rates were studied on fused silica substrates up to 600°C (where some film-substrate reaction was first observed). The films were deposited in the center of the discharge. The results showed no reduction in deposition rate to within the measurement error (±15%) up to 600°C.

X-Ray Diffractometry (XRD) indicated some crystallinity in films deposited on MgO substrates at temperatures above 550°C, but the patterns indicated that it was primarily impurity phases being formed. One film deposited at 650°C had a relatively low resistance, but its behaviour upon cooling was semiconducting. Cooler substrates had high resistance amorphous films (See Figure 3). Other groups[11] have seen crystallinity at temperatures as low as 400°C.

Using the microprobe, the composition of the films on silica (up to 600°C) and on MgO (up to 730°C) was studied as a function of substrate temperature. The results (Figure 4) indicate that the relative concentrations of copper, strontium, and calcium are not significantly affected by the substrate temperature. However, the amount of bismuth in the deposited films decreases drastically above about 350-400°C. This has serious implications for producing as-deposited superconducting films if a high substrate temperature is required during deposition.

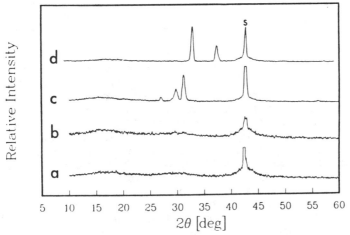

Fig. 3. Cu Kα X-ray spectrum of films deposited at different temperatures: a) 21°C b) 450°C c) 550°C d) 730°C. 'S' labels the substrate line. The other peaks are associated with impurity phases.

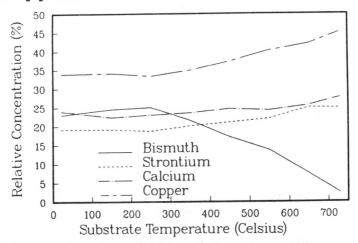

Fig. 4. Composition of the film at different substrate temperatures. The values are normalized to 100% so the rise in Ca, Cu, and Sr is largely an artifact of the decrease in Bi.

Annealed films:

The film properties were very sensitive to the annealing conditions. A difference in temperature or annealing atmosphere could lead to a significant difference in the relative proportions of the 2201, 2212, and 2223 phases as well as strongly affect the degree of film orientation, the texture and the adhesion of the film. The melting point of the films appeared to vary, presumably with the proportion of oxygen in the tube furnace. A film could be annealed as high as 895°C for 16h in an oxygen flow without significant loss of film material. On the other hand, in air the temperature would have to be closer to 860°C to prevent the evaporation of the film. This divided annealing conditions into two regimes: high oxygen and high temperature vs low oxygen and lower temperature. In both regimes, the films produced were highly oriented with the c-axis of the $Bi_2Sr_2Ca_{n-1}Cu_nO_x$ structure perpendicular to the substrate (001). They were also predominantly 2212 in phase (as seen by XRD) and longer anneal times near melting point temperatures tended to promote the 2223 (and often the 2201) phase. The most striking difference between the two regimes, however, was the degree of the film orientation. The high oxygen films tended to have very much stronger 001 orientation for the corresponding anneal times than the low oxygen ones. Furthermore, they tended to have smoother texture and better adhesion. On the other hand, the films with the highest proportions of 2223 were produced at 860°C in air (75h anneal time). See Figure 5a and b for XRD pattern and resistivity curves.

One radial effect that was observed in the annealed films was the tendency for films deposited at less than 1.5 cm from the center of the discharge to flake or evaporate much more readily.

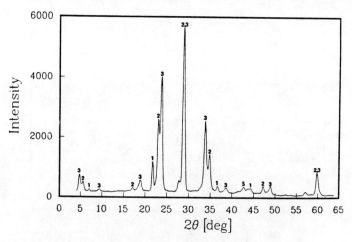

Fig. 5a. XRD scan of a film annealed at 860°C for 75 hours in air. The 001 lines of $Bi_2Sr_2Ca_{n-1}Cu_nO_x$ are labelled by n=1,2,3.

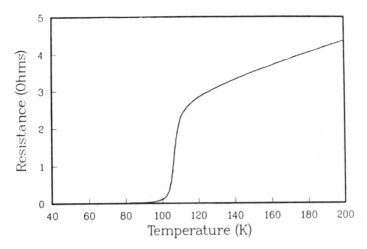

Fig. 5b. Resistance behaviour of a film annealed at 860°C for 75 hours in air. Four point current was 0.1 mA.

SUMMARY

In summary, thin films of Bi-Sr-Ca-Cu-O have been prepared by RF magnetron sputtering of superconductor oxide targets. These films show some evidence of radial variation and resputtering. The concentration of bismuth was observed to vary with substrate temperature. Upon annealing, multiphase superconducting films are formed. These films are highly oriented and their properties are strongly dependent on annealing conditions.

ACKNOWLEDGEMENTS

The authors would like to thank Dr. J. Carolan for his useful advice and the Natural Sciences and Engineering Research Council (NSERC) of Canada for their financial support.

REFERENCES

1. J. Bednorz and A. Müller, Z. Phys., B64, 189(1986).
2. C.W. Chu, P.H. Hor, R.L. Meng, L. Gao, Z.T. Huang and Y.K. Wang, Phys. Rev. Lett., 58, 405(1987).
3. Z.Z. Sheng, A.M. Hermann, Nature, 332, 58-59; 138-139(1988).
4. H. Maeda, Y. Tanaka, N. Fukutomi, T. Asano, Jpn. J. Appl. Phys. Lett., 27, L209(1988).
5. R.M. Hazen, C.T. Prewitt, R.J. Angel, N.L. Ross, L.W. Finger, C.G. Hadidiacos, D.R. Veblen, P.J. Heaney, P.H. Hor, R.L. Meng, Y.Y. Sun, Y.Q. Wang, Y.Y. Xue, Z.J. Huang, L. Gao, J.Bechtold and C.W. Chu, Phys. Rev. Lett., 60(12), 1174(1988).
6. R. Ramesh, C.J.D. Hetherington, G. Thomas, S.M. Green, C. Jiang, M.L. Rudee and H.L. Luo, Appl. Phys. Lett., 53(7), 615(1988).

7. J.M. Tarascon, W.R. McKinnon, P. Barboux, D.M. Hwang, B.G. Bagley, L.H. Greene, G. Hull, Y. LePage, N. Stoffel and M. Giroud, submitted to Phys. Rev. B.
8. B. Sullivan, N. Osborne, W. Hardy, J. Carolan, B. Yang, P. Micheal, R. Parsons, Appl. Phys. Lett., 52(23), 1992(1988).
9. C.E. Rice, A.F.J. Levi, R.M. Fleming, P. Marsh, K.W. Baldwin, M. Anzlowar, A.E. White, K.T. Short, S. Nakahara and H.L. Stormer, Appl. Phys. Lett., 52(21), 1828(1988).
10. M. Fukutomi, J. Machida, Y. Tanaka, T. Asano, H. Maeda and K. Hoshino, Jpn. J. Appl. Phys. Lett., 27(4), L632(1988).
11. H. Adachi, Y. Ichikawa, K. Setsune, S. Hatta, K. Hirochi and K. Wasa, Jpn. J. Appl. Phys. Lett., 27(4), L643(1988).
12. C.R. Guarnieri, R.A. Roy, K.L. Saenger, S.A. Shivashankar, D.S. Yee and J.J. Cuomo, submitted to Appl. Phys. Lett.
13. D. Burbidge, P. Mulhern, S. Dew, R. Parsons, AIP Conf. Proc. 165, 87(1988).
14. S.M. Rossnagel and J.J. Cuomo, AIP Conf. Proc., 165, 106(1988).

SINGLE-PHASE HIGH T_c SUPERCONDUCTING $Tl_2Ba_2Ca_2Cu_3O_{10}$ FILMS

M. Hong, J. Kwo, C. H. Chen, A. R. Kortan, D.D.Bacon

AT&T Bell Laboratories, Murray Hill, New Jersey 07974

and

S. H. Liou

Department of Physics, University of Nebraska-Lincoln, NE 68588

ABSTRACT

We have routinely produced single-phase high T_c superconducting $Tl_2Ba_2Ca_2Cu_3O_{10}$(2223) thin films 0.2 to 1.0 μm thick on MgO(100) and $SrTiO_3$(100) substrates by dc diode sputtering. A single compound target with a composition of $Tl_{2.3}Ba_2Ca_2Cu_3O_x$ was used. Pure argon sputtering has been employed. The as-deposited films need subsequent heat treatment under air or O_2 with an inclusion of a Tl-Ba-Ca-Cu-O powder compact to become superconducting. The $Tl_2Ba_2Ca_2Cu_3O_{10}$ films have a T_c onset at 125K and a T_c(R=0) at 116K. Critical current densities J_c's measured by the transport method are as high as $10^5 A/cm^2$ at 100K. X-ray diffraction studies show that the films are of single phase, and have a strong preferred orientation with the c-axis perpendicular to the film plane. Film morphology and microstructure are studied by scanning electron microscopy (SEM) and transmission electron microscopy (TEM).

INTRODUCTION

Following the initial discovery of superconductivity in the Tl-Ba-Ca-Cu-O system,[1,2] the highest T_c(R=0) at 125K has been observed in the compound of $Tl_2Ba_2Ca_2Cu_3O_{10}$ (2223).[3,4] Thin film research on the Tl-based superconductors is particularly important because their T_c(R=0)'s are much higher than those of the rare earth based $Y_1Ba_2Cu_3O_{7-x}$ (123) compounds. Furthermore, it is likely to obtain more stable high critical current densities in excess of $10^5 - 10^6 A/cm^2$ at 77K in the Tl-based system.

Recently, unoriented polycrystalline films 0.7 μm thick containing predominantly $Tl_2Ba_2Ca_1Cu_2O_8$ (2212) phase with T_c(R=0) at 97K and a transport

J_c of $1.1 \times 10^5 \text{A/cm}^2$ at 76K have been obtained by Ginley et al[5] using electron beam evaporation. Highly oriented films 0.4 μm thick (with the c-axis perpendicular to the film plane) containing nearly single phase of 2212 as well as having a $T_c(R=0)$ at 102K and a transport J_c of $1.2 \times 10^5 \text{A/cm}^2$ at 77K have been prepared by Ichikawa et al[6] using RF magnetron sputtering and a single complex oxide target. Highly textured c-axis oriented films 2.0 to 4.0 μm thick containing both 2223 and 2212 phases with a $T_c(R=0)$ at 120K and a J_c by the magnetic measurement of $1.5 \times 10^4 \text{A/cm}^2$ at 77K have been reported by Lee et al[7] using two identical oxide targets in a symmetrical RF diode sputtering system.

The ability to produce thin films with the pure high T_c $Tl_2Ba_2Ca_2Cu_3O_{10}$ phase not only can further improve J_c's at higher temperatures but also may provide a good material for a fundamental study on the Tl-based superconducting oxides. Here, we report the preparation and characterizations (crystal structures, transport properties, and microstructures) of single-phase superconducting $Tl_2Ba_2Ca_2Cu_3O_{10}$ thin films 0.2 to 0.4 μm thick on both MgO(100) and $SrTiO_3$(100) substrates. These films have a T_c onset at 125K, a $T_c(R=0)$ at 116K, and transport critical current densities of 10^5A/cm^2 at 100K measured at zero magnetic field. The films are highly textured with the c-axis normal to the film plane. The typical grain size is around 10 μm from the SEM study. A superlattice structure similar to that of the Bi-Sr-Ca-Cu-O system[8] has been revealed by a transmission electron diffraction study.

EXPERIMENTAL

The film deposition was carried out by a simple DC diode getter sputtering method. Pure argon sputtering was used with the pressure kept at 80 mTorr during deposition. No partial oxygen pressure was added in this work. The detailed experimental procedure was given elsewhere[9,10]. The film thickness is in the range of 0.2 to 0.4 μm. The substrate temperatures were varied from room temperature to 500°C. However, in this study, we have found that the substrate temperature has not affected the measured superconducting properties. The film deposition area is confined within a liquid-nitrogen cooled stainless steel can 7.5 cm in diameter and 10 cm in length, therefore eliminating the contamination of Tl oxides or others over the rest of the vacuum chamber. The targets were prepared from thallium oxide, copper oxide, barium carbonate, and calcium carbonate powders. Appropriate amounts of powders were mixed, cold pressed into a disk-shape plate 0.3 thick and 2.5 cm in diameter under a pressure of 1.5 tons/cm^2. The plate was wrapped in a gold foil, reacted at 880°C for 3 hours in a sealed quartz tube, and then cooled to room temperature in 2 hours. The nominal target composition (metal cation ratio) is 2.3:2:2:3 of Tl:Ba:Ca:Cu, which was obtained by assuming that the loss of the material is due to the Tl oxide. The actual film composition was determined by Rutherford backscattering spectrometry (RBS) on unreacted samples 0.03 μm thick to be 2:1.86:2.08:3.04:x of Tl:Ba:Ca:Cu:O.

Single crystal x-ray diffraction measurements were made on a 12 kW rotating Cu anode source equipped with a triple-axis four-circle diffractometer. A pair of perfect, flat Ge(111) monochromator and analysis crystals were used to provide a spatial resolution of 8×10^{-4} Å^{-1} for the parallel scans and 3×10^{-5} Å^{-1} for the transverse or rocking scans. Planar-view samples for TEM studies were prepared by mechanical polishing followed by ion-milling from the substrate side. After the appearance of pinholes in the samples, a brief cleaning by 3 kV Ar ion from the thin-film side was carried out to remove any unwanted deposit on the film surface during the substrate thinning process.

RESULTS AND DISCUSSION

The as-deposited films are amorphous and insulating, and need post-anneal at temperatures above 800°C to become superconducting. The heat treatment procedures are critical because of the loss of Tl or Tl oxide during the annealing. In order to minimize the loss, the films were wrapped in a gold foil with pellets of compressed composite Tl-Ba-Ca-Cu-O oxide powders and sealed in a quartz tube. The quartz tube was filled with 1.0 atm of air or O_2. We found that the films heat-treated under air have the same superconducting properties and crystal structures as the films under pure O_2.

The ability to obtain pure single-phase $Tl_2Ba_2Ca_2Cu_3O_{10}$ films depends very much on the initial as-deposited film composition, and more importantly, on the final highest annealing temperature and the heat treatment duration. For films heat treated at 870°C for 10 - 30 min, films containing the pure 2223 phase have been routinely produced. In Figure 1, the periodical sharp peaks observed in the X-ray diffraction pattern (θ - 2θ scan) are the (00ℓ) peaks of the $Tl_2Ba_2Ca_2Cu_3O_{10}$ phase with the lattice constant c = 35.647Å. Note that the unit of the x-axis is normalized based on the lattice constant c = 35.647Å, and the x-axis is plotted in terms of (00ℓ). This $Tl_2Ba_2Ca_2Cu_3O_{10}$ film was deposited on MgO(100) substrate. From the X-ray diffraction study, we have observed no trace of second phase(s) down to 1% level as clearly shown in Figure 1. The rocking curve of the (0014) peak, which measures the alignment of the crystallinity of the film normal to the substrate is 0.22° full width at half maximum (FWHM). The rocking curve data is shown in Figure 2. When the films were heat-treated under 860°C, the major phase is the $Tl_2Ba_2Ca_1Cu_2O_8$ with $T_c(R=0)$ at 102K. For films annealed over 880°C or at 870°C over 2 hours, some unknown phases appeared.

The films grown on MgO(100) were also studied by transmission electron microscopy (TEM). Again, the TEM study shows that the films are free of any second phase or inclusions. The grain size is, on the average, of 10 μm. The electron diffraction pattern obtained along the [001] zone-axis (Figure 3(a)) does not reveal any superlattice spots in the zero-order Laue zone. However, weak satellite reflections along both [100] and [010] directions are clearly visible in the first-order Laue zone. This observation suggests that the superlattice reflections must have a component along the c direction. The appearance of the superlattice is,

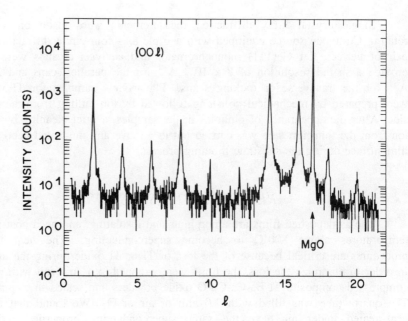

Figure 1 X-ray scan along (00ℓ) in a $Tl_2Ba_2Ca_2Cu_3O_{10}$ film grown on MgO.

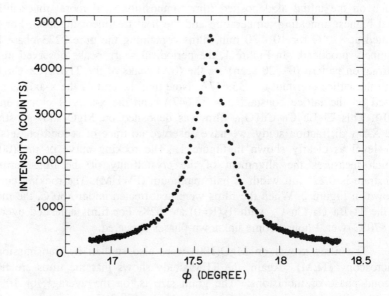

Figure 2 Rocking curve of (0014) plane of the $Tl_2Ba_2Ca_2Cu_3O_{10}$ film on MgO.

therefore, best demonstrated from an area that is oriented with [100]-axis parallel to the electron beam. Figure 3(b) shows the electron diffraction pattern obtained along the [100] zone axis. We see immediately that the superlattice is characterized by an incommensurate periodicity approximately six times of the lattice parameter along the a-axis (3.86Å) and a commensurate periodicity of the c lattice parameter (35.6Å) along the c-axis.

This incommensurate superlattice is similar to that observed in the Bi-based oxide superconductors, in which the incommensurate periodicity is 4.76 times of the lattice parameter along the b-axis.[8] We note that the incommensurate modulation in the Tl-based superconductors is much weaker than that observed in the Bi-based superconductors. Nevertheless, the incommensurate superlattice

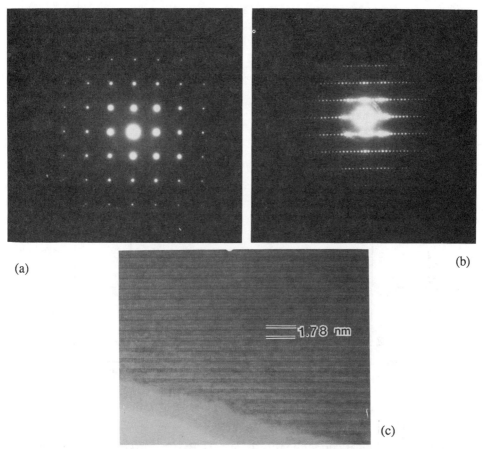

Figure 3 (a) [001] zone-axis electron diffraction pattern, (b) [100] zone-axis electron diffraction pattern, and (c) high resolution lattice images obtained along the [100] zone-axis.

reflections we observed from our thin film samples are much sharper than those reported for a bulk ceramic sample of the $Tl_2Ba_2Ca_1Cu_2O_8$ phase[11]. High resolution lattice image obtained along the [100] zone axis is also shown in Figure 3(c). Our thin film sample is very clean and free of defects such as intergrowth of perovskite layers with different layer-thickness which are commonly observed in the Bi-based superconductors.

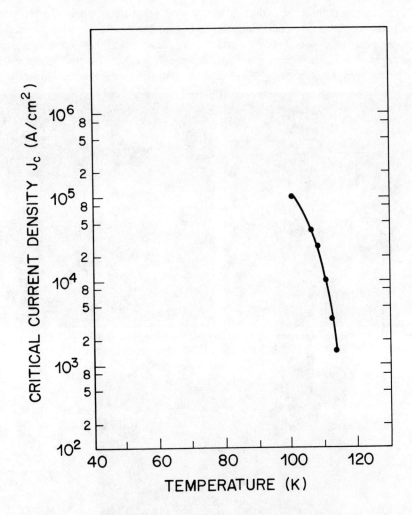

Figure 4   Critical current density J_c of a $Tl_2Ba_2Ca_2Cu_3O_{10}$ film on $SrTiO_3$ as a function of temperature.

The transport properties were measured by the standard four-point measurement using a dc method by switching the polarization of the applied current, which is usually about 2.5-100 µA. The room temperature resistivity of the $Tl_2Ba_2Ca_2Cu_3O_{10}$ films deposited on MgO or $SrTiO_3$ is low, around 100-300 µΩ-cm. The superconducting transitions are T_c onsets at 125K and $T_c(R=0)$ at 116K. For measuring the transport critical current density J_c, the films were scribed to form a narrow conducting path 100-200 µm wide. The critical current was taken to be the current at which the potential difference across the voltage leads (spaced 1 cm apart) exceeds 10 µV, and the J_c was taken to be the critical current divided by the total cross-sectional area of the narrowed restriction. Figure 4 shows the data of transport J_c versus temperature for a typical $Tl_2Ba_2Ca_2Cu_3O_{10}$ film grown on $SrTiO_3$. At zero magnetic field, the J_c's are as high as $10^5 A/cm^2$ at 100K. J_c's carried by the films deposited on $SrTiO_3(100)$ are slightly higher than those by the films on MgO(100), presumably due to a different morphology as studied by a scanning electron microscopy (SEM) study[12]. We believe that the J_c value in the present films is the lower bound of the critical current density carried by films with a pure $Tl_2Ba_2Ca_2Cu_3O_{10}$ phase, considering the degree of porosity present in our films. A typical example is shown in Figure 5 for the morphology of a $Tl_2Ba_2Ca_2Cu_3O_{10}$ film on $SrTiO_3$. With the further improvement on the film morphology, a transport J_c can be enhanced to reach $10^6 A/cm^2$ at 100K in the present Tl-based superconducting oxide films.

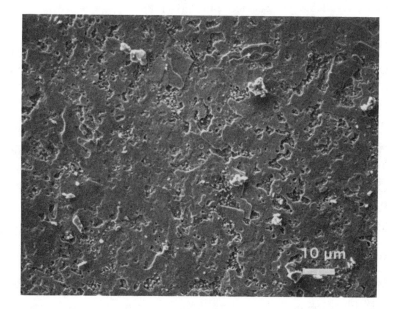

Figure 5   Scanning electron micrograph of a $Tl_2Ba_2Ca_2Cu_3O_{10}$ film on $SrTiO_3$[12].

CONCLUSION

In summary, we have routinely prepared Tl-based superconducting thin films with the pure 223 phase by using a simple dc diode sputtering technique and employing a single composite oxide target. The films on both MgO(100) and SrTiO$_3$(100) are strongly textured with the c-axis normal to the film plane. The films on MgO are free of second phase(s) and other defects as shown by the X-ray diffraction and TEM studies. Full width at half maximum (FWHM) of the rocking curve at (0014) is as narrow as 0.22°. An incommensurate superlattice is observed in our films. The superconducting transition is T_c onset at 125K and T_c(R=0) at 116K. The transport J_c is as high as 10^5 A/cm^2 at 100K. This value can be further enhanced when the film morphology is optimized.

REFERENCES

1. Z. Z. Sheng and A. M. Hermann, Nature **332**, 138 (1988).
2. R. M. Hazen, L. W. Finger, R. J. Angel, C. T. Prewitt, N. L. Ross, C. G. Hadidiacos, P. J. Heaney, D. R. Veblen, Z. Z. Sheng, A. El Ali, and A. M. Hermann, Phys. Rev. Lett. **60**, 1657 (1988).
3. C. C. Torardi, M. A. Subramanian, J. C. Calabrese, J. Gopalakrishnan, K. J. Morrissey, T.Sleight, Science, **240**, 631 (1988).
4. S. S. P. Parkin, V. Y. Lee, E. M. Engler, A. I. Nazzal, T. C. Huang, G. Gorman, R. Savoy, and R. Beyers, Phys. Rev. Lett. **60**, 2539 (1988).
5. D. S. Ginley, J. F. Kwak, R. P. Hellmer, R. J. Baughman, E. L. Venturini, and B. Morosin, Appl. Phys. Lett. **53**, 406 (1988).
6. Y. Ichikawa, H. Adachi, K. Setsune, S. Hatta, K. Hirochi, and K. Wasa, Appl. Phys. Lett. **53**, 919 (1988).
7. W. Y. Lee, V. Y. Lee, J. Salem, T. C. Huang, R. Savoy, D. C. Bullock, and S. S. P. Parkin, Appl. Phys. Lett. **53**, 329 (1988).
8. C. H. Chen, D. J. Werder, S. H. Liou, H. S. Chen, and M. Hong, Phys. Rev. B **37**, 9834 (1988).
9. M. Hong, S. H. Liou, D. D. Bacon, G. S. Grader, J. Kwo, A. R. Kortan, and B. A. Davidson, Appl. Phys. Lett. (to be pulished in a November issue, 1988)
10. M. Hong, E. M. Gyorgy, and D. D. Bacon, Appl. Phys. Lett. **44**, 706 (1984).
11. J. D. Fitz Gerald, R. L. Withers, J. G. Thompson, J. S. Anderson, and B. G. Hyde, Phys. Rev. Lett. **60**, 2797 (1988).
12. S. H. Liou, M. Hong, A. R. Kortan, J. Kwo, D. D. Bacon, C. H. Chen, R. C. Farrow, and G. S. Grader, (to be published).

Oriented Growth of $Y_1Ba_2Cu_3O_x$
Thin films by Dual Ion Beam Sputtering

J.P. Doyle, R.A. Roy, D.S. Yee, and J.J. Cuomo
IBM Research Division, T.J. Watson Research Center, P.O. Box 218
Yorktown Heights, New York 10598

ABSTRACT

We report the deposition of crystalline as-deposited ion beam sputtered $Y_1Ba_2Cu_3O_x$ thin films with orientations of (100), (001), and (110)/(103) and a method for controlling the orientation of the films. Films deposited on Al_2O_3 Yittria (9%) stabilized ZrO_2 typically have a near random orientation, while the films on MgO show a high degree of orientation. In addition, post deposition heat treatments at 500 C were found to reduce the room temperature resistivity significantly, although no superconductivity was observed. High temperature annealing produced superconducting material, maintaining a specular surface. RBS analysis shows a variation in the composition as a function of substrate position on the platen, which indicates a re-sputtering mechanism due to highly energetic neutrals from the target surface.

INTRODUCTION

$Y_1Ba_2Cu_3O_x$ (123) thin films have been prepared using a variety of methods [1-4]. Initially, high temperature processing of the film and substrate similar to that used in bulk ceramic synthesis [5] was used to post process the films in order to achieve High Tc characteristics. Associated with such temperatures are undesirable interactions between film and substrate, placing a severe restriction on the type of substrates available for film deposition [6]. Recent results show superconducting films can be obtained after deposition on substrates whose temperature is maintained at 500 °C to 700 °C, followed by annealing at lower temperatures [7]. In addition to reducing the processing temperature, several groups have demonstrated the ability to produce highly oriented films on substrates such as $SrTiO_3$. In this paper we report the oriented growth of 123 films by ion beam sputter deposition in the temperature range of 600 °C to 700 °C. By changing depostion conditions such as substrate temperature, substrate type, and ion bombardment of the growing film, the degree and type of orientation can be changed. Highly oriented (100),(001), (110), and (103) films can be deposited at the same temperature by varying other parameters. This type of control of orientation of 123 films by psuedo-epitaxial growth on MgO has not been reported and opens up the possibility for controlling the texture on other substrates. The more highly oriented films are found to display superior T_c over the randomly oriented films.

EXPERIMENTAL

Film depositions were carried out in a turbomolecular-pumped dual ion beam deposition system that acheived base pressures below 3×10^{-7} Torr. As shown in Figure 1, the deposition system consisted of a 2.5 cm dual carbon grid source for target bombardment, while a 3 cm source was used for substrate bombardment during film growth. The target ion beam (2.5 cm) was composed of Ar gas while an inlet independent of the target source was used for oxygen. The oxygen background pressure

© 1989 American Institute of Physics

ranged up to 0.4 mTorr. In some cases, a substrate ion beam (3 cm) was directed at the substrates during film growth and consisted of either an Ar/O_2 gas mixture or pure O_2. The substrates could be heated to about 800 °C using a radiant heater. The target composition used in this work was $Y_1Ba_{2.4}Cu_{3.8}O_x$ to compensate for Ba and Cu loss found in previous investigations [8]. Deposition temperatures were varied between 600 °C and 750 °C. Films were selected from each deposition and annealed in a variety of atmospheres at temperatures ranging from 400 °C to 950 °C. All Al_2O_3 and Yittria (9%) stabilized ZrO_2 (YsZrO$_2$) substrates were cleaned by detergent scrubbing followed by a DI water rinse and spin drying. This preparation method will be referred to as a standard cleaning. The MgO substrates were prepared in one of two ways: In the first method, the substrates were standard cleaned. In the second method, a chemical polish using hot phosphoric acid was used.

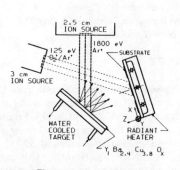

Figure 1. System Configuration

RESULTS

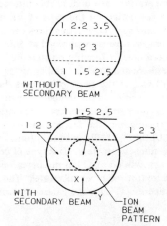

Figure 2. Composition as a function of position on the platen with and without substrate bombardment.

The composition of films deposited at various positions on the substrate platen are shown in Figure 2. At 650 °C to 710 °C, substrates in the center of the platen are slightly Ba- and Cu-poor with respect to to the target composition, but close to the desired 1-2-3 stoichiometry. Films at the bottom of the platen are found to be more deficient in Ba and Cu. The selective removal of Ba and Cu may be caused by high energy neutrals created by Ar^+ ions incident on the target, whose effects in ion beam sputtering have been extensively documented, and are often comparable to that of a secondary beam directed at the substrate [9]. In addition, a more severe Ba and Cu deficency was observed when a secondary ion beam of Ar/O_2 ions with energy of 125 eV was used, as shown in Figure 2. The effect was not as pronounced when the energy was lowered to 75 eV and pure O_2 was used.

Films deposited at 650 °C and above with O_2 in the background during deposition were found to have moderate conductivity as-deposited and XRD analysis confirmed that the films were crystalline. Films deposited on YsZrO$_2$ and Al_2O_3 typically displayed a near random orientation, but were not examined over an extensive matrix of conditions. However, films deposited on MgO (100) were, in general, highly oriented. The X-Ray Diffraction patterns of these films showed reflections characteristic of the tetragonal 123 phase with an (100) parameter of 3.86 Å to 3.9 Å and an (001) paramenter of 12.0 Å to 12.2 Å.

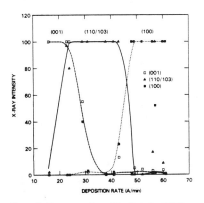

Figure 3. Texture vs. Deposition Rate for 123 films deposited on standard cleaned MgO (100) at 675 °C

Figure 3 shows the effect of orientation as a function of deposition rate for MgO substrates that were standard cleaned. When deposited above 48 Å/min the films showed a strong (100) orientation with a few crystallites of (001) orientation; however, below about 42 Å/min a mixture of the (110) and (103) orientation is seen. When the deposition rate is lowered to below 20 Å/min by decreasing the ion beam current incident on the target, the films on MgO exhibited a strong (001) texture. Table 1 additionally shows that the predominant texture goes according to the following sequence as the rate is lowered: (100) - (110) - (103) - (001).

Preliminary TEM analysis shows that under all deposition conditions some c-oriented grains grow from the substrate interface; however, beyond the interface, the orientation of the majority of the crystallites is found to depend on the deposition rate [10]. The effect of secondary bombardment on the orientation of the deposited films has also been studied. When films were deposited at a rate less than 20 Å/min, and bombarded with an Ar/O_2 ion beam of 125 eV with a current density of 0.025 mA/cm^2, the texture of the films is found to be highly (100) oriented, similar to that observed for films deposited above 50 Å/min without a secondary beam. On the other hand, for films deposited at 50 Å/min with concurrent bombardment from an oxygen ion beam with 75 eV, the orientation is (100), the same as unbombarded films deposited at that rate.

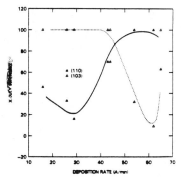

Figure 4. Texture vs. Deposition Rate for 123 films deposited on acid polished MgO (100) at 675 °C

Figure 4 shows the orientation of 123 films versus deposition rate on chemically polished MgO (100) substrates. Again an orientation dependence on rate is observed. The (103) and (110) orientations dominate over the entire deposition range investigated although the amount of (001) and (100) crystallites increases at the low and high extremes. The (103) orientation is seen to be stronger below 45 Å/min while the (110) is stronger above 55 Å/min.

The room temperature resistivity of as-deposited films ranged from 20 milliohm-cm to about 1 ohm-cm for crystalline films. However, even in the films showing the greatest conductivity, four-point probe measurements down to 4.2 °K showed an increase in resistance with decreasing temperature and no superconductivity. Therefore, a portion of the samples were subjected to treatment under various annealing schedules ranging from 400 to 950 °C in order to determine if the films would show superconductivity.

Films were first exposed to a series of low temperature anneals in the range of 400 to 500 °C. After a 500 °C anneal for five minutes in flowing O_2, the resistivity fell to several milliohm-cm or less on the MgO (100) substrates. The $YsZrO_2$ and Al_2O_3 substrates had a somewhat higher value. These films were again measured to 4.2 °K in order to observe any superconducting transitions. The results showed an increase in resistance, but much less steeply than before annealing. XRD analysis showed that the films were still tetragonal. A vacuum anneal was then performed in the range of 550 to 650 °C and a significant increase in resistivity is seen. These films were then re-annealed at 500 °C in flowing O_2. The result is a drop in resistivity back to the value observed after the first 500 °C anneal. Further annealing for up to 16 hours in flowing O_2 produced no change in resistivity on MgO.

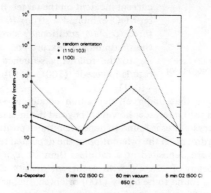

Figure 5. Room Temperature Resistivity vs. Annealing conditions on MgO (100).

Figure 5 shows the results for films deposited on MgO with different textures which suggest that there is a rapid diffusion of oxygen into and out of the films at 500 °C. At the vacuum annealing stage, the randomly oriented films incurred a much larger resistance increase, possibly indicative of more oxygen loss among other things. The (100) oriented film, in contrast, only rose to about 40 milliohm-cm. Although the films in this set did not show any signs of superconductivity after the low temperature processing, when annealed for several minutes in oxygen above 700 °C, followed by a slow cooling, superconductivity is observed. The quality of the transition improved with temperature up to 900 °C, where zero resistance is observed at 60 °K as shown in Table 1.

Substrate	Tonset	Tcompletion	Annealing Temperature
Al2O3	67 K	< 10 K	900 C
YsZrO2	70 K	< 10 K	750 C
YsZrO2	68 K	Broad	800 C
YsZrO2	75 K	48 K	900 C
MgO (100)	50 K	12 K	750 C
MgO (100)	77 K	48 K	800 C
MgO (100)	90 K	60 K	900 C

Table 1. Results of High Temperature Annealing of Films on various substrates.

DISCUSSION

The growth of the a-axis texture on MgO is interesting, inasmuch as solid phase recrystallization of amorphous films has shown c-axis texture on MgO (100) substrates. The fact that the texture can be changed continuosly from (100) to (110)/ (103) to (001) by lowering the deposition rate raises the question as to the mechanism responsible for the texture changes. Oriented growth on $SrTiO_3$ generally has been shown to be epitaxial, with (100) and (001) texture seen on $SrTiO_3$ (100) and (110) texture on $SrTiO_3$ (110). Fujita et al [11] have shown a change from (001) to (100) texture at a temperature in which the lattice mismatch between (100) film plane and the $SrTiO_3$ becomes larger than the mismatch between (001) film plane and the substrate, in accordance with behavior expected for true epitaxial growth. In the case of MgO the lattice mismatch is about 10% between (100) film and substrate directions. Therefore the changes in orientation observed are unlikely to be due to small changes in the degree of film-substrate mismatch, but rather to other factors, possibly both at the nucleation stage and during film growth. The fact that acid polishing the surface can produce (110) texture under conditions where (100) texture is seen on unpolished substrates, demonstrates that the MgO substrate interface can play a role in controlling the subsequent texture. Furthermore, the fact that the use of a low energy Ar/O_2 ion beam incident on the film during growth causes the texture to change from (001) to (100), demonstrates the role that energetic bombardment during growth can play in controlling film orientation. Since increasing the deposition rate by increasing the primary beam current changes the texture in a similar manner, one might infer that an increase in energetic substrate bombardment is indirectly incurred by increasing the ion flux to the target, possibly by producing a larger amount of reflected high energy neutrals. The fact that (001) crystallites are invariably found to some degree at the interface suggests that crystallite orientation in non (001) directions also occurs during growth in higher rate deposition or when secondary ion bombardment is used.

The fact that energetic substrate bombardment may be an important mechanism is further highlighted by the observed composition-orientation relations. It is interesting to note the relation between film composition and orientation for several depositions on MgO substrates. The films deposited at high rate without direct ion bombardment of the substrates have composition close to 123 and show strong (100) texture. The films deposited at lower rate generally had excess Cu and showed (001) texture. The films deposited under direct substrate ion bombardment at low rate were moderately Cu and Ba-deficient due to selective re-sputtering, but nevertheless showed a strong (100) texture. The coincidence of Cu- and Ba-deficient compositions with (100) texture might thus reflect the effect of high energy substrate bombardment, which would tend to selectively re-sputter Ba and Cu. Although the occurence of substrate bombardment is well established, the mechanism by which this controls orientation is not yet certain.

It is interesting to compare the orientations at high and low rates on polished and unpolished substrates with respect to the the internal arrangement of atomic planes for the different orientations. On the unpolished substrates, the low rate c-axis texture contains Cu-O planes parallel to the surface, while in the a-axis texture, the Cu-O planes are perpendicular to the surface. At the low rate on the polished substrates, the (103) is dominant and has the Cu-O planes at 45° with respect to the substrate surface, while the (110) at the high rate has Cu-O planes at 90 ° planes. This tendency for the 90 ° arrangement of Cu-O planes at high rates suggests that subtle structural differences in the unit cell tend to favor such orientation under ion or energetic neutral bombardment. Alternatively, the bombardment may simply produce more nucleation on the MgO substrate of non (001) oriented grains, which grow at the expense of the (001).

The behavior of films subject to oxygen annealing at low temperatures contrasts to that observed for polycrystalline bulk materials. In the bulk, the reversible transition between semiconducting tetragonal phase and superconducting orthorhombic phase is easily facilitated by annealing in oxygen at 450 °C 500 °C [12]. In this study no films annealed below 700 °C were found to become superconducting or exhibit orthorhombic structure. Although oxygen appears to diffuse rapidly in and out of the films at 500 °C, as evidenced by the large and reversible changes in resistivity and the temperature coefficient of resistance, the phase transition does not seem to occur. The fact that moderately high superconducting transitions can be produced by subsequent high temperature annealing tends to rule out substrate contamination as a factor contributing to the inability to produce superconducting films at low temperature. However, the large lattice parameter values observed in as-deposited films may be an important factor in preventing superconductivity. Further x-ray analysis is being conducted to monitor lattice parameter changes as a function of annealing conditions. Additional work on controlled ion bombardment of substrates at fixed deposition rates is also under way to establish the mechanism by which ion bombardment changes the texture.

CONCLUSION

We have observed highly oriented growth of $Y_1Ba_2Cu_3O_x$ on MgO (100). The crystallographic textures that have been produced are (100), (001), and (110)/(103). By controlling the deposition rate, secondary ion bombardment, and substrate preparation we are able to produce the desired texture. In addition, secondary ion bombardment during deposition has been found to change the orientation of the films. Reflected, energetic particles from the target appear to play a role in this preferred orientation and lattice distortion. Low temperature annealing enhanced the room temperature resistivity of the films, but did not produce any superconducting transitions.

ACKNOWLEDGEMENTS

We would like to thank Grant Coleman for the RBS analysis and Don Mikalsen and Steve Rossnagel for many helpful comments.

REFERENCES

1. M. Naito, R.H. Hammond, B. Oh, M.R. Hahn, J.W.P. Hsu, P. Rosenthel, A.F. Marshall, M.R. Beasley, T.H. Geballe, A. Kapitulnik, J.Mater.Res. 2 (6) 713 (1987)
2. R.M. Silver, J. Talvacchio, and A.L. de Lozanne, (Preprint)
3. J.J. Cuomo, M.F. Chisholm, D.S. Yee, D.J. Mikalsen, P.B. Madakson, R.A. Roy, E. Geiss, and G. Scilla, presented at the 1987 AVS Annual Symposium, Anaheim, CA, 1987
4. D. Dijkkamp, X.D. Venkatesan, X.D. Wu, S.A. Shaheen, N. Jisrawi, Y.H. Min-Lee, W.L. McLean, and M. Croft, Appl. Phys. Let., 51 619 (1987)

5. M.K. Wu, J.R. Ashburn, C.J. Torng, P.H. Hor, R.L. Meng, L. Gao, Z.J. Huang, Y.Q. Wang, and C.W. Chu, Phys. Rev. Let., 58 908 (1987)
6. J.J. Cuomo, M.F. Chisholm, D.S. Yee, D.J. Mikalsen, P.B. Madakson, R.A. Roy, E. Giess, and G. Scilla, A.I.P. Conf. Proc. 165, 141 (1988)
7. T. Terashima, I. Iijima, K. Yamamoto, Y. Bando, and H. Mazaki, Jpn. J. Appl. Phys. 27 L91 (1988)
8. P.B. Madakson, J.J. Cuomo, D.S. Yee, R.A. Roy, and G. Scilla, J. Appl. Phys. 63 (6), 2046 (1988)
9. E. Kay, F. Parmigiani, and W. Parrish, J. Vac. Sci. Technol. A 5 (1), 44 (1987)
10. M.F. Chisholm (Private Communication)
11. J. Fujita, T. Yoshitake, A. Kamijo, T. Satoh, and H. Igarashi, MRS Extended Abstracts High Tc Superconductors II pp. 109-112 (1988)
12. P.K. Gallagher, H.M. O'Bryan, S.A. Sunshine, and D.W. Murphy, (Preprint)

SUPERCONDUCTING Bi-Sr-Ca-Cu-O FILMS BY SPUTTERING USING A SINGLE OXIDE TARGET

M. Hong, J.-J. Yeh, J. Kwo, R. J. Felder,
A. Miller, K. Nassau, and D. D. Bacon

AT&T Bell Laboratories, Murray Hill, NJ 07974

ABSTRACT

We have prepared superconducting thin films of Bi-Sr-Ca-Cu-O with $T_c(R=0)$ at 82K and a 20% sharp drop in resistivity around 115K by RF magnetron sputtering using a single oxide target. The nominal target composition is (metal cations) 2:1.56:1.09:1.86 of Bi:Sr:Ca:Cu. The as-deposited films have a compositional ratio of $Bi_2Sr_{1.9}Ca_{1.3}Cu_{2.6}O_x$ ($\pm 5\%$) as determined by Rutherford backscattering spectrometry (RBS). The substrates used here are single-crystal $SrTiO_3(100)$ and $MgO(100)$. With heat treatments at temperatures above 860°C, we have obtained superconducting films with an almost single-phase $Bi_2Sr_2Ca_1Cu_2O_8$ as revealed by both X-ray diffraction and transmission electron microscopy (TEM). The films are highly textured with the c-axis perpendicular to the film plane. The room-temperature resistivity of the films is low around 100-200 $\mu\Omega$-cm. High critical current densities, in zero magnetic field, are of 10^5 A/cm^2 measured by the transport method at temperatures 10K below $T_c(R=0)$. The superconducting properties such as T_c's and J_c's are independent of the film thickness in the range between 0.07 to 0.2 μm.

INTRODUCTION

The discovery of superconductivity above 100K in copper oxide systems containing (Bi, Sr, and Ca)[1] and (Tl, Ba, and Ca)[2] has again caused a tremendous research activity of bulk ceramics as well as thin films on these new oxide systems. Thin film studies of the high temperature superconductors are important for a fundamental understanding of the high T_c phenomena and furthermore, for a realization of potential electronic device applications. Both the Bi- and Tl-bearing superconducting phases possess similar crystal structures as a layered perovskite oxide arrangement.[3,4] In the Tl-based system, several superconducting phases have been observed and some are identified as $Tl_2Ba_2Ca_{n-1}Cu_nO_{4+2n}$ for n = 1, 2, and 3.[4] The corresponding T_c's are 80, 110, and 125K. Similarly, for the

Bi-based system, two superconducting phases are identified: $Bi_2Sr_2Ca_1Cu_2O_8$ (2212 phase) of $T_c(R=0)$ about 80K and $Bi_2Sr_2Ca_2Cu_3O_{10}$ (2223 phase) of 107K.[5]

For the Tl-based system, the highest T_c 2223 phase has been stabilized and purified both in bulk[4] and thin-film forms[6]. For the Bi-based system, there has been certain degree of difficulty in obtaining the pure 2223 phase either in bulk ceramics or in thin films. In this paper, we report the preparation, and the characterizations of both structural and transport properties of superconducting Bi-Sr-Ca-Cu-O thin films.

We have prepared superconducting Bi-Sr-Ca-Cu-O thin films with $T_c(R=0)$ at 82K by RF magnetron sputtering. A single oxide target with a nominal composition (metal cations) 2:1.56:1.09:1.86 of Bi:Sr:Ca:Cu was used. The films need post oxygen annealing to become superconducting. X-ray diffraction studies show that the reacted films consist of nearly pure $Bi_2Sr_2Ca_1Cu_2O_8$ (2212) phase with a lattice constant c = 30.6 Å. The films are strongly textured with the c-axis perpendicular to the film plane. For films heat treated at 860°C for 20 min, only one superconducting transition with $T_c(R=0)$ in the range of 77-82K was obtained. However, for films heat treated at higher temperature of 885°C for 5-10 min, a 20% sharp drop in resistivity around 115K followed by a $T_c(R=0)$ at about 75-80K has always been observed. A cross sectional transmission electron microscopy (TEM) study[7] on the latter films showed that the stacking sequence along the c-axis contains a few isolated regions with a c-spacing of 36Å in addition to the major area with the c-spacing of 30Å. Transport critical current densities carried by these films deposited on both $SrTiO_3$ and MgO single-crystal substrates are as high as $10^5 A/cm^2$ at temperatures 10K below $T_c(R=0)$. The criterion for defining J_c is that the potential difference across the voltage leads (spaced 1 cm apart) exceeds 1.0 μV.

FILM PREPARATION AND CHARACTERIZATION

The film deposition was carried out by a RF magnetron sputtering technique using a single oxide target. Pure argon sputtering was used with the pressure kept at 30 mTorr during deposition. This Ar pressure is relatively high for magnetron sputtering. Film deposition using a lower argon pressure of 5 mTorr is now in progress to study the effect of the argon pressure. In this work, the substrate temperature was kept at room temperature. Experiments using a mixture of Ar/O_2 and with a higher substrate temperature are also in progress. A background pressure of 5×10^{-8} Torr was achieved prior to the introduction of argon. Each target was presputtered for one hour to ensure that the target surface had reached an equilibrium condition before the first film deposition. Single-crystal substrates of cleaved MgO(100) and polished $SrTiO_3$(100) were used. Earlier we found that the best results were obtained with the films deposited on

SrTiO$_3$.[8] However, with the improvement in the film composition and the heat treatment, we have found that there is not much difference in the superconducting properties of T_c's and J_c's between the films deposited on MgO and SrTiO$_3$.

The substrate to target distance was chosen to be 3 cm. This distance could vary with different Ar pressures. The deposition rate is about 15 nm/min, which certainly depends on the input power, the Ar pressure, and the substrate-to-target distance. We have prepared films with thickness ranging from 0.02 to 1.0 µm. Due to the high Ar sputtering pressure (30 mTorr) and the short target-to-substrate distance used in this work, the film thickness is uniform only within a 1.6 cm diameter area as indicated by the film composition analysis using Rutherford backscattering.

The Bi-Sr-Ca-Cu-O oxide targets with various compositions were prepared from bismuth oxide, copper oxide, strontium carbonate and calcium carbonate powders (of at least a 4N purity). Among the various targets, we have found that the target with a nominal composition of $Bi_2Sr_{1.56}Ca_{1.08}Cu_{1.86}O_x$ has produced superconducting films with a nearly pure single-phase $Bi_2Sr_2Ca_1Cu_2O_8$. The appropriate amounts of powders were mixed, reacted at 800-850°C for 6 hours, thoroughly reground and cold-pressed into a disk-shape plate 5 cm in diameter at a pressure of 1.5 tons/cm^2. The plates were sintered at 840°C in air and then furnace-cooled to room temperature.

RESULTS AND DISCUSSION

Previously, we have prepared superconducting thin films of Bi-Sr-Ca-Cu-O with T_c onsets at 84K and full superconductivity at 72.5K by RF magnetron sputtering using a single oxide target with a composition of $Bi_1Sr_1Ca_1Cu_2O_x$.[8] The films, after a post oxygen annealing of 800°C for 30 min followed by a slow cooling to 500°C within 2 hr, and then a furnace-cooling to room temperature, contain a mixture of the superconducting phase $Bi_2Sr_2Ca_1Cu_2O_8$ and the semiconducting phase $Bi_2(Sr, Ca, Bi)_2CuO_6$.[9] Structural analysis by x-ray diffraction shows that the films have a strong preferred orientation with the c-axis perpendicular to the film plane. The resistivity of the films deposited on single-crystal SrTiO$_3$(100) shows a linear dependence on temperature with $\rho(300K) = 450$ µΩ-cm, and a ratio $\rho(300K)/\rho(100K) \approx 2.4$. High critical current densities of 10^5 A/cm^2 measured by the transport method at 61K in zero field and of 10^7 A/cm^2 measured by the magnetic method at 6K have been obtained.

In this work, using our present target, we have obtained a different film composition $Bi_2Sr_{1.9}Ca_{1.3}Cu_{2.6}O_x$ as determined by Rutherford backscattering spectrometry (RBS) using 1.6 MeV ^{4}He$^+$ ions. The as-deposited films are insulating and amorphous in structure. A post-deposition oxygen anneal for recrystallization is needed to make films superconducting.

Two different heat treatments were used in this work: (1) under a flow of oxygen, films were heated from room temperature to 450°C in about 1 hr, then quickly heated up to 860°C in 45 min and remained at this temperature for 20 min, and then cooled to 500°C in 2 hr. (2) the same as heat treatment procedure (1) with the only difference in the stage of soaking at the highest temperature, 885°C for 5-10 min. The crystal structures of the annealed films were characterized by a powder x-ray diffractometer. A Philips θ-2θ diffractometer with Cu K$_\alpha$ radiation was used. The samples are of nearly single-phase (2212) for films under both heat treatment conditions (1) and (2).

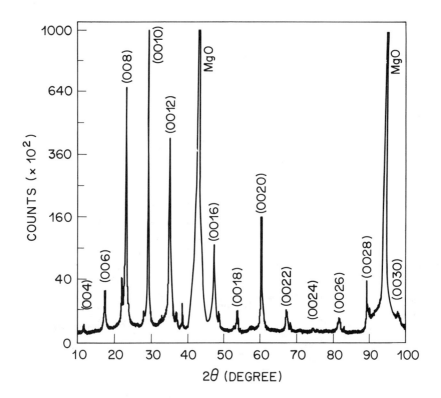

Figure 1   X-ray θ-2θ scan of a film 0.2μm thick containing almost pure $Bi_2Sr_2Ca_1Cu_2O_8$ phase annealed at 860°C. The film was grown on MgO(100).

Figures 1 and 2 are the X-ray diffraction patterns for the reacted films 0.2 and 0.07 μm thick, respectively. The films were deposited on MgO substrates. The film in Fig. 1 was annealed under heat treatment (1), while the film in Fig. 2 was reacted under (2). Both films show a strong texture with the c-axis normal to the film plane. The observed Bragg peaks of both films follow the selection rule of scattering that only (00ℓ) reflection, $\ell = 2n$, are allowed for the space group Fmmm[9]. The best fits among (00ℓ) peaks give a c-axis lattice parameter of 30.6 A for the superconducting phase $Bi_2Sr_2Ca_1Cu_2O_8$. We have noted that in Fig. 2 five peaks marked with asterisks are situated evenly between the peaks (00ℓ) where $\ell = 18$ to 28. At this time, we are still unable to identify the phase(s) associated with these peaks.

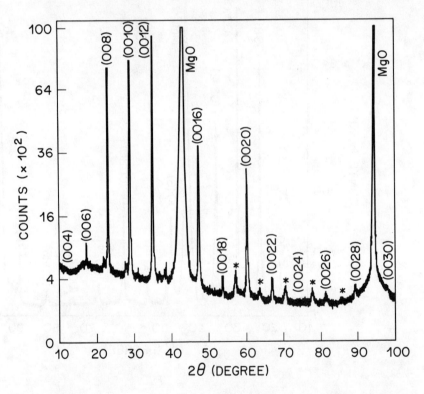

Figure 2　X-ray θ-2θ scan of a film 0.07μm thick with the same as-deposited composition as the film in Figure 1, but with a different annealing temperature of 885 °C.

The superconducting and transport properties were measured by the standard 4-point measurement using both an ac method and a dc method by switching the polarization of the applied current during the measurement. Figure 3(a) is a typical result for films annealed under Condition (1) with $T_c(R=0)$ at 82K. Usually the $T_c(R=0)$'s are around 78-82K. However, for films annealed under Condition (2), we have always observed a drop in resistivity around 110 to 115K as shown in Figure 3(b), regardless of the film thickness from 0.07 to 0.2 μm. Note that the $T_c(R=0)$'s for the latter films are somewhat lower around 75K. More recently, similar results were also obtained in an even thinner film 0.02 μm[10]. From our X-ray diffraction studies on these films, we have not seen any set of peaks corresponding the $Bi_2Sr_2Ca_2Cu_3O_{10}$ phase with a c-axis lattice constant of $\sim 36\mathring{A}$ even for thin films 0.02 to 0.07 μm thick. Interestingly, a cross sectional transmission

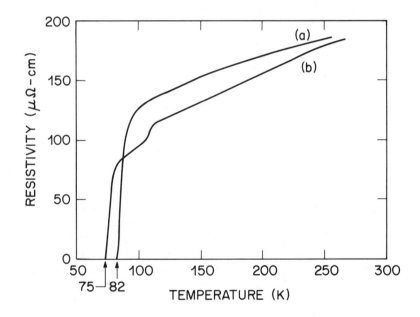

Figure 3.   (a) Resistivity vs. temperature of the $Bi_2Sr_2Ca_1Cu_2O_8$ film annealed at 860°C, and (b) resistivity vs. temperature of the $Bi_2Sr_2Ca_1Cu_2O_8$ film annealed at 885°C, showing a 20% drop in resistivity around 115K.

electron microscopy (TEM) study on a film 0.07 μm thick annealed under Condition (2) showed that along the c-axis stacking sequence, there are a few isolated regions with a c-spacing of 36Å in contrast to the rest area with the c-spacing of 30Å of the $Bi_2Sr_2Ca_1Cu_2O_8$ phase. The room temperature resistivity for both types of films is usually low around 100-200 $\mu\Omega$-cm, which is close to the value 130 $\mu\Omega$-cm reported in a superconducting $Bi_2Sr_2Ca_1Cu_2O_8$ single crystal[9]. This value is much lower than 450 $\mu\Omega$-cm obtained in our previous films of mixed-phase[8].

The transport critical current density was measured directly by passing currents through a restricted section of 1mm wide with 1.0 μV as a criterion. J_c increases rapidly with the decreasing temperature in a manner similar to that observed in epitaxial $Y_1Ba_2Cu_3O_7$ films of essentially a pure phase.[11] At temperatures 10K below the $T_c(R=0)$'s, a J_c of 10^5 A/cm^2 in zero field is obtained in these nearly pure single-phase Bi-Sr-Ca-Cu-O films. Previously, we found that films grown on MgO generally show a resistivity ρ(300K) about 2-5 times higher than those grown on $SrTiO_3$ with a similar composition, and a much lower J_c of 10^3 A/cm^2 at 4.2K.[8] Now with the present target and the two new heat treatment conditions, we have found that there is very little difference in the superconducting properties such as T_c's and J_c's, and the normal-state resistivity between films grown on $SrTiO_3$ and MgO substrates.

CONCLUSION

In this work, highly-oriented single-phase superconducting $Bi_2Sr_2Ca_1Cu_2O_8$ films with the c-axis normal to the film plane have been prepared by RF magnetron sputtering using a single oxide target. With heat treatments at 860°C, we have obtained superconducting films with $T_c(R=0)$ in the range of 77 to 82K. For films heat treated at a higher temperature of 885°C, we have observed a 20% sharp drop in resistivity at 115K followed by a $T_c(R=0)$ at a lower temperature of 75-80K. In a separate effort of our high T_c superconducting oxide thin film research, we have routinely produced films with pure 2223 phase in the Tl-Ba-Ca-Cu-O system with $T_c(R=0)$ at 116K and a high transport J_c of 10^5A/cm^2 at 100K.[6] For the Bi-Sr-Ca-Cu-O system, the pure highest T_c 2223 may be more difficult to prepare. However, with the adjustment in film-composition, substrate-temperature, and the final annealing condition, it may now be possible to obtain superconducting films with $T_c(R=0)$ over 100K in the Bi-based system.

REFERENCES

1. H. Maeda, Y. Tanaka, M. Fukutomi and T. Asano, Jpn. J. Appl. Phys. **27**, L209 (1988).

2. Z. Z. Sheng and A. M. Hermann, Nature **332**, 55 and 138 (1988).

3. R. M. Hazen, L. W. Finger, R. J. Angel, C. T. Prewitt, N. L. Ross, C. G. Hadidiacos, P. J. Heaney, D. R. Veblen, Z. Z. Sheng, A. El Ali, and A. M. Hermann, Phys. Rev. Lett. **60**, 1657 (1988).

4. C. C. Tordrdi, M. A. Subramanian, J. C. Calabrese, J. Gopalakrishnan, K. J. Morrissey, T. R. Askew, R. B. Flippeht, U. Chowdhry, and A. W. Sleight, Science **240**, 631 (1988).

5. M. Takano, J. Takada, K. Oda, H. Kitaguchi, Y. Miura, Y. Ikeda, Y. Tomii, and H. Mazaki, Jpn. J. Appl. Phys. **27**, No. 6, June, (1988).

6. M. Hong, S. H. Liou, D. D. Bacon, G. S. Grader, J. Kwo, A. R. Kortan, and B. A. Davidson, Appl. Phys. Lett. November (1988).

7. S. Nakahara and M. Hong (unpublished).

8. M. Hong, J. Kwo, and J.-J. Yeh, J. Crystal Growth **91**, 382 (1988).

9. S. A. Sunshine, T. Siegrist, L. F. Schneemeyer, D. W. Murphy, R. J. Cava, B. Batlogg, R. B. van Dover, R. M. Fleming, S. H. Glarum, S. Nakahara, R. Farrow, J. J. Krajewski, J. V. Waszczak, J. H. Marshall, P. Marsh, L. W. Rupp, Jr., and W. F. Peck, Phys. Rev. **B** (1988).

10. J.-J. Yeh and M. Hong (unpublished).

11. P. Chaudhari, R. H. Koch, R. B. Laibowitz, T. R. McGuire, and R. J. Gambino, Phys. Rev. Lett. **58**, 2684 (1987)

EFFECTS OF OXYGEN PARTIAL PRESSURE ON PROPERTIES OF Y-Ba-Cu-O FILMS PREPARED BY MAGNETRON SPUTTERING

T. Miura, Y. Terashima, M. Sagoi, K. Kubo,
J. Yoshida, K. Mizushima
Toshiba Research and Development Center, 210, Kawasaki, JAPAN

ABSTRACT

Dependence of orientation in Y-Ba-Cu-O films on oxygen partial pressure has been investigated for films prepared by magnetron sputtering on $SrTiO_3$ (100) planes. Orientations were closely related to the oxygen partial pressure in the 530 to 680 °C substrate temperature range. The dependence on substrate temperature was not observed clearly below 680 °C. At low oxygen pressure, the c axis was oriented perpendicular to the film plane. With an increase in oxygen pressure, the films tended to have a axis normal to the film surface. When the gas pressure during sputtering was 0.65 Pa with oxygen at 50 %, zero resistivity was obtained at 70 K for many films without post annealing and 80 K for the best film. At higher oxygen pressure, the a axis was oriented with the expanded lattice parameter and superconducting films were not obtained without post annealing. At 0.32 Pa of oxygen, the oxygen uptake during film growth was most effective. At the higher substrate temperature of 730 °C, c axis oriented films were formed.

INTRODUCTION

Various methods have been reported for fabricating thin films of the high Tc superconductor Y-Ba-Cu-O system. These include sputtering[1-3], electron beam evaporation[4-6], laser ablation[7,8], molecular beam epitaxy[9,10] and chemical vapor deposition[11,12], by which excellent superconducting films have been prepared. In order to obtain films which show zero resistivity at a temperature above 80 K, post annealing at high temperature (>800 °C) has been extremely effective. However, it causes interdiffusion between film and substrate and deterioration in surface morphology. Recently, as grown superconducting films have been successfully prepared with a low temperature process by many groups[13-15]. The low temperature process has made it possible to control preferential film orientation by appropriately selecting the preparation condition. It has been reported that the films have different orientations in depending on substrates and preparation conditions[4,16,17]. Factors controlling film orientations, however, have not yet been clarified. This paper reports effects of oxygen partial pressure on the orientations of as grown Y-Ba-Cu-O films and post annealing effects on superconductivity.

EXPERIMENTAL

An multi-target rf magnetron sputtering technique was used for preparing superconducting films. Yttrium metal, copper metal and sintered Ba_2CuO_3 ceramics targets were simultaneously sputtered. As summarized in Table 1, the sputtering gas pressure was between 0.65 and 1.0 Pa, mixed with oxygen ranging from 5 to 100 %. As a esult, the oxygen partial pressure varied from 0.032 Pa to

Table 1   Sputtering Conditions

total pressure	0.65-1.0 Pa
oxygen partial pressure	0.032-1.0 Pa
target	Y, Ba_2CuO_3, Cu
substrate temperature	530-730 C
rate	3 nm/min.
thickness	500 nm
substrate	$SrTiO_3$ (100)

1.0 Pa. The substrates were $SrTiO_3$ single crystals with (100) face, and were heated at temperatures ranging from 530 to 730 °C to obtain superconducting films without post annealing at high temperature. The separation between substrates and targets was 9 cm. The film thickness was about 500 nm and deposition rate was about 3 nm/min. Film composition was sensitive to oxygen partial pressure. At higher oxygen pressure, barium and copper tended to decrease in the composition. Then, the composition was controlled so that it would be within the 5 % of the stoichiometric composition, by regulating the power applied to the targets. After deposition, 1 atm oxygen was introduced into the chamber, and the films were cooled to 200 °C in presence of the sputtering gas. The film composition was analyzed by inductively coupled plasma emission spectroscopy (ICP). The surface morphology and the crystal structure were examined by scanning electron microscopy (SEM) and X-ray diffraction. The electrical resistivity was measured by the standard four probe method.

RESULTS AND DISCUSSIONS

Several films with a or c axis oriented perpendicular to the film plane were obtained on $SrTiO_3$ (100) substrates. Figure 1 shows the oxygen partial pressure dependence of c and a axis parameters for as grown films. Some groups reported that the orientation depended on the composition[4], substrate temperature[16,18] and oxygen pressure[18] in evaporation methods. In the present sputtering method, the orientation was closely related to the oxygen partial pressure for as grown films with stoichiometrical composition within 5 % deviation. No substrate temperature dependence of the orientation was clearly observed in the range from 530 °C to 680 °C. However, c axis orientation was observed for films sputtered at 730 °C.

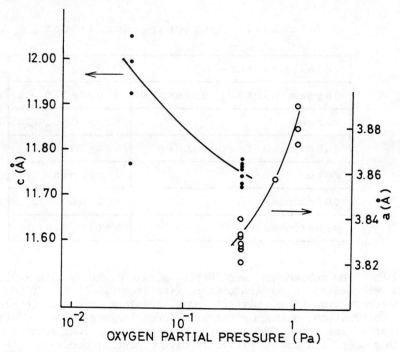

Fig. 1. Lattice parameters versus oxygen partial pressure for as grown films.

At 0.032 Pa oxygen pressure, all films had an orientation with the c axis normal to the film plane and the lattice parameter extremely expanded. These films exhibited zero resistivity below 40 K, except one which showed zero resistivity at 60K. Figure 2(a) shows a typical X-ray diffraction pattern for c axis oriented films.

At 0.32 Pa, two different orientations were mixed. Almost all films exhibited preferential orientation with the a axis perpendicular to the film surface. The lattice parameters for the c axis and a axis were 11.72 to 11.78 Å and 3.82 to 3.84 Å. Figure 2(b) shows a typical X-ray diffraction pattern for as grown films, with highly preferred orientation of a axis. These orientations were observed for films deposited below 680 °C substrate temperature. However, films composed of only c axis oriented grains were formed at 730 °C. In a sputtering method, the c axis seems to be oriented normal to the film surface at higher substrate temperature than in evaporation methods.

At higher oxygen partial pressure, the films exhibited an orientation with the a axis normal to the film plane. The X-ray diffraction pattern showed no peaks from the c axis oriented grains and the lattice parameter a expanded to more than 3.86 Å,

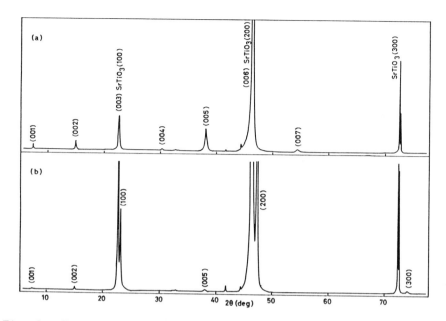

Fig. 2. X-ray diffraction patterns for as grown films.
(a) c axis normal to the film surface. Oxygen pressure, 0.032 Pa.
(b) a axis normal to the film surface. Oxygen pressure, 0.32 Pa.

indicating the growth of a tetragonal phase. The resistivity change was semiconductive and zero resistivity was not observed for any as grown films. An increase in the oxygen partial pressure was not effective in improving superconductivity in as grown films.

Figure 3 shows the oxygen partial pressure dependence for the lattice parameters c and a, after post annealing in oxygen of 1 atm at 800 °C for 1 hour. The orientation dependence did not change after annealing. In the entire range of oxygen partial pressure, both lattice parameters c and a of all films contracted, indicating progress in oxygen uptake. However, the contraction of lattice parameter a was not sufficient for films prepared at high oxygen partial pressure, 1.0 Pa. Zero resistivity was not observed above 80 K for these films. On the contrary, zero resistivity was often observed for films formed at or below oxygen partial pressure of 0.32 Pa. Throughout the experiments, films with excellent characteristics were most frequently obtained at this oxygen pressure, whose total pressure was 0.65 Pa. Figure 4 shows a scanning electron micrograph of a film surface with a axis orientation after post annealing. The surface was smooth and shiny, although grains below 0.3 µm in size were observed on the dense film with no texture.

Figure 5 shows the resistivity versus temperature relation for the as grown film, prepared at 0.32 Pa of oxygen at a substrate temperature of 560 °C. Zero resistivity was obtained at 80 K. However, the lattice parameter c expanded more than that for the ideal

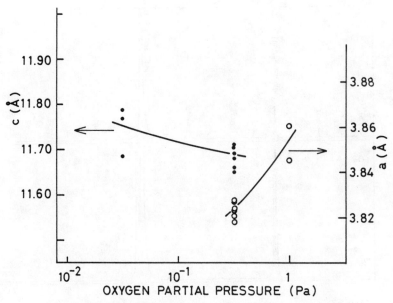

Fig. 3. Lattice parameters versus oxygen partial pressure for annealed films.

Fig. 4. SEM micrograph of an annealed film with the a axis oriented grains.

orthorhombic structure. On the other hand, a axis oriented grains with the lattice parameter a, which corresponded to the orthorhombic structure, were formed for this film. The lattice parameters c and a were 11.74 and 3.82 Å. Thus, zero resistivity is thought to be obtained at 80 K through a axis oriented grains with the orthorhombic structure, although the film includes low Tc phase. Post annealing

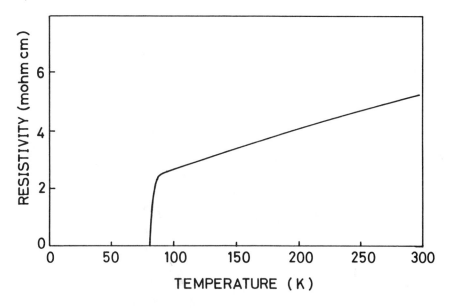

Fig. 5. Resistivity versus temperature for an as grown film, prepared at 560 C.

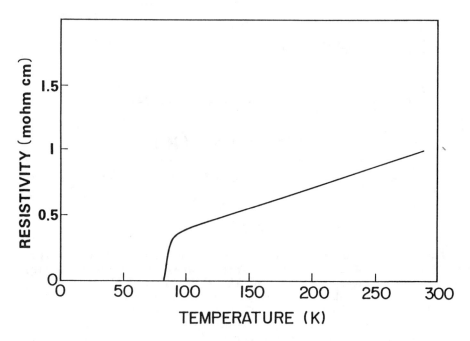

Fig. 6. Resistivity versus temperature for an annealed film.

at high temperature is expected to improve film characteristics. Figure 6 shows a typical resistivity versus temperature relation after annealing at 800 °C in atmospheric oxygen for 1 hour. The X-ray diffraction analysis indicated that the film crystal structure was orthorhombic with 3.82 Å for the lattice parameter a and 11.68 Å for the lattice parameter c. The resistivity decreased linearly with the temperature and zero resistivity was obtained at 85 K. The improvement was confirmed by magnetization measurements, as shown in Fig. 7. Curves (a) and (b) indicate magnetic susceptibility versus temperature relation, before and after annealing, respectively. This shows that the film is composed of a high Tc phase after annealing.

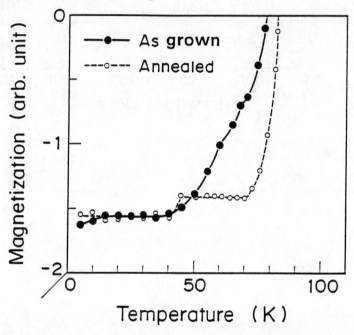

Fig. 7. Magnetic susceptibility (a) For an as grown film (b) For an annealed film.

CONCLUSIONS

For as grown Y-Ba-Cu-O films, prepared by magnetron sputtering, the orientations of grains were closely related to the oxygen partial pressure in sputtering gas in the 530 to 680 °C substrate temperature range. At low oxygen pressure, grains with c axis normal to the surface grew in the film. With an increase in oxygen partial pressure, the a axis tended to be normal to the surface. However, the c axis was oriented normal to the surface at 730 °C. Superconductive films with zero resistivity at 80 K were successfully obtained at 0.32 Pa oxygen partial pressure. Further increases in the oxygen pressure were not effective in stimulating oxygen uptake into the films. At 1.0 Pa, the a axis was oriented perpendicular to the film surface and its lattice parameter a was larger than 3.86 Å, indicating oxygen deficiency.

REFERENCES

1) K. Char, A. D. Kent, A. Kapitulnik, M. R. Beasly and T. H. Geballe: Appl. Phys. Lett. 51 (1987) 1370.
2) Y. Enomoto, T. Murakami, M. Suzuki, and K. Moriwaki: Jpn. J. Appl. Phys. 26 (1987) L1248.
3) M. Komuro, Y. Kozono, Y. Yazawa, T. Ohno, M. Hanazono, S. Matsuda and Y. Sugita: Jpn. J. Appl. Phys. 26 (1987) L1907.
4) P. Chaudhari, R. H. Koch, R. B. Laibowitz, T. R. McGuire and R. J. Gambino: Phys. Rev. Lett. 58 (1987) 2684.
5) B. Oh, M. Naito, S. Arnason, P. Rosenthal, R. Barton, M. R. Beasley, T. H. Geballe, R. H. Hammond and A. Kapitunik: Appl. Phys. Lett. 51 (1987) 812.
6) M. Mukaida, M. Yamamoto, Y. Tazoh, K. Kuroda and K. Hohkawa: Jpn. J. Appl. Phys. 27 (1988) L211.
7) J. Narayan, N. Brunno, R. Singh, O. W. Holland and O. Auciello: Appl. Phys. Lett. 51 (1987) 1845.
8) X. D. Wu, D. Dijkkamp, S. B. Ogale, A. Inam, E. W. Chase, P. F. Miceli, C. C. Chang, J. M. Tarascon and T. Venkatesan: Appl. Phys. Lett. 51 (1987) 861.
9) J. Kwo, T. C. Hsieh, R. M. Fleming, M. Hong, S. H. Liou, B. A. Davidson and L. C. Feldman: Phys. Rev. B36 (1987) 4039.
10) C. Webb, S. L. Lang, J. N. Eckstein, N. Missert, K. Char, D. G. Schlom, E. S. Hellman, M. R. Beasley, A. Kapitulnik and J. S. Harris, Jr.: Appl. Phys. Lett. 51 (1987) 1191.
11) J. V. Mantese, A. H. Hamdi, A. L. Micheli, Y. L. Chen, C. A. Wong, J. L. Johnson, M. M. Karmarkar and K. R. Padmanabhan: Appl. Phys. Lett. 52 (1988) 1631.
12) H. Abe, T. Tsuruoka and T. Nakamori: Jpn. J. Appl. Phys. 27 (1988) L1473.
13) H. Adachi, K. Hirochi, K. Setsune, M. Kitabatake and K. Wasa: Appl. Phys. Lett. 51 (1987) 2263.
14) T. Terashima, K. Iijima, K. Yamamoto, Y. Bando and H. Mazaki: Jpn. J. Appl. Phys. 27 (1988) L91.
15) X. D. Wu, A. Inam, T. Venkatesan, C. C. Chang. E. W. Chase, P. Barboux, J. M. Tarascon and B. Wilkens: Appl. Phys. Lett. 52 (1988) 754.
16) S. W. Chan, L. H. Green, W. L. Feldman, P. F. Miceli and B. G. Bagley: Proceedings of Thin film Processing and Characterizations of High Temperature Superconductors, American Vacuum Society, Anaheim, CA, 1987.
17) D. K. Lathrop, S. E. Russek and R. A. Buhrman: Appl. Phys. Lett. 51 (1987) 1554.
18) J. Fujita, T. Yoshitake, A. Kamijo, T. Satoh and H. Igarashi: J. Appl. Phys. 64 (1988) 1292.

EVAPORATION

A Flash Evaporation Technique for Oxide Superconductors 140

Sequentially Evaporated Thin Y–Ba–Cu–O Superconductor Films:
Composition and Processing Effects 147

Epitaxial Growth of Dy–Ba–Cu–O Superconductor on (100)$SrTiO_3$ Using
Reactive and Activated Reactive Evaporation Processes 155

Rate and Composition Control by Atomic Absorption Spectroscopy for the
Coevaporation of High T_c Superconducting Films 163

A FLASH EVAPORATION TECHNIQUE FOR OXIDE SUPERCONDUCTORS

Matthew F. Davis, Jaroslaw Wosik, J. C. Wolfe,
and Christopher L. Lichtenberg
Department of Electrical Engineering,
University of Houston, Houston, TX 77004

ABSTRACT

An electron beam flash evaporation technique is presented which can reproduce the stoichiometry of $YBa_2Cu_3O_x$ and $Bi_2CaSr_2Cu_2O_y$ source powders. Process details such as source powder degassing and electron beam operating parameters are described. Zero resistance temperatures near 80 K were obtained for both materials. The $Bi_2CaSr_2Cu_2O_y$ films deposited on MgO are highly textured with the c-axis normal to the substrate.

INTRODUCTION

Superconducting thin films have been deposited by a number of evaporation techniques, including multisource coevaporation[1,2], sequential evaporation[3,4], and pulsed laser ablation[5,6]. Recently, a novel flash evaporation technique has been developed independently by ourselves[7] and NRL[8]. This technique differs from conventional flash evaporation by the use of an electron beam evaporator. This variation permits the evaporation of refractory materials, including yttrium, and, as we show below, has an important influence on the fractionation of the source pellets. As with conventional flash evaporation, the main objective is to exactly reproduce the stoichiometry of the source material in the film. This eliminates the need for precise control of multiple sources and permits the coverage of large substrate areas. The main concern of this paper is to verify that exact reproduction of source stoichiometry is possible for both $YBa_2Cu_3O_x$(123) and $Bi_2CaSr_2Cu_2O_y$(2122). The superconducting properties of the resulting films are described.

EXPERIMENTAL TECHNIQUE

Evaporation was carried out in a cryopumped vacuum system with a typical base pressure of $5*10^{-6}$ torr. The electron beam evaporation source is a commercial 14 kW, 40 cm^3, e-gun fitted with a water cooled copper hearth insert. The hearth insert facilitates cleaning, prevents accidental damage to the e-gun, and permits viewing of the evaporation process. Substrates are mounted on a Monel block which is heated by internal quartz lamps to 900 °C. The temperature of the block is monitored by a thermocouple. A programmable temperature controller is used to control the substrate thermal cycle. Substrates are mounted to the block with Pt-Rh spring clips without any heat sink material. Therefore, the temperature of the substrates may be significantly below that of the block(see below). The source to substrate distance is 30 cm and the substrate elevation is 40 degrees. A quartz crystal microbalance is used to monitor the evaporation process. The system is vented with pure oxygen gas.

The electron beam flash evaporation technique (EBFE) is a two-step process. The first step is to degas and melt the source powder to form evaporation pellets. These pellets are then rapidly evaporated to completion in the second step. Degassing is accomplished by placing 0.3 grams of powder in a line approximately 0.4 cm wide and 1.8 cm long on the beam axis. The line of powder is then scanned at a 2 Hz rate while the beam power is gradually increased. Degassing is terminated when the system pressure returns to its base level and the deposition rate monitor shows that 2 nm of film has been deposited. The powder melts during the degassing process and agglomerates into pellets which are approximately 1 mm in diameter. The diameter of the pellets is determined by the scan rate. Evaporation is then accomplished by making one pass over the line of pellets with a 3 kW beam in approximately 10 s. The beam is rectangular in shape with a width(normal to the beam axis) of 1 cm and a length of 3 mm. Each pellet contributes approximately 20 nm to the thickness of the film.

Powders for evaporation were obtained from Davison Chemical Division of W.R. Grace & Co. The composition of the powders was determined by inductively coupled plasma spectroscopy(ICP) to be within 2 mole percent of the nominal composition for the 123 material. Thin film composition was determined by energy dispersive spectroscopy, calibrated by ICP. The accuracy is estimated to be 10 mole percent.

A summary of the complete deposition process for 123 films is as follows. Films are deposited by 4 sequential cycles of flash evaporation. Each cycle adds approximately 240 nm to the thickness of the film. The evaporation source is cleaned and polished between evaporation sequences. The nominal substrate temperature during each evaporation cycle is 850 °C. The system is slowly(i.e. 500 Torr is achieved over a period of 10 minutes) vented with pure oxygen after deposition is complete. The nominal temperature then rises to 880 °C, indicating that the temperature of the heater block is approximately this value during deposition. One can assume that the substrate temperature also approaches this value during venting. The nominal temperature returns to 850 °C within 4 minutes after venting is begun. The nominal substrate temperature is then reduced to 700 °C over a period of 15 minutes. This temperature is maintained for 40 minutes. The sample is then cooled to room temperature in 1.5 hours. The following post-deposition heat treatment in flowing oxygen is applied to the films after all four cycles of flash evaporation are complete: 850 °C for 15 minutes, furnace cool over 30 minutes to 650 °C, soak at 650 °C for two hours, and slow cool to room temperature.

A summary of the formation process for 2122 films is as follows. Films are again deposited by 4 sequential cycles of flash evaporation. Deposition is carried out with a substrate temperature of 200 °C and the system vented with oxygen. The following post-deposition heat treatment is applied to the finished films: load into an 850 °C tube furnace in air, flowing oxygen for 15 minutes, furnace cool(6 hours) to room temperature in oxygen.

Resistances were measured with a standard low frequency(1 kHz) 4-point probe technique. The noise limit of the lock-in voltmeter is 0.1 microvolts. With the typical excitation current of 10 microamperes, the corresponding resistance noise limit is 1.0 mohm. Contacts were made by back-etching and depositing sputtered gold dots through a shadow mask. Silver paste was used to glue silver wires to the dots.

RESULTS

Stoichiometry: Comparisons have been made between the stoichometry of the deposited films and that of the source material. A set of six substrates, distributed over the 6 by 8 cm heater block was coated simultaneously from a nominal $YBa_2Cu_3O_x$ source powder. The mean composition of the resulting films was $Y_{1.0}Ba_{2.1}Cu_3O_x$. The corresponding experiment with $Bi_{1.9}Ca_{0.8}Sr_{1.9}Cu_2O_y$ source powder yielded a film with composition $Bi_{2.0}Ca_{1.1}Sr_{2.0}Cu_2O_y$. In spite of these encouraging results, quite wide variations(e.g 20 mole percent) have been observed between runs. In all cases where stoichiometry of 123 films is poor, they are Ba rich. The most probable cause of these variations is preferential reevaporation of the high vapor pressure components from that portion of the pellets which lands on the hearth. This effect can be controlled to some extent by polishing the hearth between deposition cycles and by minimizing the amount of material evaporated during each cycle. In this way, thermal contact with the hearth is maximized and reevaporation minimized.

Superconducting properties: The transition curves of the best 123 and 2122 superconductors on MgO are shown in figures 1 and 2, respectively. The 123 films are superconducting when they are removed from the evaporator. However, performance improves after post-deposition annealing. The diamonds in fig. 1 show the transition for an as-deposited film, while the squares show the transistion of the post-deposition annealed film. Typical zero resistance temperatures (defined as 0.1% of the normal resistance above the transition) are above 60 K for 123 and above 75 K for 2122.

Structural properties: Figure 3 is an X-ray diffraction pattern for the 2122 film shown in fig. 2 above. The film clearly shows a strong c-axis texture.

Fractionation: Fractionation is observed in the evaporation of single particles. The Rutherford backscattering spectroscopy(RBS) profile of an 80 nm thick film deposited at room temperature on a silicon substrate from a single evaporation pellet is shown in figure 4. The profile shows excess copper at the silicon interface and excess barium at the surface, with yttrium distributed uniformly through the film. While the high surface concentration of barium could have a variety of causes including reaction with the atmosphere, the copper distribution indicates that it evaporates preferentially. This is confirmed by the green color of copper which dominates the evaporation plume during the initial evaporation stage of each pellet. Thus, the pellets do not fractionate in the ratio of the vapor pressures of the elements. This may be due to the evaporation of intermetallic compounds or possibly because the rate limiting step for evaporation is the diffusion of the components to the surface of the liquid pellet. Copper would be expected to have the highest diffusion constant and therefore the highest arrival rate at the surface of the particle.

SUMMARY AND CONCLUSIONS

Electron beam flash evaporation has been shown to be effective in reproducing the stoichiometry of starting powder and of achieving uniform composition over large substate areas. We have shown, in addition, that high quality, oriented films can be deposited. The run-to-run reproducibility requires improvement.

One might expect, based on the order of elemental vapor pressures, a strong similarity between EBFE and sequential evaporation with a Ba-Cu-Y evaporation sequence. Instead, Cu is preferentially evaporated during the initial stages of pellet evaporation and the degree of fractionation much less extreme. A full explanation of this result requires further investigation.

ACKNOWLEDGEMENTS

This material is based upon work supported by the Texas Center for Superconductivity at the University of Houston under prime grant MDA 972-88-G-0002 to the University of Houston from the Defence Advanced Research Projects Agency and the State of Texas, by the National Science Foundation under grant MSM 8718776, and by the National Aeronautics and Space Administration under grant NAG9 -27. Any opinions, findings and conclusions or recommendations expressed in this publication are those of the authors and do not necessarily reflect the views of the Texas Center for Superconductivity at the University of Houston, the Defense Advanced Research Projects Agency, the National Science Foundation, or the National Aeronautics and Space Administration.

The authors are grateful to Paul Chu and Pei Hor for useful discussions and X-ray diffraction services, Joe Keenan and Tom Shaffner of Texas Instruments, Inc. for RBS services and to John Bear and Micky Puri for source material during the initial stages of these studies.

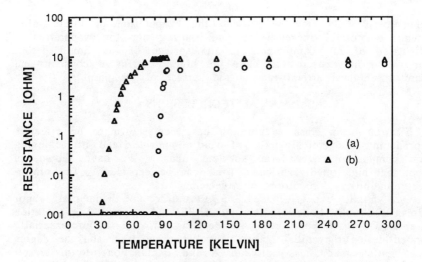

Fig. 1. Resistance versus temperature for a 123 film deposited (a) with, and (b) without post-deposition furnace oxygen annealing.

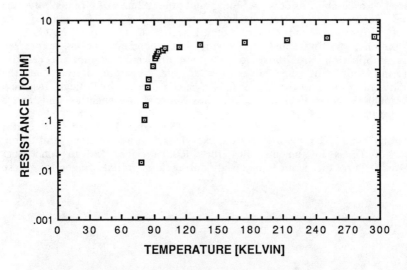

Fig. 2. Resistance versus temperature for a 2122 film on an MgO substrate, annealed at 860 °C for 15 minutes in flowing oxygen.

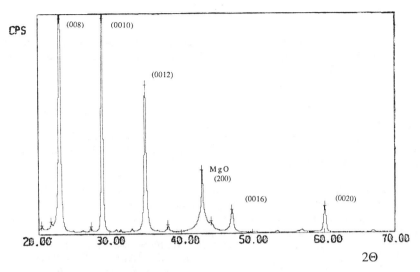

Fig. 3. X-ray diffraction pattern (counts versus 2Θ) for the 2122 film in fig. 2. The dominance of (00L) peaks indicate that the film is highly oriented with c-axis perpendicular to the film plane.

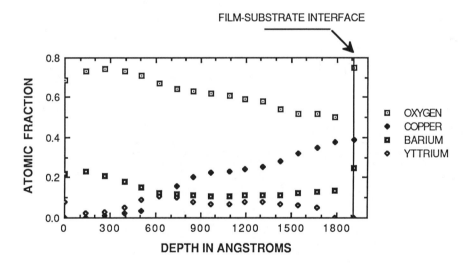

Fig. 4. RBS depth profile for a 123 thin film deposited on a Si substrate at room temperature.

REFERENCES

1. M. Naito, R.H. Hammond, B. Oh, M.R. Hahn, J.W.P. Hsu, P. Rosenthal, A.F. Marshall, M.S. Beasley, T.H. Geballe, and A. Kapitulnik, J. Mater. Res.**2**, 713 (1987).
2. D.K. Lathrop, S.E. Russek, and R.A. Buhrman, Appl. Phys, Lett. **51**, 1554 (1987).
3. C-A. Chang, C.C. Tsuei, C.C. Chi, and T.R. McGuire, Appl. Phys, Lett. **52**, 72 (1988).
4. M. Nastasi, P.H. Arendt, J.R. Tesmer, C.J. Maggiore, R.C. Cordi, D.L. Bish, J.D. Thompson, S-W. Cheong, N. Bordes, J.F. Smith, and I.D. Raistrick, J.Mater. Res., **2**, 726 (1987).
5. B.F. Kim, J. Bohandy, T.E. Phillips, W.J. Green, E. Agostinelli, F.J. Adrian, K. Moorjani, L.J. Swartzendruber, R.D. Shull, L.H. Bennett, and J.S. Wallace, NBS, Gaithersburg Md 20899, unpublished.
6. D. Dijkkamp, T. Vekatesan, X.D. Wu, S.A. Shaheen, N. Jisrawi, Y.H. Min-Lee, W. L. Mclean, and M. Croft, Appl. Phys. Lett. **51** (1987).
7. J.C.Wolfe, Proceedings of the NSF Grantees Meeting on Processing of High-Temperature Superconducting Materials, April 26-27,1988.
8. M.S. Osofsky, P. Lubitz, M.Z. Harford, A.K. Singh, S.B. Qadri, E.F. Skelton, W.T. Elam, R.J. Soulen Jr., W.L. Lechter and S.A. Wolf, Naval Research Laboratory, Washington DC 20375, unpublished.

SEQUENTIALLY EVAPORATED THIN Y–Ba–Cu–O SUPERCONDUCTOR FILMS: COMPOSITION AND PROCESSING EFFECTS

George J. Valco and Norman J. Rohrer
Ohio State University, Columbus, Ohio 43210

Joseph D. Warner and Kul B. Bhasin
NASA Lewis Research Center, Cleveland, Ohio 44135

ABSTRACT

Thin films of $YBa_2Cu_3O_{7-\delta}$ have been grown by sequential evaporation of Cu, Y, and BaF_2 on $SrTiO_3$ and MgO substrates. The onset temperatures were as high as 93K, while T_c was 85K. The Ba/Y ratio was varied from 1.9 to 4.0. The Cu/Y ratio was varied from 2.8 to 3.4. The films were then annealed at various times and temperatures. The times ranged from 15min to 3 hours, while the annealing temperatures used ranged from 850°C to 900°C. There was found a good correlation between transition temperature (T_c) and the annealing conditions; the films annealed at 900°C on $SrTiO_3$ had the best T_c's. There was a weaker correlation between composition and T_c. Barium poor films exhibited semiconducting normal state resistance behavior while barium rich films were metallic. The films were analyzed by resistance versus temperature measurements and scanning electron microscopy. The analysis of the films and the correlations are reported.

INTRODUCTION

With the widespread effort to investigate and develop electronic applications of the superconducting oxides, a large variety of techniques have been used to form high temperature superconducting thin films. One such technique consists of the sequential evaporation of a multi-layer stack containing the constituents of the superconductor followed by annealing in an oxygen ambient. When performed by electron beam evaporation from a multi-hearth gun, this technique allows deposition of films with little spatial variation of stoichiometry as all components of the film are evaporated from the same point in space. The stoichiometry of the films is also easily adjusted by controlling the thickness of the individually deposited layers. This technique has been employed with a variety of starting materials; Y, Ba and Cu metals themselves[1-3], a combination of the metals and oxides[2,4,5], a combination of metals and BaF_2[3] and a combination of oxides and BaF_2[6].

In the work reported here, we have performed sequential evaporation of Cu, Y and BaF_2 to study the formation of superconducting films on $SrTiO_3$ and MgO substrates. We have varied the stoichiometry of the films by adjusting the thicknesses of the individual layers. Films of fixed composition have also been differently annealed to assess the influence of annealing conditions. The fabricated films were characterized through measurement

© 1989 American Institute of Physics

of their resistance as a function of temperature and through scanning electron microscopy. The results of these characterizations are presented here.

DEPOSITION OF FILMS

Deposition of the films was performed in a CHA Industries electron beam evaporator. The system is equipped with a four hearth gun, allowing deposition of the multi-layer stack without breaking vacuum. Thickness of the layers was controlled via an Inficon XTC thickness monitor and rate controller. The depositions were calibrated by measurements of step heights using a surface profilometer.

A cross sectional drawing of the structure of a typical as deposited film is shown in Figure 1. First approximately 510Å of copper was deposited on the substrate. This was followed by an approximately 480Å thick layer of yttrium which was followed by an approximately 1920Å thick layer of barium fluoride. For most of our depositions, this multi-layered sequence was repeated four times for a total of twelve layers. For the thicknesses listed above, the film is characterized by a barium/yttrium atomic ratio of 2.25 and a copper/yttrium atomic ratio of 3.01, which has produced our best results on $SrTiO_3$ substrates. We have investigated the properties of films with barium/yttrium ratios ranging from 1.9 to 4.0 and copper/yttrium ratios ranging from 2.8 to 3.5. We have used barium flouride rather than elemental barium since barium flouride is less reactive.

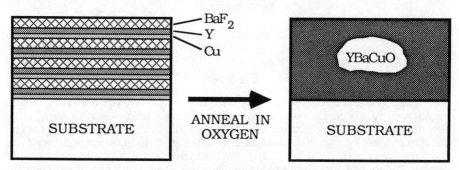

Fig. 1. On th left: Typical twelve layer structure of the as deposited film. On the right: $YBa_2Cu_3O_{7-\delta}$ formed after annealing in oxygen.

ANNEALING

The deposited films were annealed in a hot wall, programable, quartz tube furnace. The furnace was preheated to the annealing temperature and purged with oxygen prior to inserting the samples. Temperatures ranged from 850°C to 900°C. The samples were pushed into the furnace with either a fast push of approximately 30sec or a slow 5min push. The duration of the anneals ranged from 15min to

3hr. The temperature was then ramped to 450°C at a rate of -2°C/min. The samples were held at 450°C for 6hr and then the temperature was ramped to room temperature at -1°C/min. During the high temperature portion of the anneal the ambient consisted of ultra high purity oxygen bubbled through room temperature water to assist in removal of fluorine from the films. Dry oxygen was used during all other portions of the annealing process.

CHARACTERIZATION

Ohmic contacts were formed on the films to allow measurement of their resistance as a function of temperature. Most of the samples were rectangular in shape with widths of approximately 5mm and lengths of approximately 1cm. The contacts for these samples were deposited by evaporation of $1\mu m$ of silver through shadow masks to produce four stripes across the width of the samples. For some irregularly shaped samples, shadow masks which produced four dots were used. The contacts were annealed in dry oxygen at 500°C for 1hr. The temperature was ramped to 250°C at a rate of -2°C/min and then to room temperature at -1°C/min.

The samples were mounted to a sample holder and gold ribbon bonds were made between the silver contacts and bonding posts. A four probe DC measurement was employed to determine the resistance. The samples were cooled in a closed cycle helium refrigerator. Measurements were performed from room temperature to well below the transition temperature for superconducting films or to approximately 10K for non-superconducting films.

Scanning electron microscopy (SEM) was employed to observe the morphology of the films. In addition, several of the films on $SrTiO_3$ substrates were analyzed by x-ray diffraction spectroscopy (XDS) to observe orientation and the presence of other phases.

RESULTS

Figure 2 is a plot of the distribution of composition for several films. Each point on the plot represents the combination of Cu/Y ratio and Ba/Y ratio used for a deposition. Each of these films was made with a 12 layer deposition and was between 1.0 and $1.2\mu m$ thick before annealing. All of the samples shown in this plot were annealed with the procedure described above for 45min at 900°C. A few films with composition well outside the range plotted here (larger Ba/Y ratios) were also deposited. They are not plotted since they had a small residual resistance below the "transition" temperature even though they showed a fairly sharp onset of superconductivity and transition. Films of many other compositions were also deposited but were not annealed at the same conditions and are thus not plotted here.

Associated with each point on the plot are two numbers and a letter. The first number is the onset temperature in Kelvin. The second number is the temperature below which the resistance of the sample was zero. The letter "m" refers to a metallic behavior of the normal state resistance with temperature while the letter "s"

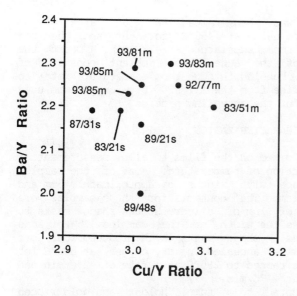

Fig. 2. Composition of several films annealed at 900°C for 45min. The numbers give onset/T_c. m: metallic normal state, s: semiconducting normal state.

indicates that the normal state resistance of that sample showed a semiconducting behavior with temperature.

The data in this plot show that there is a minimum Ba/Y ratio below which the samples have a semiconducting normal state resistance-temperature (R-T) characteristic. Those films deposited with a Ba/Y ratio of greater than approximately 2.2 have a metallic normal state R-T characteristic. This was also true of the films, mentioned above, with the large Ba/Y ratios (≈3.0 and ≈4.0).

Over the ranges studied, the Cu/Y ratio does not affect the qualitative character of the normal state resistance. As would be expected, adjusting the thicknesses of the individual layers to make the films copper rich or poor does degrade the onset temperature and critical temperature of the resulting films.

Our best films on $SrTiO_3$ are those deposited with a Ba/Y ratio of approximately 2.25 and a Cu/Y ratio of ≈3.0. The R-T characteristic below room temperature for one of these films is plotted in Figure 3. This film had a room temperature resistivity of approximately 2mΩ-cm. The onset temperature of this film was 93K and the critical temperature was 85K. The 90% to 10% transition width for this sample was 3.6K. As can be seen from Figure 2, there is a range of ratios near these values which produce films with nearly the same onset and critical temperatures.

Films of several compositions were annealed at different temperatures or for different durations. Data for one of these experiments, involving four samples, is listed in Table I. These samples were annealed for 45min at three temperatures. Two samples were annealed at 850°C. One was quickly pushed into the furnace while the other was pushed into the furnace over a period of 5min. The other two samples were annealed at 875°C and 900°C, both with slow pushes. The last of these samples is included in Figure 2. The onset temperature, critical temperature, 90%-10% transition width and 99%-1% transition width are tabulated. The transition widths are referenced to the resistance just above the onset.

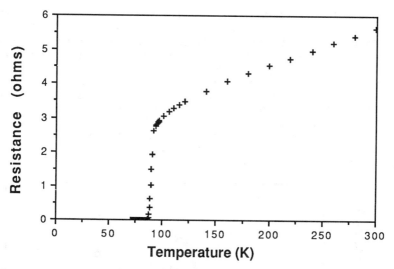

Fig. 3. Resistance as a function of temperature for a Y-Ba-Cu-O thin film with composition Ba/Y=2.25 and Cu/Y=3.01. Annealed at 900°C for 45min. Onset temperature=93K. T_c=85K.

Table I. Transition widths for three annealing temperatures.

Annealing Temperature (°C)	Onset (K)	T_c (K)	ΔT_c 90%-10% (K)	ΔT_c 99%-1% (K)
850 fast push	55	21	14.7	25.1
850 slow push	69	41	10.3	18.4
875 slow push	83	49	13.0	21.5
900 slow push	83	51	12.6	20.2

The tabulated data shows that both the onset temperature and critical temperature are improved with the higher temperature anneals. In addition, for the same onset temperature, a higher temperature anneal results in a slightly sharper transition. The push rate also has a dramatic effect on the R-T characteristic. The sample which was slowly pushed into the furnace had significantly higher onset and critical temperatures.

For several compositions of films on $SrTiO_3$, two samples were annealed for 45min or 60min respectively at 850°C. No significant or consistent differences in their R-T characteristics were observed.

One of the films with a critical temperature of 85K was remeasured at two later times. The resistance as a function of temperature from 70K to 120K from these measurements is shown in Figure 4. The initial measurement is marked by the squares. The second measurement, plotted with crosses, was made 14 days later. The normal state resistance had increased by approximately .8Ω from the first measurement. The transition had degraded only slightly over this period. The sample was measured a third time 33 days after the first measurement, shown by the circles in Figure 4. The normal state resistance was seen to have increased very slightly from the second measurement. There was no additional change in the superconducting transition temperatures.

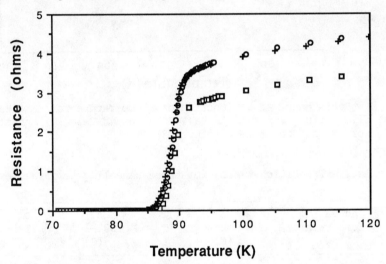

Fig. 4. Three measurements of the resistance as a function of temperature for a Y-Ba-Cu-O film. Squares: Initial measurement. Crosses: Second measurement, 14 days. Circles: Third measurement, 33 days.

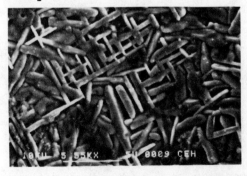

Fig. 5. Scanning electron micrograph of a 1.2μm film deposited in 12 layers.

A scanning electron micrograph of the film of Figure 3 is shown in Figure 5. This film was deposited in twelve layers and had a thickness of approximately 1.2μm prior to annealing. The film is polycrystalline and can be seen to have both ordered and unordered grains. Observations of this and other areas of the film as well as other samples appear to indicate that the film consists of a layer of ordered grains in contact with the substrate and an overlayer of unordered grains. The x-ray diffraction spectrum for this sample also shows incomplete ordering.

Fig. 6. Scanning electron micrograph of a 0.9μm film deposited in 18 layers.

In addition, superconducting films have been formed with depositions consisting of both more and less layers. Figure 6 shows a scanning electron micrograph of a film deposited in 18 layers which had a thickness of 0.9μm prior to annealing. This film appears to be more completely ordered than the thicker films formed with twelve layers.

Superconducting films on MgO substrates are not as readily formed. When the same annealing conditions used for $SrTiO_3$ are used for MgO, films with semiconducting normal state resistance temperature characteristics result. These films still show sharp onset temperatures in the vicinity of 90K, but the transitions have long resistive tails and zero resistance is achieved only below 15K, if at all. Better results are achieved by annealing at 850°C for times of two to three hours, yielding metallic normal state resistance characteristics, an onset of 93K and a critical temperature of 50K.

SUMMARY AND CONCLUSIONS

The technique of multi-layer sequential evaporation of BaF_2, Y and Cu layers has been employed for the formation of high temperature superconducting films. The effects of varying the composition and annealing conditions have been studied. For films on $SrTiO_3$ annealed at 900°C for 45min, variation of the Ba/Y ratio significantly altered the resistance-temperature characteristics. Barium poor films exhibited semiconducting normal state resistance-temperature characteristics while barium rich films were metallic. The transition temperatures varied with both the Ba/Y and Cu/Y ratios. Films which were superconducting at 85K were formed with Ba/Y=2.25 and Cu/Y=3.0. For a given composition, films on $SrTiO_3$ had higher transition temperatures when annealed at 900°C than at 850°C. Conversely, films on MgO were better when annealed at 850°C for longer times. The normal state resistance of the films increased over a period of two weeks but the transition temperature degraded only slightly.

REFERENCES

1. B-Y. Tsaur, M.S. Dilorio and A.J. Strauss, Appl. Phys. Lett., 51, 858 (1987).

2. C.X. Qiu and I.Shih, Appl. Phys. Lett., 52, 587 (1988).

3. A. Mogro-Campero and L.G. Turner, Appl. Phys. Lett., 52, 1185 (1988).

4. C-A. Chang, C.C. Tsuei, C.C. Chi and T.R. McGuire, Appl. Phys. Lett., 52, 72 (1988).

5. Z.L. Bao, F.R. Wang, Q.D. Jiang, S.Z. Wang, Z.Y. Ye, K. Wu, C.Y. Li and D.L. Yin, Appl. Phys. Lett, 51, 946 (1987).

6. N. Hess, L.R. Tessler, U. Dai and G. Deutscher, Appl. Phys. Lett., 53, 698 (1988).

EPITAXIAL GROWTH OF Dy-Ba-Cu-O SUPERCONDUCTOR ON (100)SrTiO$_3$
USING REACTIVE AND ACTIVATED REACTIVE EVAPORATION PROCESSES

R. C. Budhani, H. Wiesmann, M. W. Ruckman, and R. L. Sabatini
Brookhaven National Laboratory, Upton, NY 11973

ABSTRACT

Epitaxial and highly oriented thin films of Dy-Ba-Cu-O superconductor have been grown by thermal evaporation of Dy, Ba and Cu metals onto MgO and SrTiO$_3$ substrates. Oxidation of the evaporated metal species is realized by releasing oxygen gas near the substrate during evaporation. The films deposited on (100)SrTiO$_3$, when annealed at 900°C, crystallize epitaxially with c-axis on the plane of the substrate. These films are metallic with superconducting transition width ~1.5 K. On magnesium oxide substrates, the film crystallize into a polycrystalline material which undergoes a strong c-axis reorientation upon further annealing at 950°C. Activated Reactive Evaporation process has been successfully used for in situ growth of the superconducting phase on Al$_2$O$_3$ substrates. These films reach a zero resistance state at 60 K.

INTRODUCTION

The normal state transport and superconducting properties of Y-Ba-Cu-O films are highly sensitive to oxygen concentration and the degree of crystallographic orientation in the material. C-axis oriented and epitaxial structures of Y$_1$Ba$_2$Cu$_3$O$_{7-x}$, for example, show metallic character in the normal state and a sharp transition to the superconducting (SC) phase. The deposition of epitaxial and high T$_c$ thin films is facilitated by three-source thermal evaporation of Cu, Ba and Y in low partial pressures of oxygen.[1-3] The high reactivity and low vapor pressure of molten yttrium however, requires the use of electron beam gun for evaporation. The e-beam evaporation becomes difficult if the reactive gas pressure increases above a certain value (~10^{-4} Torr). This feature of the e-beam evaporation process hampers the in situ growth of the SC phase in high oxygen partial pressures or in an oxygen plasma environment. In contrast to yttrium, lanthanides such as Dy, Er, Ho, and Yb, have a higher vapor pressure and a tendency to sublime. Both these features make them ideal for evaporation from Knudsen type thermal sources, thus allowing deposition in higher partial pressures of oxygen.

In this paper we will discuss the deposition of epitaxial thin films of Dy$_1$Ba$_2$Cu$_3$O$_{7-x}$ oxide on strontium titanate and magnesium oxide substrates by coevaporation of Dy, Ba and Cu from Knudsen type sources. Preliminary results on the use of Activated Reactive Evaporation (ARE) for in situ growth of the SC phase are also reported.

EXPERIMENTAL

Films, typically 800-1000 nm thick, were prepared by coevaporation of Dy, Ba and Cu onto heated (~400°C) MgO(100) and SrTiO$_3$(100) substrates. Resistively heated alumina crucibles were used for all the metals. The rates of evaporation were monitored and controlled with three separate quartz crystal monitors placed above the sources. Intermixing of the evaporated species in the vicinity of the crystals was minimized by partitioning the chamber into three sections. Oxidation of the evaporated metal species was realized by releasing oxygen gas just beneath the substrates. A perforated copper tube was used for this purpose. The oxygen partial pressure in the chamber during evaporation was maintained at 4×10^{-5} Torr.

In a separate series of runs, the compound forming reaction at the substrate was enhanced by letting the evaporated species pass through a rf excited oxygen plasma. The plasma was created by coupling 100 watts of rf power (13.6 MHz) to a copper mesh (80 percent open area) placed 6 cm below the substrates. The partial pressure of oxygen and the substrate temperature during these runs were 5×10^{-4} Torr and ~650°C respectively. The superconducting behavior of the films was characterized by four probe dc resistivity measurement. Sputter deposited gold pads were used to provide contacts for the current and voltage leads. Typical measuring currents ranged from 50 µ to 100 µA. The crystal structure and the surface topography of the samples were determined using x-ray diffraction and scanning electron microscopy techniques respectively. The cation concentration in the films was measured by electron microprobe.

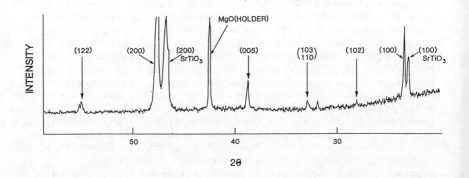

Fig. 1. X-ray diffraction pattern taken from a film deposited on SrTiO$_3$(100) and annealed at 900°C for 18 min.

RESULTS AND DISCUSSION

Films on SrTiO3 and MgO Using Reactive Evaporation

As desposited films on both $SrTiO_3$ and MgO substrates were amorphous and electrically insulating. A well-developed crystalline phase exhibiting the metallic conduction was obtained by annealing the films at 900 and 950°C for 18 minutes in a flowing oxygen environment. Figure 1 shows the x-ray diffraction pattern of a film deposited on $SrTiO_3$ and subsequently annealed at 900°C for 18 minutes. The diffraction pattern is dominated by (100) and (200) reflections of the orthorhombic $Dy_1Ba_2Cu_3O_{7-x}$ lattice, which are adjacent to (100) and (200) reflections of the strontium titanate substrate. Unlike powder diffraction results from polycrystalline Y-Ba-Cu-O, the intensity of the reflections with nonzero k and l indices is

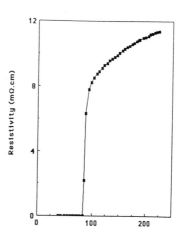

Fig. 2. Variation of electric resistivity of a a-axis oriented film on $SrTiO_3(100)$.

negligible in the present case. This observation indicates that the film is growing epitaxially with "a" axis normal to the plane of the substrate. The lattice parameters for the orthorhombic $Dy_1Ba_2Cu_3O_{7-x}$ are a=3.81 Å, b=3.89 Å and c=11.67 Å. Since the substrate is cubic with a_0=3.91 Å, the mismatch between a and a_0 is ~2.5 percent. This relatively small mismatch facilitates alignment of the bc plane of the SC lattice parallel to the substrate. The observed growth pattern on (100)$SrTiO_3$ can be compared to the behavior of Y-Ba-Cu-O system which generally grows with c-axis normal to the plane of the substrate. Webb et al.[4] have discussed the growth of Dy-Ba-Cu-O films by using Molecular Beam Epitaxy (MBE). Their films show varying degrees of a, b and c axis orientation. In Fig. 2 we show the temperature dependence of electrical resistivity of one of the a-axis oriented films. The normal state resitivity of the film is strongly metallic. The T_c onset and the transition width, which is defined as the temperature range in which the resistivity decreases by 10 and 90 percent of its value at the onset, are 90.0 and 1.5 K respectively. One interesting feature of these a-axis oriented films is the magnitude of the normal state resistivity. The room temperature resitivity for example, is higher by a factor of 15~25 when compared with the data for the "c" axis oriented Dy-Ba-Cu-O films grown on MgO (as described in the following section).The out-of-plane and in plane resistivity, ρ_c and ρ_{ab} respectively, data for single crystals of Y-Ba-Cu-O show anisotropy (ρ_c/ρ_{ab}) in the range 50 to 110.[5] Hence, the results for our films are consistent with the single crystal data. Films deposited on MgO(100), when annealed at

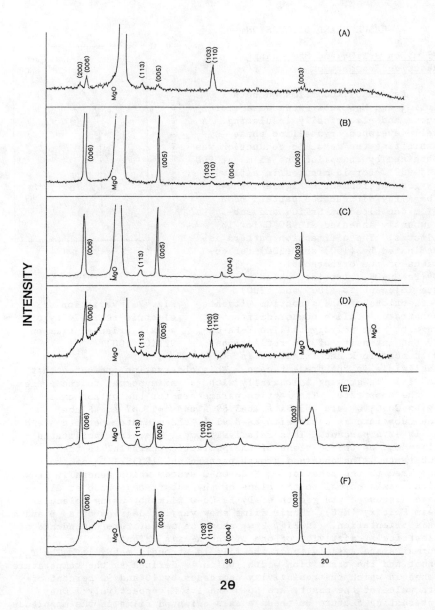

Fig. 3. X-ray diffraction pattern of some films deposited on MgO substrate. Curves A and B are for the same film annealed at 900 and 950°C respectively. C, D, E and F are all annealed at 950°C.

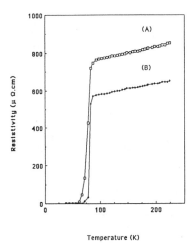
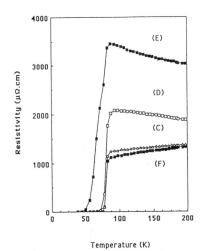

Fig. 4. Effect of annealing temperature on the resistivity vs temperature of film B.

Fig. 5. Electrical resistivity of films C, D, E, and F plotted as a function of temperature.

900°C, result in a randomly oriented polycrystalline material (Fig. 3 curve A). Further annealing at 950°C for 18 min. however, causes a significant enhancement of the intensity of (001) reflections and the reflections with the nonzero h and k indices diminish into the background. These changes in the x-ray diffraction pattern are seen in Fig. 3 (curve B). Results for the electrical resistivity measurements of these films, annealed at 900 and 950°C, are compared in Fig. 4. In the random polycrystalline state, the residual resistance ratio (ρ_{RT}/ρ_o) and the transition width are 1.38 and 12 K respectively. On annealing at the higher

Table I. Properties of the films deposited on MgO.

Sample	ρ_{RT} ($\mu\Omega$ cm)	$T_{c1}(10\%)$ K	$T_c(90\%)$ K	ΔT_c K	Metal Ion Concentration Dy:Ba:Cu	ρ_{RT}/ρ_o
B	689	83.0	80.0	3.0	1:1.93:2.99	1.37
C	1456	83.0	79.0	4.0	1:2.12:3.3	1.35
D	1890	83.0	75.0	8.0	1:2.36:3.74	semi conducting
E	2898	81.5	57	24.5	1:1.93:4.26	"
F	1543	83.0	78	5.0	1:2.08:3.32	1.65

temperature, the transition width decreases to 3 K but the resistance ratio does not change. The lack of any change in the resistance ratio and the difference in the transition width of the two samples highlights the role of phonon scattering and the intergrain Josephson coupling in deciding (ρ_{RT}/ρ_o) and ΔT_c respectively. The identical slope of the normal state resistivity of the two samples indicates that phonon scattering dominates over the entire temperature range. Alternatively, the temperature independent scattering times originating from structural imperfections and grain boundaries, are large as compared with the phonon scattering time even in the random polycrystalline sample. The polycrystallinity and the grain boundaries in the sample annealed at 900°C however, significantly reduce the intergrain Josephson coupling which results in higher ΔT_c.

The effects of composition on the structure and superconducting response of the films deposited on MgO are shown in Figs. 5 and 6 respectively. Some important features of the films, as derived from these figures, are listed in Table I. The table also lists the cation concentration in the films. Film B, whose cation concentration is closest to 1:2:3 stoichiometry shows the sharpest transition (ΔT_c=3 K) and the lowest resistivity (ρ_{RT}= 690 µΩ cm). Electron microprobe data for film C and F sug-gest that a large increase in the copper concentration from 1:2:3 stoichiometry does not have a significant influence on ρ_{RT} and ΔT_c. However small excess barium concentration, as in film D, leads to a semiconductor like transport behavior in the normal state.

Activated Reactive Evaporation (ARE) Process

The use of high temperature superconductors in microelectronics requires the development of a deposition technique that would permit *in situ* growth of the superconducting phase on substrate materials such as Al_2O_3, SiO_2, Si and GaAs at low substrate temperatures. Such a technique should also permit easy controlling of the generation, transport and condensation processes involved in deposition of a film. The use of reactive gas plasma in the deposition environment has been found to enhance the kinetics of compound forming reaction at the substrate. This type of activated reactive evaporation has been very successful in deposition of a large number of refractory compounds.[6] In general, the presence of ionic or the highly excited atomic form of the reactive gas in the ion assisted deposition processes significantly improves the structure and morphology of the compound films.[7] Considering the well documented benefits of ion assisted deposition, we have explored the use of ARE to grow Dy-Ba-Cu-O films on Al_2O_3 substrates.

Figure 6 shows results of electrical resistivity measurements on a 600 nm thick film. The temperature of the copper block holding the substrate was 650°C. The resistivity of the sample shows semiconducting behavior down to 93 K and then the onset of the SC state at 87 K. The sample reaches a zero resistance state at 60 K. Figure 7 shows x-ray diffraction pattern of this sample in 2θ range 20 to 55 degrees. The pattern is dominated by (103) and (005) reflections of the orthorhombic $Dy_1Ba_2Cu_3O_{7-x}$. The broad width of these peaks suggests that the material is not completely crystallized to the orthorhombic phase, and this may, in turn, affect the temperature dependence of the resistivity and the superconducting transition width. At this point it is worth noting that the ARE technique has been successfully used by Terashima et al.[8] for in situ growth of epitaxial $Y_1Ba_2Cu_3O_{7-x}$ on $SrTiO_3$. The advantage of using atomic oxygen in molecular beam epitaxy of Dy-Ba-Cu-O has been reported by Spah et al.[9] In this process the

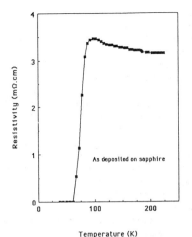

Fig. 6. Temperature dependence of electrical resistivity of a as deposited film on sapphire.

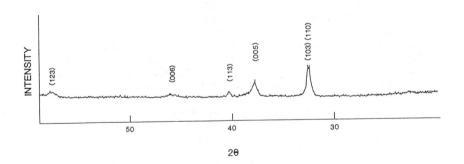

Fig. 7. X-ray diffraction pattern of a film deposited on Al_2O_3 substrate using ARE processes. No post deposition annealing treatment was given to the film.

atomic oxygen is extracted from a cold cathodic rf discharge. The films deposited on Al_2O_3 at 580°C show granular superconducting behavior with zero resistance at 48 K. In ARE however, since the plasma extends over the entire source - substrate space, every vapor species present in that region is ionized or excited. The high energy associated with these ionic/excited species can be utilized to improve the film properties.

In summary, reactive thermal evaporation of Dy, Ba, and Cu in low oxygen partial pressure leads to the formation of amorphous and insulating deposits. Upon furnace annealing at 900°C the material deposited on $SrTiO_3$ crystallized epitaxially along the orientation of the substrate. These films are strongly metallic with superconducting transition width ~1.5 K. Similar films deposited on MgO are polycrystalline when annealed at 900°C. On further annealing at 950°C, the material undergoes a strong c-axis oriented growth.

The feasibility of using ARE technique for in situ growth of the superconducting phase has been established. Preliminary results on these films show that they exhibit granular superconductivity. The large parameter space of the process allows for further experimentation to improve the films.

ACKNOWLEDGEMENTS

This research was performed under the auspices of the U.S. Department of Energy, Division of Materials Science, Office of Basic Energy Sciences under Contract No. DE-AC02-76CH00016.

REFERENCES

1. P. Chaudhari, R. H. Koch, R. B. Laibowitz, T. R. Macguire, and R. J. Gambino, Phys. Rev. Lett. $\underline{58}$, 2684 (1987).
2. J. Kow, T. C. Hsieh, R. M. Fleming, M. Hong, S. H. Liou, B. A. Davidson, and L. C. Feldman, Phys. Rev. B $\underline{36}$, 4039 (1987).
3. R. H. Hammond, M. Naito, B. Oh, M. Hahn, P. Rosenthal, A. Marshall, N. Missert, M. R. Beasly, A. Kapitulnik, and T. H. Gaballe, Proc. Mater. Res. Soc., M. Schulter and D. U. Gubser, Editors, Vol. EA-11, MRS, Pittsburgh, 1987.
4. C. Webb, S. L. Weng, J. N. Eckstein, N. Missert, K. Char, D. G. Schlom, E. S. Hellman, M. R. Beasly, A. Kapitulnik, and J. S. Harris Jr., Appl. Phys. Lett. $\underline{51}$, 1191 (1987).
5. S. J. Hagen, T. W. Jing, Z. Z. Wang, J. Horvath, and N. P. Ong, Phys. Rev. B $\underline{37}$, 7928 (1988).
6. R. F. Bunshah and A. C. Raghuram, J. Vac. Sci. Technol. $\underline{9}$, 1385 (1972).
7. J. M. E. Harper, J. J. Cuomo, R. J. Gambino, and H. R. Kaufman, in Ion Beam Modification of Surfaces; Fundamentals and Applications, O. Auciello and K. Kelly, Editors, p. 127, Elsevier, Amsterdam, 1984.
8. T. Terashima, K. Iijima, K. Yamamoto, Y. Bando, and H. Mazaki, Jpn. J. Appl. Phys. $\underline{27}$, L91 (1988).
9. R. J. Spah, H. F. Hess, H. L. Stormer, A. E. White, and K. L. Short, Appl. Phys. Lett. $\underline{53}$, 441 (1988).

RATE AND COMPOSITION CONTROL BY ATOMIC ABSORPTION SPECTROSCOPY
FOR THE COEVAPORATION OF HIGH Tc SUPERCONDUCTING FILMS

C. Lu
Xinix, Inc., 3500 Thomas Road, Santa Clara, CA 95054

N. Missert, J. E. Mooij*, P. Rosenthal, V. Matijasevic,
M. R. Beasley, and R. H. Hammond
Applied Physics Department, Stanford University,
Stanford, CA 94305

ABSTRACT

Atomic absorption spectroscopy has been used to control the deposition rates during coevaporation processes with multiple electron-beam sources. This technique is material specific and thus allows the deposition rate of each component to be controlled independently. Because only a light beam is needed to interact with the vapor stream, the sampling region can be selected to be very close to the substrate for precise control of the film composition. With its high sensitivity and no limitations on operation pressure, this technique offers some unique advantages for the preparation of high Tc superconducting films by coevaporation in a high oxygen partial pressure environment. The performance of a multi-source deposition controller and the resultant film properties are presented.

INTRODUCTION

Since the discovery of the new high Tc superconductors, numerous techniques have been developed for depositing thin films of these materials. Recently there has been considerable effort dedicated to growing thin films of these superconductors in situ at lower temperatures and on a wider variety of substrates. It is now clearly demonstrated that superconducting thin films of these new materials can be prepared in situ by a number of techniques such as coevaporation from multiple sources[1-5], sputtering from a composite-target [6], and laser ablation [7,8]. The substrate temperature can be as low as 400°C without post-annealing at temperatures higher than the growth temperature [8]. The elimination of post-annealing at high temperature and in oxygen ambient is considered essential for the fabrication of multi-layer superconducting devices and for the eventual merging of the new superconductor device technology with the existing semiconductor fabrication technology.

In these fabrication processes for in situ superconducting films, the main challenges at the present appear to be the incorporation of a sufficient amount of oxygen into the films and the control of film composition during deposition. In order to accomplish the former task, all processes reported so far inevitably

*On leave from Delft University of Technology, Netherlands.

require the presence of high oxygen flux at the substrates during deposition. On the quest to control the film composition, there have been two approaches. In processes which employ a single source only, such as sputtering and laser ablation from a composite target, the resulting film composition is determined in part by that of the source material, as well as by various process parameters. Because of the existence of these process variables, it is not easy to reproduce the correct stoichiometry of the source material when it is being transformed into thin films. The other approach to control the film composition is to use multiple sources in the forms of elements, alloys, or compounds, and controlling the deposition rate of the individual components independently as in coevaporation and cosputtering processes. This approach is unquestionably the most versatile method for the synthesis of new superconducting thin films, allowing optimization of film stoichiometry. Furthermore, films with graded or modulated compositions may be prepared under controlled conditions. Unfortunately, the difficulties involved in controlling the deposition rate of each component precisely and independently has limited the full potential of this approach.

Although quartz crystal microbalances have been used in most codeposition processes for rate control, they are far from ideal for the present purpose. The quartz crystal microbalance measures the deposited mass only, thus it is material non-specific. To use it for controlling multi-component codeposition processes, multiple sensors must be installed with proper shieldings to prevent cross-talk. The resultant configuration may not be satisfactory because of the non-ideal sensor/substrate relation and the directionality of the sensors. The presence of oxygen causes additional problems to rate controlling by quartz crystal microbalance because the sensor can not distinguish whether the mass deposited is pure metal or its oxide. The sensitivity and response time of quartz crystal rate monitors are also marginal for closed-loop control of electron-beam sources at low deposition rates. The ion gauge type rate controllers have higher sensitivity and faster response time than those of quartz crystal monitors but they cannot be operated satisfactorily above 10^{-5} Torr. Other types of material-specific deposition rate monitors, such as mass spectrometer and electron impact emission sensor, require only a single sensor to control multiple components, but all employ a thermionic emitter in the sensor. Thus they suffer the same pressure limitations as the ion gauge type detectors.

In this paper we describe the application of atomic absorption spectroscopy to control in situ growth of the high-temperature oxide superconductors by reactive electron-beam coevaporation processes. This technique is material specific and highly sensitive to most metal atoms. It has no limitations on operating pressure and thus it is ideally suited for reactive processes with a high partial pressure of oxygen or other gases. We feel that this rate controlling technique can significantly expand the utility of codeposition processes for the preparation of high Tc superconducting thin films.

PRINCIPLE OF OPERATION

The atomic absorption spectroscopy is a very sensitive and proven technique for the detection of metal atoms in the vapor phase. The feasibility of using this technique for rate monitoring in physical vapor deposition processes has been demonstrated [9,10]. Light emitted from a hollow cathode lamp made of the same metal as the source to be controlled is allowed to pass through the evaporant vapor. For a specific resonance line, the amount of radiation absorbed by the atomic vapor is a function of the absorption path length l, and the number density of the atom N. Normally, absorption phenomena can be described by the simple Beer's law, which is given by

$$I = I_o e^{-\mu l N} \qquad (1)$$

where I_o and I are the intensities of radiation before and after absorption, and μ is the absorption cross section of the atoms. Equation (1) in the form shown here is valid only if the absorption line profile, or bandwidth, is equal to or broader than the emission line profile. However, in a typical evaporation system the absorption line profile [Fig. 1(b)] is narrower than the emission line profile of a hollow cathode lamp source [Fig. 1(a)], owing to the fact that the mean velocity component of evaporant atoms is nearly perpendicular to the direction of observation. Therefore, the Doppler broadening of the absorption line is relatively small. Consequently, only a portion of the atomic emission within the bandwidth of the absorption line profile can be absorbed by the evaporant vapor [Fig. 1(c)]. The intensity of radiation after absorption thus deviates significantly from that described by Eq. (1). For practical purposes, the relationship between light absorption (I/I_o) and atomic vapor density N typically needs to be determined by experimental methods.

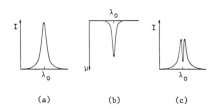

Fig. 1. Spectral profiles of (a) emission line intensity, (b) absorption line, and (c) intensity after absorption.

Strictly speaking, the measurement of the number density N alone is insufficient to determine the mass deposition rate D. This can be seen in the equation

$$D = kmNv \qquad (2)$$

where k, m, and v are the sticking coefficient, atomic mass, and the average velocity of atoms towards the substrate. In typical deposition processes, the sticking coefficient depends on the substrate temperature and generally is a constant. Equation (2) thus indicates that a one-to-one correspondence between D and N can only exist under the condition of a constant v. The criterion needs more careful examination because in common evaporation processes the source temperature is always varied to keep the deposition rate constant. Fortunately, for most metals, a 10% change in absolute source temperature will cause one order of magnitude change in evaporation rate but results in only about a 5% changes in v . Thus for a specific evaporation source, the average velocity of atoms can be assumed as constant over a limited range of deposition rate. This justifies the use of atomic number density N measurements to control the mass deposition rate D. From a practical point of view, this assumption is necessary because a simple method for measuring v has yet to be developed.

The collisions between the evaporant atoms and residual gases are another factor which can change the atomic velocity. Collisions reduce the average velocity of evaporant atoms and broaden the absorption line width. The net effect is an increase in light absorption as compared to the same deposition rate under high vacuum emvironment, or an increase in deposition rate sensitivity.

The above discussions indicate that an absolute calibration function relating the photodetector output signals and deposition rates may be difficult to derive theoretically because it depends on many parameters. This problem is avoided in practice by using an empirical calibration function F to relate the two quantities as shown in the following equation.

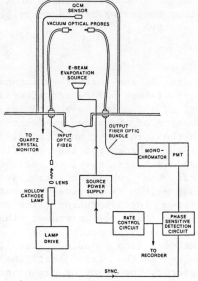

Fig. 2. Block diagram of a single-channel atomic absorption deposition controller.

$$I = F(D,p) \cdot D \qquad (3)$$

where F, for a specific system, is a function of deposition rate D and ambient pressure p . The calibration function can be determined by using a quartz crystal microbalance or other film thickness measuring device for absolute measurements.

ATOMIC ABSORPTION DEPOSITION CONTROLLER

Atomic absorption deposition controllers developed by Xinix are used in all experiments. For the purpose of testing the general performance of atomic absorption deposition controllers, a single-channel instrument was installed in a standard bell-jar type, electron-beam evaporation system. A block diagram of the controller is shown in Fig. 2. A hollow cathode lamp made of the metal of interest was used as source of specific atomic emission. Light from the hollow cathode lamps is guided into the vacuum chamber by a UV optic fiber (1 mm in diameter). A quartz lens is used to collimate the light exiting from the end of the input optic fiber. The collimated light beam is then allowed to pass through the evaporant stream at a region very close to the substrate. Another optical probe similar to the one described above is used to collect the light and focus it onto the end of an optic fiber bundle (5x600 micron fibers). Optional thin glass microchannel plates can be placed in front of the quartz lenses to reduce the amount of condensable materials reaching the lenses but at the expense of reduced transmitted light intensity. The output optic fiber bundle couples to a monochromator which acts as a narrow bandpass filter. The monochrmator is tuned to the wavelength of the atomic emission line to be detected. The intensity of the selected atomic emission line after absorption is then detected by a photomultiplier tube (Hamamatsu R928). In order to discriminate the radiation from ambient and the evaporation sources, the hollow cathode lamp is electronically modulated at about 400 Hz and phase sensitive detection circuitry is used to detect the photo-signals. The system has a response time of better than 40 ms which is sufficiently fast to control electron-beam sources.

The output signal from the photomultiplier is inverted and offset so that at zero rate (no absorption and maximum detector output) the output indicates zero while at 100% absorption (with the light source turned off or completely blocked off) the output reads 10 volts. Thus the zero level stability is affacted by small light intensity fluctuations of the hollow cathode lamp. We found that most hollow cathode lamps require a warm-up period of at least one hour. The light output from hollow cathode lamps is also slightly temperature sensitive. But after warm-up and by keeping the ambient temperature relatively constant, the zero drift can often be kept to less than 1% over a period of one hour.

A typical calibration curve is shown in Fig. 3 for Cu deposition in high vacuumm ($<1 \times 10^{-5}$ Torr). This curve is obtained using a quartz crystal microbalance for absolute calibration. The absorption path length is 13 cm with a source-to-sensor distance of 25 cm. As one can see, the calibration curve is nonlinear and becomes saturated at high deposition rates. As mentioned earlier, the calibration curve will be different for a different absorption path length, ambient pressure, type of evaporation source, and system geometry.

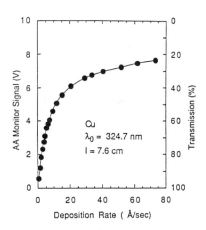

Fig. 3. Atomic absorption monitor output as a function of Cu deposition rate.

DEPOSITION OF SUPERCONDUCTING THIN FILMS

To investigate the applicability of the atomic absorption deposition controller for codeposition of superconducting thin films, a three-channel monitoring and control system was installed in a large coevaporation system which employs three electron-beam sources. A schematic of this deposition system is shown in Fig. 4. The three channels of deposition controller are independent and each channel is similar to that shown in Fig. 2, except for the input and output optic fibers. The optic fibers form a single bundle inside the vacuum chamber but split into three channels externally for coupling to individual controllers. The length of input and output optic fiber bundles is 3 m each. Hollow cathode lamps made of Y, Ba, and Cu are used as light sources. Each monochromator is tuned to the strongest resonant emission line from the corresponding lamp. The wavelengths are 410.2, 553.5, and 327.4 nm for Y, Ba, and Cu, respectively. To avoid possible inter-channel crosstalk due to interfering emission lines, the modulation frequencies of light sources and the corresponding frequencies for the phase sensitive detection systems, are different for the three channels.

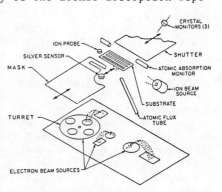

Fig. 4. Schematic of the deposition chamber for superconducting thin films.

For in situ growth of superconducting YBaCuO thin films, we use the atomic absorption deposition controller to control the metal stoichiometry (123). Typical deposition conditions are 2Å/sec for the total deposition rate, 3500Å for film thickness, and a substrate temperature of 600°C. Films were grown on $Al_2O_3 \langle 1\bar{1}02 \rangle$, $SrTiO_3$ $\langle 100 \rangle$, $\langle 110 \rangle$, and MgO $\langle 100 \rangle$. A flux of oxygen was supplied to the substrates during and after the deposition. The films were not subjected to post-anneal above or at the growth temperature, but a programmed cool-down procedure was implemented with the presence of oxygen flux[11].

Typical signals from individual channels during deposition with sources under closed-loop control are shown in Fig. 5 illustrating the excellent signal-to-noise ratio at

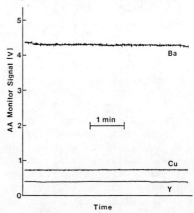

Fig. 5. Output signals from atomic absorption monitors during reactive coevaporation of Ba, Y, and Cu.

low deposition rates. With the metal stoichiometry under control, we are able to compare the effectiveness of different types of oxygen sources for the incorporation of oxygen into the films. Three oxygen sources (a modified Kaufman ion source, an electron-beam activated source, and an atomic oxygen source) were evaluated under similar deposition conditions. Details of these oxygen sources are described in Ref. 11.

The resistive transitions of films grown in a background pressure of 2×10^{-4} Torr with an additional flux of 3×10^{15} atoms/cm^2-sec from the atomic oxygen source are shown in Fig. 6. These films show the onset of the resistive transition at about 88 K and zero resistance anywhere from 77 to 80 K. Critical current densities (at 4.2 K) of 2×10^5 A/cm^2 on Al$_2$O$_3$, 5×10^5 A/cm^2 on SrTiO$_3$ <110>, and 1×10^6 A/cm^2 on MgO, indicate the high quality of these films.

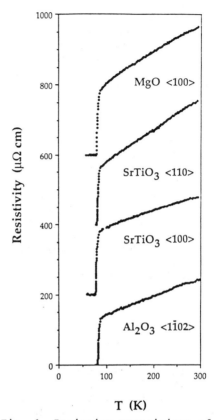

Fig. 6. Resistive transitions of films on different substrates.

Although the metal stoichiometry of these films reported here was found to be 1:2:4 by electron microprobe analysis, a fine tuning of the calibration function of the atomic absorption deposition controller has since allowed routine production of films with the 1:2:3 metal ratios. This demonstrated that once the calibration function has been determined, the atomic absorption technique provides a reproducible, sensitive control of the metal deposition rates.

CONCLUSIONS

We have used the atomic absorption spectroscopic technique to control the metal stoichiometry during the growth of YBaCuO thin films by reactive electron-beam coevaporation. With a proper oxygen source, _in situ_ growth of superconducting thin films has been successfully demonstrated. The atomic absorption spectroscopic technique for deposition rate control offers the following advantages:

(1) This technique is material-specific thus allowing the simultaneous control of multiple sources in a coevaporation system.

(2) This method is non-intrusive so that the sampled region can be selected to be adjacent to the substrate for precise control of the composition of evaporant flux.

(3) There are no limitations on operating pressure for this method thus it is ideally suited for reactive evaporation and sputtering processes.

ACKNOWLEDGEMENTS

The in situ growth of YBaCuO films at low temperatures discussed above was reported at the 1988 Applied Superconductivity Conference, San Francisco, California, August 21 - 25, 1988. The other authors are T. H. Geballe, A. Kapitulnik of Stanford University, S. S Laderman of Hewlett-Packard Company, E. Garwin of the Stanford Linear Accelerator Center, and R. Barton of Conductus, Inc. The Stanford authors wish to acknowledge the support of the Office of Naval Research, the National Science Foundation, and the Air Force Office of Scientific Research. Additional support was provided through a DuPont Research Grant to the Department of Applied Physics at Stanford University.

REFERENCES

1. D.K.L. Lathrop, S.E. Russek, and R.A. Buhrman, Appl. Phys. Lett. 51, 1554 (1987).
2. T. Terashima, K. Iijima, K. Yamamoto, Y. Bando, and H. Mazaki, Jpn. J. Appl. Phys. 29, L91 (1988).
3. R.M. Silver, A.B. Berezin, M. Wendman, and A.L. de Lozanne, Appl. Phys. Lett. 52, 2174 (1988).
4. R.J. Spah, H.F. Hess, H.L. Stormer, A.E. White, and K.T. Short, Appl. Phys. Lett. 53, 441 (1988).
5. P. Berberich, J. Tate, W. Dietsche, and H. Kinder, Appl. Phys. Lett. 53, 925 (1988).
6. H.C. Li, G. Linkes, F. Ratzel, R. Smithey, and J. Greek, Appl. Phys. Lett. 52, 1098 (1988).
7. D. Dijkkamp, T. Venkatesan, X.D. Wu, S.A. Shakeen, N. Jisrani, Y.M. Min-Lu, W.L. McLean, and M. Croft, Appl. Phys. Lett. 51, 619 (1987).
8. S. Witanachchi, H.S. Kwok, X.W. Wang, and D.T. Shaw, Appl. Phys. Lett. 53, 234 (1988).
9. P.E. Cade, master's thesis (University of Vermont, 1971) (unpublished).
10. E.I. Fazekas and M. Mezey, J. Vac. Sci. Technol. 9, 1119 (1972).
11. N. Missert, R. Hammond, J. E. Mooij, V. Matijasevic, P. Rosenthal, T. H. Geballe, A. Kapitulnik, M.R. Beasley, S.S. Laderman, C. Lu, E. Garwin, and R. Barton, Proceedings of the 1988 Applied Superconductivity Conference, San Francisco, California, 1988.

TRANSPORT

High Critical Currents and Flux Creep Effects in *e*-Gun Deposited Epitaxially 001 Oriented Superconducting $YBa_2Cu_3O_{7-\delta}$ Films 172

Factors Influencing Critical Current Densities in High T_c Superconductors 180

Limitations on Critical Currents in High Temperature Superconductors 194

HIGH CRITICAL CURRENTS AND FLUX CREEP EFFECTS IN
E-GUN DEPOSITED EPITAXIALLY 00L ORIENTED
SUPERCONDUCTING $YBa_2Cu_3O_{7-\delta}$ FILMS

B. Dam, G.M. Stollman
Philips Research Laboratories,
P.O. Box 80.000, NL-5600 JA Eindhoven, The Netherlands.

P. Berghuis, S.Q. Guo
Kamerlingh Onnes Laboratorium, faculteit Wiskunde en
Natuurwetenschappen
der Rijksuniversiteit, NL-2311 SB Leiden, The Netherlands

C.F.J. Flipse, J.G. Lensink, R.P. Griessen
Natuurkundig Laboratorium der Vrije Universiteit,
NL-1081 HV Amsterdam, The Netherlands.

ABSTRACT

Thin films of $YBa_2Cu_3O_{7-\delta}$ have been made by codeposition of Cu, Y and BaF_2 in a UHV system. Annealing in humid oxygen at 835°C produces high quality single phase preferentially c oriented films. In 1000 Å thin films the transport current density is larger than $10^6 A/cm^2$ at 86 K. Even 500 Å films prove to be superconducting above 77 K. Extrapolation to zero field of the magnetization data obtained from torque experiments are consistent with the current transport data. This indicates that the sample is homogeneous and the critical state model can be applied. Both magnetic relaxation measurements and flux creep resistivity in a magnetic field indicate the importance of thermally activated processes.

INTRODUCTION

To increase the range of possible substrate materials suited for the high − T_c oxide superconductors a lowering of the processing temperature is of the utmost importance. Furthermore, for applications a high J_c is wanted, which requires a film of high perfection.
In a short time a lot of progress has been made in the deposition of superconducting $YBa_2Cu_3O_{7-\delta}$ films. By now most deposition techniques have proven to be successful in depositing a stoichiometric $YBa_2Cu_3O_{7-\delta}$ film. However, to induce the transition from the amorphous state into the crystalline superconducting state, usually a post-deposition anneal treatment in O_2 at temperatures up to 900 °C is needed. Alternatively, in situ growth techniques have been developed using an increased substrate temperature and an enhanced oxygen pressure. In this way the processing temperature is lowered, which reduces the substrate film interaction and generally improves the critical current density[1]. It has been shown that the formation temperature can be lowered down to 400°C [2] when using reactive oxygen during deposition.
A method which improves the film properties was introduced by Mankiewich et al.[3]. The electrical film properties appear to bene-

fit from the use BaF$_2$ instead of Ba as one of the constituents of the deposited layer. This approach is not only very useful in UHV deposition, but it has also shown its merits in laser ablation[4], sputtering[5] and thick film technology[6], all using the two step growth mode with a high temperature anneal step.

Here, we report on the preparation of high J_c films taking the BaF$_2$ approach, with a UHV triple e-gun deposition system. As compared to our triode sputtered layers[7] film characteristics such as surface flatness, chemical inertness and J_c all benefit from this deposition technique.

Apart from the structural charaterization using Rutherford Back Scattering (RBS), Electron Probe Micro Analysis (EPMA) and X-ray diffraction (XRD) the electrical properties of the layers are considered in some detail. The critical current density is measured both by a transport current (in magnetic fields up to 8T) and by magnetization measurements in a torque magnetometer, both giving consistent results.

DEPOSITION

Amorphous films were co-deposited in a UHV triple e-gun system, taking Cu, Y and BaF$_2$ as source materials. The rate of the individual sources is monitored by three quartz crystals. A feedback system controls the deposition rate. The standard total rate of 8Å/s is made up from 5.1 Å/s BaF$_2$, 1.5 Å/s Cu and 1.4 Å/s Y. The film thickness is varied between 500 and 5000 Å. Throughout this paper the standard thickness is 2500 &angstrom, unless stated otherwise.

Oxygen is supplied by a nozzle ending with a glancing angle close to the rotating substrate holder. The oxygen partial pressure during deposition is controlled by a needle valve to a value of 10^{-7} mbar. As the pressure is monitored at a height roughly 30cm lower than the substrate the oxygen pressure near the substrate will be much higher. RBS and electron probe micro analysis on as deposited BaF$_2$ layers confirmed the 2:1 ratio of F with respect to barium.

As substrates we used polished (100) SrTiO$_3$ sawn with high precision (within ±0.03° along (100)) from a single crystalline boule. For these substrates RBS typically shows a [100] channeling minimum yield of 1.5%, while X-ray rocking curves give a FWHM for the 200 reflection of 0.15°.

ANNEALING AND ITS CHEMISTRY

The deposited films were annealed ex situ under an oxygen flow at temperatures up to 835°C. The temperature ramps proved to be less critical as compared to those used for our triode sputtered films[7]. We preferred a heating ramp of 300°C/h, while a cooling ramp of 100°C/h was chosen. The addition of water is essential for the hydrolysis of BaF$_2$ into bariumhydroxide, which then reacts with the metal oxides to form YBa$_2$Cu$_3$O$_{7-\delta}$. The necessity of water is also shown by a film partly covered by gold strips prior to the anneal step. After annealing, EPMA indicates that Ba and F are still present in the covered regions in a 1:2 ratio, whereas in the uncovered regions no fluorine is detected. The optimal procedure in our anneal system was found by checking the powder X-ray diffraction for the disappearance of the BaF_2 diffraction lines. We introduce additional water by passing the O_2 through a water slot. Maintaining the maximum anneal temperature at 835°C for 1 hour, a 15 minute soak with H_2O is sufficient. Contrary to Mogro-Campero[8] after this procedure we do not detect any fluorine (neither by EPMA nor by RBS), as long as the films have the correct metal stoichiometry.

BaF_2 is advantageous because of its stability towards the formation of $BaCO_3$, which is thought to be the main reason for the degradation of films containing Ba(0). BaF_2 not only protects the as-deposited layers. The hydrolysis apparently stops as soon as the bariumhydroxide cannot react any further. The fact that we could not reproduce the hydrolysis in a powder mixture of BaF_2 and the metal oxides treated in the same way as the films, indicates that the equilibrium constant of the hydrolysis reaction is very small. No $YBa_2Cu_3O_{7-\delta}$ was found by XRD, not even after heating the powders at 950°C in a flow of oxygen and water steam. It is clear that the equilibrium can be shifted to the right by manipulating the partial pressures of HF and H_2O and the concentration of bariumhydroxide.

Summarizing, the effect of taking BaF_2 as one of the constituents of the as-deposited film is primarily to protect Ba against the formation of unwanted by-products such as $BaCO_3$. In addition the formation of $YBa_2Cu_3O_{7-\delta}$ is facilitated.

STRUCTURAL CHARACTERIZATION

The annealed $YBa_2Cu_3O_{7-\delta}$ films are semi-transparent, sometimes with shining surfaces. The thickness variations as measured by a Talley step profiler are generally of the order of ± 200Å.

On the epitaxial nature of the BaF_2 type films various results are reported. Powder X-ray diffraction (XRD) using Cu Kα radiation shows the strong tendency of our films to grow with the c-axis parallel to the $SrTiO_3$ substrate (100) surface normal. This is in conflict with the results of others[4,5] who found rather large quantities of pref a oriented material. Only when we decrease the amount of BaF_2 small h00 peaks appear, otherwise the XRD spectrum only shows 001 diffraction lines. In contrast to our sputtered films[7] we did not observe any tendency of an increase of the pref a and random type of growth neither upon increasing the layer thickness from 500 to 5000 Å nor by lowering the anneal temperature. The X-ray rocking curves show that the width (FWHM) of a 005 diffraction line of a 2500 Å film is 0.40°. In contrast to results e.g. of Levi et al.[9] the misorientation of our pref c oriented film with respect to the $SrTiO_3$ 100 substrate reflection is less than 0.05°. The varying results of groups on the misorientation of the $YBa_2Cu_3O_{7-\delta}$ films may be due to the degree of misorientation of the substrate surface with respect to the 100 direction, which is very small in our case.

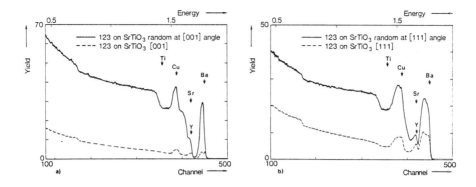

FIG.1: RBS of a 500 Å $YBa_2Cu_3O_{7-\delta}$ film on (100) $SrTiO_3$ a) Channeled along the [100] direction. b) Channeled along the [111] direction.

The epitaxial nature of the films is confirmed by the results of Rutherford Back Scattering (RBS) experiments. Applying 2MeV He^+ ions a very low minimum yield χ_{min} (the ratio of the back scattered yields of aligned versus random incidence) of 9 % is found when channeling a 500 Å film along the [100] normal of the $SrTiO_3$ substrate (fig.1a). This compares favourably with the 'single crystal' film values by Meyer et al.[10]. The splitting of the channeled Ba signal may indicate a small enrichment of Ba at the interfaces, which is amorphous or random oriented crystalline. Whereas the [100] channeling indicates the degree of texture, the epitaxiality of the 123 ab-plane with respect to the $SrTiO_3$ main axes is better examined by a [111] channeling experiment. In this case we find χ to be $\simeq 40\%$ (fig.1b) for each of the equivalent body diagonals. Compared with our sputtered layers these values are 20 and 10% respectively better. In the sputtered layers already a high degree of epitaxiality was observed by TEM, hence we conclude these layers to be at least as good in this respect.

Lowering the anneal temperature down to 650°C does not change the nature of the preferential orientation. However, apart from the 001 $YBa_2Cu_3O_{7-\delta}$ reflections an increased amount of preferentially oriented phases with lattice plane distances of $n \times 13.6$ and $n \times 10.6$Å (the first phase being related to the $Y_2Ba_4Cu_8O_x$[11] phase) could not be avoided.

ELECTRICAL CHARACTERIZATION

Resistance measurements R(T) were performed by a conventional DC 4-point probe method. Indium spheres pressed on sputtered patches of gold served as contacts. The 1000 Å film's R(T) can be approximated by $\rho(T) = \beta T$. We estimate β to be approximately $0.56 \mu\Omega cm/T$. Both in the 500 and the 1000 Å film no extra phases are found by X-ray diffraction. However, the thinner layer will be more sensitive to small amounts of impurities, which might influence both the zero resistance temperature T_{cf} and the curvature around onset temperature of the transition T_{co}.

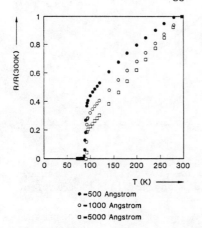

• = 500 Angstrom
o = 1000 Angstrom
□ = 5000 Angstrom

A larger value for β is reported by Levi et al.[9]. On the other hand the 100 K resistivity of our 1000 Å film approaches the single crystal value for the resistivity in the ab-plane measured by Ong et al.[12].

The behaviour of the 5000 Å film corresponds in two aspects to a mixed $YBa_2Cu_3O_{7-\delta}/Y_2Ba_4Cu_8O_x$ behaviour as described by Kapitulnik[11].

Fig.2 Normalized R(T) plots of a 500, 1000 and 5000 Å $YBa_2Cu_3O_{7-\delta}$ film deposited on (100) $SrTiO_3$.

R(T) extrapolates to zero at about 45 K and due to a small foot the zero resistivity temperature is reduced to 83.0 K. Indeed, the XRD of this film shows both the $YBa_2Cu_3O_{7-\delta}$ and the $Y_2Ba_4Cu_8O_x$ phase equally well developed and both pref c oriented.

While the T_{cf}'s of the 5000 and 500 Å films in fig.2 are reduced both to about 83 K the 90-10% width of the transition is only 1.5 K and 1.1 K respectively. The 1000 Å film has an even sharper transition with ΔT_c(90-10% %) =0.5 K and a zero resistance temperature T_{cf}=92 K.

A sharp transition often indicates a high critical current density. For the 1000 Å film we find at 86.7 K a J_c of $1.3 \times 10^6 A/cm^2$. Due to the fact that the current transport through the wires of the cryostat is limited, measurements could be extended only down to 81 K, where we find a critical current density of $3 \times 10^6 A/cm^2$. The critical current density is in this case measured by current transport along a small constriction 30 μm wide and 1mm long isolated from the rest of the film by a laser treatment. Instead of laser ablating a track, the isolating strips around the track are made by laser heating. The width of the track was checked by optical microscopy and SEM. The critical current is defined as the current at which a voltage drop across the strip reached 0.05 μV (i.e. $\simeq 0.5 \mu V/cm$).

In order to get a better insight in the temperature dependence of the critical current density a smaller 10μm track was made in a series of films with a thickness of 500, 600, 700, 800 and 900 Å Though apart from the 900 Å film all films show a foot in their R(T) plots, the temperature dependence of R above T_c is quite linear (Fig.3a.). The temperature dependence of the critical current density of the films is plotted in Fig.3b. as a function of $t = T/T_c$. Again very high J_c's are obtained. At 77 K the 900 Å film reaches $4 \times 10^6 A/cm^2$, at 40 K we measure $1.6 \times 10^7 A/cm^2$. The current density of the 500 Å film reaches $10^6 A/cm^2$ at 62 K. Comparing the critical current densities a saturation like behaviour is observed for layers thicker than 700 Å. We assume that this is due to the presence of a non superconducting film-substrate interface layer.

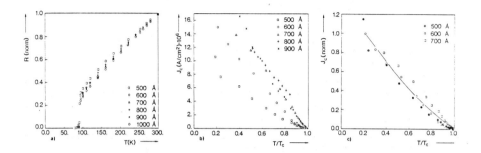

Fig.3. Electrical characteristics in zero field of 500, 600, 700, 800 and 900 Å films deposited on (100) SrTiO$_3$. a) Normalized R(T) plot. b) Temperature dependence of the critical current densities. c) Fit using eq.2 of the normalized $J_c(T)$'s.

Torque measurements[13] confirm the current densities obtained from current transport. In these experiments the magnetization of the sample is determined from the magnetic field and the resulting torque. Applying the model of Bean for homogeneous superconductors, J_c is calculated from the hysteretical behaviour of the magnetization as a function of the field. Furthermore, it is assumed that the critical currents are flowing in the plane of the sample even when the magnetic field is not perpendicular to the plane of the film. Measurements at an 0.01 T field at 77 K on a 700 Å film give a current density of $1.2 \times 10^6 A/cm^2$, which compares well with the transport value in zero field of $1.6 \times 10^6 A/cm^2$. At 4.2 K and 0.01 T torque on this 700 Å film gives a J_c of $1.8 \times 10^7 A/cm^2$. At 1 T this reduces to $J_c = 4 \times 10^6 A/cm^2$.

Analysis of the temperature dependence of the critial current in zero field shows that the $J_c(T)$ behaviour cannot be reproduced if one uses the granular model of Clem et al.[14] for a Superconductor Insulator Superconductor junction (i.e. SIS) or a Superconductor Normal-metal Superconductor junction (i.e. SNS). To demonstrate this we normalized the measured critical current densities arbitrarily at $T/T_c = 0.2$. This could only be done for the 500, 600 and 700 Å films (Fig.3c). Note, that the normalized behaviour of $J_c(T)$ is more or less the same for all films although the absolute values for T_{cf} and J_c vary considerably. The critical current in these films apparently cannot be understood in terms of a granular model. Therefore, we examined the possibility of a voltage drop due to thermally activated flux motion in the strip. In the case of high $- T_c$ superconductors where the activation energy U is comparable to kT the drift velocity of flux lines in presence of an electric current j and a magnetic field B is given by:

$$V = V_0 e^{(-U/kT)} \cdot \sinh(Aj/kT) \qquad (1)$$

where Aj represents the increase in energy of flux lines when the current density j flows through the strip. The electric field associated with the motion of flux lines is $E = V \times B$. As in the present experiment no external field is applied, $B = \gamma j$ and eq.1 is thus an implicit relation for j which can be calculated as soon as the temperature dependences of U and A are known and the detection limit E_d for the electric field is specified. In the following we take $E_d = 10^{-6}$ V/cm. Assuming that $U(t) \propto B_c^2 \xi^3(t)$ and $A(t) \propto \phi_0 \xi^2(t)$ (ϕ_0 and ξ being the elementary flux quantum and the coherence length respectively) the temperature dependences of U(t) and A(t) are $U(t) = U(0)(1+t^2)^{3/2}(1-t^2)^{1/2}$ and $A(t) = A(0)(1+t^2)/(1-t^2)$, where $t = T/T_c$. Taking $U(0)/kT_c = 10.5$ which implies an activation energy for flux motion U(0) of 83 meV, a good fit to the experimental data is obtained (fig.3c). For the 10 μm wide strip used in the experiment $\gamma = 6 \times 10^{-14}$ T m^2/A and the velocity prefactor V_0 which corresponds to the best fit is $V_0 = 1.7 \times 10^5$ cm/s. Assuming a jump length of the order of ξ which is in agreement with the assumptions made above, we find an attempt frequency for hopping over the activation barrier $\omega_0 = 1.7 \times 10^{12}$ s^{-1} which is comparable to a typical phonon frequency. It should however be pointed out that the quality of the fit depends only weakly on V_0. Consequently, the fitted values for ω_0 have a large uncertaintly.

Recently Tinkham[15] has proposed that $U(t) \propto B_c^2 \xi a_0^2$ with $a_0 = (\phi_0/B)^{1/2}$ to explain the broadening of the resistive transition in a magnetic field. Although this form for U(t) can also well be fitted to the critical current data in fig.3c it leads to too strong a magnetic field dependance at low temperatures.

Flux motion can also explain the high sensitivity of the critical current to a magnetic field. In a 0.5 T field directed normal to the film surface at 83.3 K the (transport) critical current density of the 1000 Å film (the same as in Fig.1) reduces from 2×10^6 A/cm^2 (in zero field) to 1×10^5 A/cm^2. At high magnetic fields a kind of flux flow resistance[16] is observed and no critical current can be found anymore. Within our accuracy of 20 nV, no current could be measured at 83.3 K in a field larger than 1 T with a zero voltage; i.e. ρ_{f0} defined as $(dE/dJ)_{J=0}$ is not equal to zero. ρ_{f0} depends strongly on the applied magnetic field. It increases 3 orders of magnitude when the field is increased from 2 to 5 T. As ρ_{f0} is much smaller than the extrapolated normal state resistance, the non zero onset resistance should be associated with flux creep, which we think is thermally activated.

CONCLUSION

It is clear that the BaF$_2$ method enables one to produce very thin high quality superconducting thin films. Thus, intrinsic properties of the oxide superconductors can be studied. Our results show that the J_c of thin films can reach values larger than 10^7/cm^2, however the critical current is very sensitive to magnetic fields. This is interpreted as being a result of an extremely weak pinning of flux lines resulting in thermally activated flux creep.

The fact that our films are much more sensitive to magnetic fields than other high-J_c films may be ascribed to the homogeneity of our films. E.g. Itozaki et al.[17] deposit 7000 Å films on MgO and a kind of double layer cannot be excluded. The reactivity of MgO is somewhat higher than that of $SrTiO_3$ Thus the existence of an interface layer on top of the MgO is very likely. Such an interface layer may then provide for the pinning centra needed to maintain the critical current density at high magnetic fields.

1. H. Adachi, K. Hirochi, K. Setsune, M, Kitabatake and K. Wasa, Appl. Phys. Lett. 51 2263 (1987).
2. S. Witanachchi, H.S. Kwok, X.W. Wang and D.T. Shaw, Appl. Phys. Lett. 53, 234 (1988).
3. P.M. Mankiewich, J.H. Scofield, W.J. Skocpol, R.E. Howard, A.H. Dayem, Appl. Phys. Lett., 51, 1753 (1987); G.J. Fisanick, P. Mankiewich, W. Skocpol, R.E. Howard, A. Dayem, R.M. Fleming, A. E. White, S.H. Liou and R. Moore, MRS proceedings Vol.99, 703 (Pittsburgh 1988).
4. A.M. DeSantolo, M.L. Mandich, S. Sunshine, B.A. Davidson, R.M. Fleming, P. Marsh and T.Y. Kometani, Appl. Phys. Lett. 52, 1995 (1988).
5. S.H. Liou, M. Hong, J. Kwo, B.A. Davidson, H.S. Chen, S. Nakahara, T. Boone and R.J. Felder, Appl. Phys. Lett. 52, 1735 (1988).
6. A. Gupta, R. Jagannathan, E.I. Cooper, E. A. Giess, J.I. Landman and B.W. Hussey, Appl. Phys. Lett. 52, 2077 (1988).
7. B. Dam, H.A.M. van Hal and C. Langereis, Europhys. Lett. 5, 455 (1988); A.P.M. Kentgens, A.H. Carim and B.Dam, J. Crystal Growth 91, 355 (1988).
8. A. Mogro-Campero, L.G. Turner, E.L. Hall and M.C. Burrell, Appl. Phys. Lett. 52, 2068 (1988).
9. A.F.J. Levi, J.M. Vandenberg, C.E. Rice, A. P. Ramirez, K.W. Baldwin, M. Anzlowar, A. E. White and K. Short, J. Crystal Growth 91, 386 (1988).
10. O. Meyer, F. Weschenfelder, J. Geerk, H.C. Li, G.C. Xiong, Phys. Rev. B37, 9757 (1988).
11. A. Kapitulnik, Proceedings of the Int. conf. on HTSC and materials and mechanisms of superconducivity, Interlaken, feb.-march 1988, Physica C152-155, 520 (1988).
12. N.P. Ong, Z.Z. Wang, S. Hagen, T.W. Jing, J. Clayhold, and J. Horvath, Physica C 153-155, 1072 (1988).
13. R. Griessen, M.J.G. Lee, D.J. Stanley, Phys. Rev. B16, 4385 (1977)
14. J.R. Clem, B. Bumble, S.I. Raider, W.J. Gallagher and Y.C. Shih, Phys. Rev. B35, 6637 (1987).
15. M. Tinkham, Phys. Rev. Lett. 61, 1658 (1988).
16. M. Tinkham, Introduction to Superconductivity (McGraw-Hill, New York, 1975).
17. H. Itozaki, S. Tanaka, K. Higaki and S. Yazou, Physica C152-155, 1072 (1988).

FACTORS INFLUENCING CRITICAL CURRENT DENSITIES IN HIGH T_C SUPERCONDUCTORS

G. J. Fisanick
AT&T Bell Laboratories, 600 Mountain Ave., Murray Hill, NJ 07974

ABSTRACT

A key requirement for many practical applications of the new class of high T_C superconductors is the ability to sustain high current densities in the presence of magnetic fields. The new materials present two substantially different behaviors. For $Ba_2YCu_3O_7$ critical current behavior of polycrystalline material is dominated by intergranular transport. We will review a model for transport involving the effects of the correlation of defects with grain boundary orientation, the presence of impurities at the grain boundaries, measurements of tunneling characteristics through boundaries, and the effect of grain alignment on critical current. The case is substantially different in the 84K superconducting phase $Bi_{2.2}Sr_2Ca_{0.8}Cu_2O_{8+\delta}$ where there is a very strong dependence of critical current densities at 30-84K on magnetic field *for single crystals*. There is a lack of threshold behavior in the I-V characteristics in finite applied fields larger than $H_{c1}(T)$ which is attributable to flux motion. This suggests that in the Bi material there is insufficient pinning of vortices in the *intrinsic* material to prevent them from moving in the presence of an applied transport current, causing energy dissipation. These results are in agreement with observations of flux lattice melting in single crystals of $Ba_2YCu_3O_7$ and $Bi_{2.2}Sr_2Ca_{0.8}Cu_2O_{8+\delta}$. The implications of these various factors on potential applications of these materials will be discussed.

INTRODUCTION

Virtually all potential applications of the new high T_C superconductors require high critical current densities J_C. This paper presents an overview of our present understanding of the factors influencing the magnitude of J_C. These factors can be conveniently divided into intergranular effects which dominate transport in polycrystalline ceramics and thin films, and "intrinsic" effects which can be studied using single crystals. A caveat that must be kept in mind is that all single crystal measurements are made on highly defective materials. For perfect defect-free single crystals there would be little pinning and J_C would be vanishingly small except for surface pinning effects. Therefore "intrinsic" will be used here in the sense of properties measured on materials which do not involve intergranular transport.

INTERGRANULAR EFFECTS

Work elucidating the nature of intergranular transport was an early focus of attention in $Ba_2YCu_3O_7$ because of the three to four order of magnitude decrease in J_C measured on bulk ceramics compared to thin films or single crystals. It was appreciated quite early that this difference, as well as the strong drop off of J_C in magnetic fields was consistent with weak links at the grain boundaries. A wide variety of causes were suggested including orientation dependence of transport and contamination at the grain boundary. In fact, no single factor produces the full three order of magnitude drop, but rather combinations of factors contribute. The dominant factor for any sample is a sensitive function of processing history.

GRAIN BOUNDARY DEFECT STRUCTURE

Grain boundaries in bulk sintered samples of $Ba_2YCu_3O_7$ were investigated using transmission electron microscopy (TEM).[1,2] The samples were prepared[3] using BaO_2, Y_2O_3, and CuO precursors, reacted at 900°C/6h/O_2, sintered at 975°C/40h/O_2 with a final oxygen anneal performed at 400-600°C. The dense (6.1 g-cm^{-3}), large-grained (1-40 μm) material exhibited critical current densities in the range 40-100 A-cm^{-2} in zero field at 77 K, and a resistivity ρ(300 K)=790 μohm−cm. Samples for TEM analysis were cut from regions adjacent to those used for J_C measurements, and were mechanically polished and further thinned by ion milling.

Three classes of grain boundaries were observed[1,2]: 1) coherent grain boundaries, containing only the inevitable intrinsic sheet of dislocations, 2) semicoherent grain boundaries, with an extensive region (5-150 nm) of dislocation loops and 3) completely incoherent grain boundaries, in which there are cracks and voids within a layer of disordered material. The latter two classes correlate with boundaries faced by a (001) basal plane. A typical micrograph showing the preponderance of basal faced boundaries and type 2 class defects is shown in Figure 1. This correlation is attributed[1,2,4] to local tensile stress arising from the anisotropic thermal expansion of $Ba_2YCu_3O_7$ during cooling from sintering temperatures. There is an excess 2.4% contraction along the c-axis which cannot be accomodated in a randomly oriented sample. This view is supported by the observation that thermal cycling between 650 and 850 °C leads to a decrease in J_C.[4] It is obvious that grain alignment will reduce this problem.

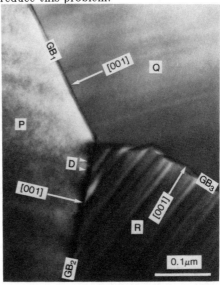

Figure 1. TEM image of three grain boundaries demonstrating the prevalence of basal plane faceting. Note the defects associated with GB2, a class 2 grain boundary, which appear only on the basal-plane-faced side of the boundary.

Electrical transport across such boundaries must be strongly impeded. It is likely that the extensively defective regions of Type 2 boundaries are semiconducting or insulating, based on studies of the effect of ion-implantation-induced disorder[5] in $Ba_2YCu_3O_7$ and measurements on amorphous material. At least at low temperatures the regions introduce a series resistance which is much greater than that of the grain, and will not pass a supercurrent. Assuming a simple tunneling model with a WKB square barrier of $\approx$5eV leads to a decay length of 0.14nm for vacuum tunneling. Thus voids as large as 1 nm will pass negligible current. $Ba_2YCu_3O_7$ has a platelet growth habit, with an average basal face-to-diameter to thickness ratio of $\sim$6 to 1. Thus $\sim$75% of the surface area of the grain is faced by a basal plane, and is therefore defective. Clearly the result will be to decrease the apparent conductivity, by roughly the ratio $(1-f_{disorder}-f_{crack})$, i. e., $\sigma_{obs} \simeq \sigma_{grain} \cdot (1-f_{disorder}-f_{crack})$, where $f_{disorder}$ is the fraction of the total area of type 2 grain boundaries, and f_{crack} is the areal fraction which is of type 3. So by the above estimate $f_{crack}+f_{disorder} \simeq 0.75$ and $\sigma_{obs} \simeq \sigma_{grain} \cdot 0.25$. Intrinsic conductivity anisotropy[6] due to the crystal structure of $Ba_2YCu_3O_7$ presents an additional complication and might affect the measured resistivity by roughly another factor of two in a random polycrystalline sample, even though the conductivity anisotropy is quite large.

The observed resistivity of our samples at 300 K is 790 μohm–cm, which implies a resistivity ρ_{grain} =200 μohm–cm by this estimate. This is remarkably close to the scaled resistivity of rho(300K)=150–190 μohm–cm inferred by Beasley[7] to explain the fluctuation conductivity in thin films. The temperature dependence of the normal resistivity of $Ba_2YCu_3O_7$ bulk samples has been observed to fit into two classes: for 500<ρ(300 K)<5000 μohm-cm, $\rho(T)/\rho$(300 K) falls on a universal curve, while samples with ρ(300 K)>5000 μohm-cm exhibit a wide variety of temperature dependences. The essential point of this model, that the resistivity is modified by a multiplicative geometrical factor, is supported by the universality of the temperature dependence of the normal resistivity in the first class. Bulk samples with grain boundary wetting phases will have a parallel channel contributing to the observed resistivity, so that the temperature dependence will no longer be universal. This is indeed observed in samples with ρ(300 K)>5000 μohm-cm.

CARBON SEGREGATION AT GRAIN BOUNDARIES

While grain boundary defects can account for the fivefold increase in apparent resistivity, they cannot explain the thousandfold decrease in J_C from values expected based on single crystal measurements or depairing estimates, or the strong dependence of J_C on magnetic field. It is necessary to invoke weak coupling between strongly superconducting regions. The source of weak coupling could be intrinsic, e. g., due to anisotropy in the conductivity of $Ba_2YCu_3O_7$ leading to inefficient coupling at high-angle grain boundaries[8,9] or it could be extrinsic, due to chemical inhomogeneity, e. g., in oxygen stoichiometry, precipitation of a second phase, or inadvertent reaction with contaminants. TEM observations imply an upper bound of $\sim$1 nm on the thickness of any intergranular inhomogeneity at tight type 1 boundaries in our large-grained sample. This upper limit makes some possibilities unlikely, such as amorphous decoration of grain boundaries, or the presence of merely semiconducting material, as such materials should present a relatively low barrier height, so that the scale length for tunneling would be

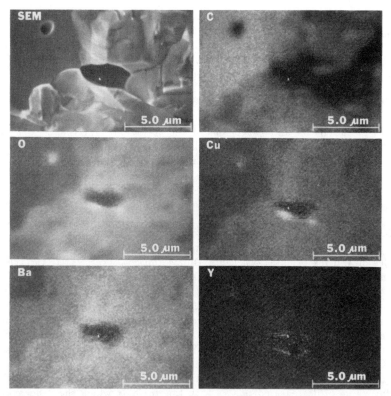

Figure 2. SEM image and Auger maps of in situ fractured surface of $Ba_2YCu_3O_7$. In the upper left corner is a region of intergranular fracture faced by a basal plane. Carbon free regions associated with transgranular fracture provide a calibration for estimating C coverage and demonstrate that the C contamination is not associated with e-beam irradiation.

correspondingly long. But it does not rule out the possibility of a large-gap insulating barrier, for which even a thickness of 1 nm can drastically attenuate the probability of electron tunneling.

Scanning Auger Microscopy was used to map a surface which was fractured *in situ* in a UHV system.[1] Figure 2 shows and SEM image of a typical fracture surface, along with maps of C, O, Ba, Y, and Cu Auger intensities. The SEM image reveals that a substantial portion of the fracture surface involves transgranular fracture, a feature also observed by Cook, *et al.*.[11] In Figure 2 it is seen that many grains are decorated by the fine step structure characterisically associated with transgranular fracture. This decoration is correlated with the absence of carbon and with increased Ba, Y, and Cu signals, as would be expected for a surface-segregated C layer. Depth profiling confirms that the C does not extend into the bulk of the grain. The thickness of the C layer was estimated to be 1-3 monolayers[1] by measuring the O, Ba and Cu Auger intensities and comparing values on inter- and transgranular fractures using estimated Auger electron escape depths.

Chemically it is plausible that Ba-based carbonates would form on the surface of $Ba_2YCu_3O_7$ by reaction with trace CO_2 in the atmosphere. (It is noteworthy that this carbon contamination is found in material which was not prepared using $BaCO_3$ reagent.) Surface reaction with CO_2 has been reported[12] and the bulk chemistry of CO_2-$Ba_2YCu_3O_7$ has been investigated by thermogravimetry,[13] explicitly revealing the strong tendency to form barium carbonates. The chemical identification of the boundary layer species can be found from XPS measurements. A particularly elegant study by Schrott and coworkers[14] tracks the presence of changes in the C, Ba and O regions of the spectrum with the areal coverage of transgranular fracture, and attributes them to the presence of carbonate.

The presence of such a 10Å insulating layer at grain boundaries is sufficient to account for most of the reduction in J_C obtained in bulk ceramic material. At 77 K this critical current density is $10-10^3$ A–cm^{-2}, dropping exponentially in an applied magnetic field with a characteristic field of 10-50 Oe.[15]

ORIENTATION EFFECTS

While carbon segregation and grain boundary defect structure in randomly aligned samples may mask orientation effects, Dimos and coworkers[9] at IBM have explored this issue in thin films of $Ba_2YCu_3O_7$ grown on large-grained polycrystalline $SrTiO_3$. By micromachining bridge structures on or near the $SrTiO_3$ boundary, and assuming that the $Ba_2YCu_3O_7$ was *exclusively* epitaxially oriented with respect to the substrate, they found that J_C across the boundary was always significantly reduced. Working on $SrTiO_3$ bicrystals, reductions of a factor of 50 were found for misorientation in the basal plane for angles greater than 15° for the assembly of $Ba_2YCu_3O_7$ grains within the bridge. As expected, the density of dislocations at the boundary varies with angle, and appear "tight" by TEM. They suggest that the limiting value of decrease correlates with the angle at which the dislocations begin to overlap.

TUNNELING BEHAVIOR

One can then ask if these junctions exhibit the ideal behavior associated with well-behaved granular superconductors.[14] In simplest form this model involves a cubic array of josephson-coupled grains, for which, in the low-temperature limit, and neglecting depairing, $j_c = \pi \Delta(0)/2e\rho_{eff}L$ where $\Delta(0)$ is the energy gap at T=0, L is the grain diameter, and ρ_{eff} is the effective normal state resistivity $\rho_{eff}=R_jL$, where R_j is the normal-state resistance of a single tunnel junction. It is assumed that $R_j \gg R_g$, the internal resistance of the grain.

To compare this result to the behavior of bulk ceramic $Ba_2YCu_3O_7$ we must first estimate R_j/R_g. For this we consider a typical grain as a rectangular prism and infer the tunneling resistance of the presumed junction at an end face, R_j, from the critical current using the full temperature-dependent Ambegaokar-Baratoff expression:

$I_c R_j = (\pi\Delta(T)/2e)\tanh[\Delta(T)/2k_B T]$.

The normal state resistance of the grain itself is simply $R_g = \rho L/A$. Then

$$\frac{R_j}{R_g} = \frac{\pi\Delta(T)\tanh[\Delta(T)/2k_B T]}{2eI_c} \cdot \frac{A}{\rho L} = \frac{\pi\Delta(T)\tanh[\Delta(T)/2k_B T]}{2ej_c\rho L}$$

To estimate $\Delta(77)$ we assume[20] $2\Delta/k_BT_c \sim 4.5$, so $\Delta(0)=18$ meV and $\Delta(77\ K) \sim 7.2$ meV. Taking $j_c=400$ A-cm^{-2}, $\rho(300K)=200$ μohm−cm and L=40 μm from our measurements on the large-grained sample, we find $R_j/R_g \simeq 17$ even at 300 K. That is, the resistivity should be dominated by the series resistance of the junction, as is necessary in the model. Second we verify that self-shielding does not dominate in the junctions: the Josephson penetration depth $\lambda_J=(h/2e\mu_0\lambda j_c)^{1/2}=18\mu$m *(for $j_c=400$ A−cm^{-2} and $\lambda(77\ K)=200$ nm)*,
so even the largest grains are only barely out of the small-junction limit required by the ideal-junction model. For ideal junctions, J_C is inversely proportional to the observed resistivity. This is not found for $Ba_2YCu_3O_7$ where we have observed $j_c=10$-1000 A-cm^{-2} in samples with $\rho(300\ K)=800\mu$ohm−cm, and $j_c \sim 1000$ A-cm^{-2} for samples with $500<\rho<1000$ μohm-cm. Additionally, for ideal junctions, ρ_{eff} is inversely proportional to grain size, and again this is not observed in our bulk $Ba_2YCu_3O_7$. It is necessary to infer that most of these weak links have a small I_cR_j product. Such a result is not unprecedented—in materials which are sensitive to disorder and which also have a short coherence length, such as the A15 compounds ($\xi_s(0) \sim 2$nm), the product I_cR_j of a tunnel junction is often orders of magnitude less than $\pi\Delta/2e$. Thus, even with the 1 nm upper bound on the junction width, we must infer the presence of low quality tunnel junctions.

The correlation of these electrical, TEM and Auger measurements suggests a picture for the (non-basal-plane-faced) tight boundaries, namely a core barium carbonate insulating layer 1-3 ML thick, sandwiched between disordered regions of $Ba_2YCu_3O_7$ roughly a coherence length thick. These observations may in part explain the difficulty of obtaining high-quality junctions for tunneling studies of the superconductor.

MELT-TEXTURED GROWTH

A significant improvement in the critical current of ceramic $Ba_2YCu_3O_7$ has been obtained by a radical change in the processing of this material: instead of conventional solid-phase sintering, the technique involves partial or complete melting and controlled solidification.[15] The resulting material is extremely dense and develops a microstructure which consists of locally aligned, high-aspect-ratio grains with short dimensions of 2-5 μm. The short grain axis is parallel to the **c** axis. Adjacent grains are only slightly misoriented with respect to each other, so low angle grain boundaries predominate.. Growth anisotropy of $Ba_2YCu_3O_7$ under quasiequilibrium conditions has been observed, as mentioned above, with the **a-b**-axis rate many times that of the **c** axis. Under those conditions (solid state growth or slow cooling from an off-stoichiometric melt) a platelet morphology obtains. The quantative details of the growth of crystallites in these $Ba_2YCu_3O_{7-x}$ supercooled melts are not yet understood.

Enhanced critical currents have been achieved by melt-textured growth following two substantially different protocols. In the first, the compound is heated only into the two-phase (liquid+solid) region, *i.e.*, to below about 1100°C. This results in decomposition into a liquid phase which is Ba- and Cu- rich, and solid-phase BaY_2CuO_5. The second protocol avoids the problem of decomposition by heating into the near fully melted region (*e. g.*, 1320°C) and cooling quickly through the peritectic range. This protocol yields the highest critical currents.[15] Figure 3 shows the transport J_C at 77 K as a function of applied magnetic field for the best melt-processed material, and compares it to the range of values obtained in conventionally sintered $Ba_2YCu_3O_7$.

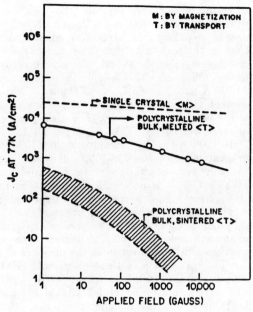

Figure 3. Plots of J_C versus magnetic field demonstrating the improvement in J_C and its reduced field dependence for melt-textured grown samples as compared to conventionally sintered $Ba_2YCu_3O_7$. Also shown are results for single crystals of $Ba_2YCu_3O_7$.

Clearly the melt-processing technique reduces the effect of weak-link connections between grains. Indeed, the signature property of those weak links, namely the extreme magnetic field dependence of J_C, is no longer evident. The field dependence seen in Figure 3 is low enough that it may in part be due to other phemonena, such as flux flow (although it should be noted that the field dependence is still stronger than that observed for single crystals.) It is not possible at this time to definitively infer the mechanism responsible for these improved properties.

The grains in this material are at least locally highly oriented so that there are many low-angle grain boundaries—avoiding the potential problem of intrinsic anisotropy in the crystal structure. This may be crucial, but it is also true that the high-temperature processing is likely to decompose any carbonates on the surface of grains, and that renucleation within the melt ensures that newly-formed grain boundaries will be made in the absence of CO_2 or any other source of C. Thus the success of the melt-processing technique does not argue exclusively for either mechanism. It is indeed fortunate that this technique yields at least two beneficial changes simultaneously, allowing the formation of clean grain boundaries in dense, highly oriented material.

"INTRINSIC" FACTORS

As in standardly prepared polycrystalline ceramic samples of $Ba_2YCu_3O_7$, $Bi_{2.2}Sr_2Ca_{0.8}Cu_2O_{8+\delta}$ exhibits substantial reduction of J_C compared to single crystals, implying that at least some of the intergranular effects described above are operative. In this section, the focus will be on single crystal measurements which demonstrate the large differences present in the intrinsic behavior of $Bi_{2.2}Sr_2Ca_{0.8}Cu_2O_{8+\delta}$ versus $Ba_2YCu_3O_7$. Furthermore, measurements of J_C by both transport and magnetization methods show substantially different field dependences for $Bi_{2.2}Sr_2Ca_{0.8}Cu_2O_{8+\delta}$, shown in Figure 4, compared to $Ba_2YCu_3O_7$.

Limitation of J_C in single crystals is related to the dynamics of vortex motion. In conventional superconductors, when a transport current J_T passes through a superconductor above H_{c1}, the transport current interacts with the magnetic flux in the vortices generating a Lorentz force $f_L = \vec{J} \times \vec{B}$. For isolated vortices, the force per unit length tending to displace the vortex laterally is given by $f_L = J_T \Phi_0$ for fields applied perpendicular to J_T. It is only due to the presence of pinning centers that impede vortex motion that the material can be considered superconducting in fields $H > H_{c1}$. In actuality only a small fraction of the vortices need to be pinned to defect sites; the others are constrained by the intervortex interactions. In currently used materials, to obtain large J_C at high fields, the presence of a flux lattice stabilized by the forces between vortices allows a relatively few strong pinning centers to pin all of the flux lines and prevent dissipation caused by vortex motion.

For conventional Type II superconductors, a vortex that is pinned will have to surmount a potential energy barrier before it is free to move under the Lorentz force. It will generate heat as it moves until it is trapped at another pinning site. This heating in turn may enhance the likelihood that nearby vortices can exceed the activation barrier, and the pinning force itself is temperature dependent leading to a highly nonlinearly coupled system with positive feedback. This thermally activated motion is termed flux creep, and is present at all finite temperatures. Thermal fluctuations are sufficient to release vortices at a constant rate. Flux creep typically exhibits a logarithmic time decay following field or current transients, and reverts to random motion as the Lorentz driving force decays.

For transport currents above a depinning threshold the statistically averaged sum of the Lorentz forces on the vortex array exceeds the total pinning force, and the entire array moves laterally at a steady velocity (flux flow). The motion of the vortex lattice results in significantly higher energy dissipation and heat generation, by inducing an electromotive force which drives current through the normal core of the vortex. In a Bardeen-Stephen model, this leads to a velocity-dependent force on the vortex $\vec{T}_v = -\eta \vec{v}$ where η is a viscosity coefficient, and $\vec{v}$ is the vortex velocity. Thus above the depinning current there is a force balance given by $\eta v = f_L - f_p$. The flux motion produces an electric field $E = nv\Phi_0 = vB$, where B is the average magnetic induction. Combining these results, a differential flux flow resistivity $\rho_f = \dfrac{dE}{dJ_T} = \Phi_0 B/\eta$ can be defined, which can be obtained from the linear portion of I-V characteristics in a field for currents above the depinning current. For an operational threshold voltage criterion in a transport measurement the onset of the linear portion of the curve defines J_C. In conventional materials $\rho_{creep} \ll \rho_f$, so that establishment of a working criterion for defining J_C is not a problem.

A second defining characterisitic of flux flow is the scaling of the resistivity ρ_f with normal state resistivity as $\rho_f/\rho_N = H/H_{c2}(T)$. Because of the high normal state resistivity of the new materials it is clear there will be substantial energy dissipation at fields near H_{c2} in this regime.

TRANSPORT AND MAGNETIZATION MEASUREMENTS

The first indications that vortex motion plays a role in limiting J_C in $Bi_{2.2}Sr_2Ca_{0.8}Cu_2O_{8+\delta}$ appear in the distinctly different behavior seen in field dependent transport and magnetization measurements compared to that observed for $Ba_2YCu_3O_7$. Measurements on BSSCO single crystals by van Dover, et al.[16]

Figure 4. J_C versus temperature for various magnetic fields for single crystal $Bi_{2.2}Sr_2Ca_{0.8}Cu_2O_{8+\delta}$.

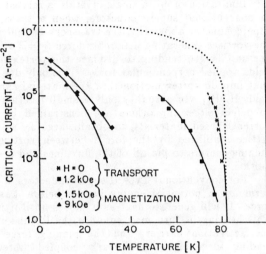

using a pulsed technique to minimize heating and a $10\mu V/mm$ threshold criterion are shown in Figure 4. Also indicated by the dotted curve is an estimate of the depairing current, based on values of $H_c(0)=0.62T$ and $\lambda(0)=140nm$ for a crystal with $T_C=82K$. This leads to $J_d(0) = 1.5 \times 10^7 A/cm^2$. The data lie within a factor of 20 of the depairing limit and are higher than values for single crystals $Ba_2YCu_3O_7$ scaled to the equivalent $(T-T_C)$. However, in a 1kOe field, J_C defined by this voltage criterion decreases by 2-3 orders of magnitude at temperatures near T_C. Examination of I-V characteristics shows, however, that there is no threshold current, i.e. an extremely linear curve. Although a formal cbut instead an current can be defined there is apparently no current for which $R \rightarrow 0$.

Data obtained at higher field at 0-30K using magnetization techniques demonstrate a marked change in behavior near 25K for $Bi_{2.2}Sr_2Ca_{0.8}Cu_2O_{8+\delta}$.[17] At lower temperatures magnetization loops had shapes predicted by the Bean model, and critical currents could be extracted using the standard formulat $J_C = 30\Delta M/D$ where D is the diameter of the disk shaped single crystal and ΔM is the magnetization difference for increasing and decreasing fields. The critical currents are quite high in this regime, demonstrating the presence of pinning in a material with no twin boundaries. With increasing temperature, the loop is significantly reduced and substantially distorted, implying a rapid decrease in J_C with applied field. The data is show as diamonds and triangles in Figure 4.

FLUX LATTICE MELTING IN SINGLE CRYSTALS

The remarkable falloff of J_C above 25K in moderate fields suggests the onset of vortex motion and hence energy dissipation. Mechanical measurement of the flux lattice melting provides the most stringent test of energy dissipation, being sensitive to the level of $10^{-14} W/cm^3$. Such experiments locate the point of phase transition in temperature and magnetic field from a static array of vortices to vortices with small amplitude (100 nm) random motion. In this regime the strongly pinned vortices are not moving; it is the vortices held only by the intervortex interactions that are initially affected as the temperature is raised.

While such techniques do not directly provide a value for J_C as in transport studies, they mark a boundary above which energy dissipation by vortx motion must be considered. Melting can be verified by Bitter decoration experiments which confirm that in the unmelted regime a vortex array with short range order is present, while in the melted regime, no decoration is achieved because vortex motion is rapid on the time scale of the decoration process. Recent measurements of flux lattice melting in single crystals of both $Ba_2YCu_3O_7$ and $Bi_{2.2}Sr_2Ca_{0.8}Cu_2O_{8+\delta}$ by Gammel, et al.[18] generate a pessimistic picture of usefulness of these materials for high field applications. The onset of vortex motion as a function of temperature and applied magnetic field was directly measured using a high-Q silicon oscillator technique. Briefly, a single crystal mounted on a fabricated oscillator is oriented with respect to the applied field as shown in the inset in Figure 5, is driven self-resonantly using a phase-locked loop. Below the flux lattice melting temperature, the vortex lattice moves rigidly with the crystal in the field because it remains structured about the strong pinning centers. This leads to an additional component to the measured elastic modulus of the system. As T_M is approached, thermal fluctuations dominate the motion of the vortices so that there is rapid relaxation on the time scale of the oscillation, and the contribution of the vortex lattice elastic modulus to the system drops to zero. A peak in the energy dissipation (the imaginary component of the oscillator response) also appears as shown in Figure 5 for a single crystal of $Ba_2YCu_3O_7$ with a T_C of 87K. This peak is used to operationally define T_M.

Figure 6 shows the variation of T_M for two orientations of the crystal with respect to the applied magnetic field fo BSSCO. The Meissner signal is also displayed to indicate the T_C and the extent of bulk superconductivity. Flux lattice melting occurs near 30K for both orientations, although bulk superconductivity is present at 75K. More recent results on Tl-based single crystals which show a T_C of 120K have a T_M of only 25K. The simplest models predict flux lattice melting above $T_M \propto H_{c2}^2 \xi^2$ where the right hand side is the energy per unit length of a vortex line. Thus, materials with smaller coherence lengths ξ and high H_{c2} will melt at lower temperatures. As expected, based on its larger ξ, the situation is very different for $Ba_2YCu_3O_7$, as shown in Figure 7. The most striking difference is that T_M is now occurs at H_{c2}, as is predicted for 3-dimensional superconductors for $\vec{H} \perp \hat{c}$. Melting occurs because the penetration depth $\lambda(T)$ diverges at T_C, so that the vortices are not well-defined. The situation is different for $\vec{H} \parallel \hat{c}$, where T_M is shifted $\approx 3.6K$ to lower temperatures than H_{c2}, while exhibiting a similar slope, for all values of field.

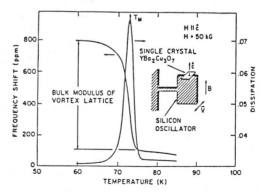

Figure 5. Changes in oscillator frequency and dissipation from mechanical measurements of motion of $Ba_2YCu_3O_7$ single crystal in a magnetic field. The peak in the dissipation is used to define the flux lattice melting temperature. The inset illustrates the experimental configuration on the micromachined Si oscillator.

Figure 6. Flux lattice melting temperature for single crystals of $Bi_{2.2}Sr_2Ca_{0.8}Cu_2O_{8+\delta}$ for both field orientations. The Meissner signal is also shown.

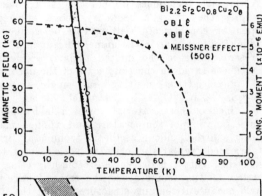

Figure 7. Flux lattice melting temperature for single crystals of $Ba_2YCu_3O_7$ for both field orientations. Also shown are the solid and dashed line indicating the dependence of H_{c2}.

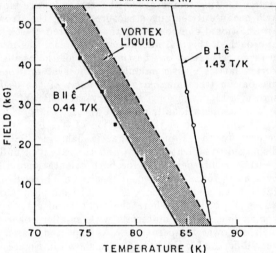

The presence of flux lattice melting substantially below T_C for large interplanar spacings mitigates the associated increase in T_C for these materials. For instance in $Bi_{2.2}Sr_2Ca_{0.8}Cu_2O_{8+\delta}$, J_C drops abruptly at $\approx$30K, as detailed above, and results on TL single crystals with T_C of 120K, melt at 25K. In YBCO ceramics, J_C is low to within 2K of T_C, while SQUID noise diverges below T_C. It is predicted that thin film devices with $\vec{H} \perp \hat{c}$ should have intrinsically lower flux noise, suggesting that growth with $\hat{c}$ in-plane would be preferred.

RESISTANCE MEASUREMENTS OF THERMAL DISSIPATION

For many applications the appropriate metric for judging the suitability of a material is a resistivity criterion. For example, low resistance interconnects may require $\rho < 10^{-8}$ Ω–cm, while persistent current magnets have more stringent requirements of $\rho < 10^{-13}$ Ω–cm. Field and temperature dependent measurements of resistivity in this regime by Palstra and coworkers[19] probe energy dissipation in the 10^{-4}W/cm^3 regime, and monitor more macroscopic flux line motion. Currents of 10mA, well within the linear I-V response regime were used, corresponding to current densities of $\approx 5 \times 10^3$ A/cm^2 based on the crystal dimensions. As first seen by van Dover, et al,[16] *using these transport current densities, there is no threshold in the I-V characteristic, indicating that the Lorentz forces have already exceeded the pinning forces in this regime of temperature and magnetic field.*

Figure 8. Arrhenius plots of the resistivity versus temperature for various values of the applied magnetic field.

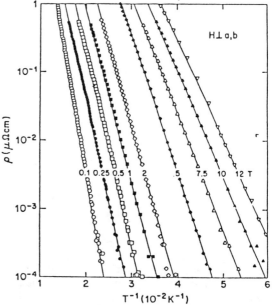

Measurements on $Bi_{2.2}Sr_2Ca_{0.8}Cu_2O_{8+\delta}$ above H_{c1} show that the resistance is current independent and follows Arrhenius type activation behavior, $\rho = \rho_0 \exp(-U_0/T)$ for resistivities in the range of 10^{-6}–10^{-2} ρ_N where ρ_N is the normal state resistivity of $140\mu\Omega$cm. Typical plots of log resistivity versus T^{-1} as a function of magnetic field are shown in Figure 8, convincingly demonstrating the Arrhenius behavior. The slope of the curve gives U_0 directly. The dependence of U_0 on H is monotonically decreasing with increasing H, varying by only a factor of 3-4 in the range 0.1-10. T, with values with H||a,b larger at all fields. The values range between 900K at 0.1 T and 300K at 10T. which are similar to those typically observed for flux creep in conventional superconductors. This thermally activated process is discussed by Palstra et al[19] in terms of flux creep.

Plotting ρ versus a normalized temperature U_0/T leads to a superposition of all curves, including those for measurements with those $\vec{H}$ || a,b. This implies a common prefactor *independent of temperature, magnetic field and orientation*. It is not yet known whether the prefactor is sample dependent.

MEASUREMENTS IN THE FLUX FLOW REGIME

Resistivity measurements in the 0.1–$10\mu\Omega$cm regime at temperatures close to T_C by van Dover, et al in low field show the characterisitics typically associated with flux flow for $R_f \ll R_n$. Data is presented in Figure 9, normalized to the normal state resistivity for several temperatures near T_C. The data is pictured as a different cut through parameter space than that in the Figure 8, and would correspond to a region above the upper left hand corner of that graph. The field axis is scaled by a temperature dependent factor determined by forcing the data to overlay. In the flux flow regime, as described above, a linear dependence of R_f/R_n on B would be expected and is indeed observed above an onset field related to $H_{c1\perp}$ when demagnetization effects are taken into account. This is quite different than the behavior observed in Figure 8, where parallel lines would be expected for a linear dependence of the normalized resistance on magnetic field. This behavior is

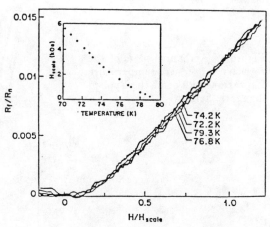

Figure 9. Low field, high temperature behavior of resistivity in single crystal $Bi_{2.2}Sr_2Ca_{0.8}Cu_2O_{8+\delta}$. The values of H_{scale} are shown in the inset. Extrapolation of the linear region of the curve allows estimation of $H_{c2\perp}$.

approached in the higher temperature regions of the graph, as would be expected when the activation barrier for depinning is easily surmounted. Instead of having flux motion limited by activation over a barrier, it is now limited by a field dependent viscosity. In comparison to the thermal activation work the energy dissipation in this regime is on the order of 3×10^3 higher.

In the simplest models of flux flow described above the scaling field is related to $H_{c2}(0)$. The value inferred from such an interpretation is consistent with that found by Palstra et al[19] by measuring R(T) for constant H. This lends further support to the view that under these conditions energy dissipation is dominated by conventional flux flow.

SUMMARY

The major point which emphasized in this review is that the new high T_C materials fall into two distinct classifications of factors limiting J_C. $Ba_2YCu_3O_7$ acts as a 3-dimensional material with strong pinning. Flux lattice melting occurs for one orientation only, and very close to H_{c2}. Transport measurements of critical currents show well-defined regions of zero resistance even close to T_C in magnetic fields. J_C in polycrystalline samples is limited by intergranular effects that are processing dependent. In contradistinction, the higher T_C compounds exhibit flux lattice melting, and energy dissipation by both thermally activated processes and flux flow in very modest fields. As T_C appears to increase with increasing interplanar separation in these newer materials, the tendency to more two-dimensional behavior and strongly field dependent pinning is enhanced, leading to poorer current carrying ability in magnetic fields. In zero field applications of polycrystalline material many of the same considerations applying to intergranular transport as occur in $Ba_2YCu_3O_7$ are expected to apply.

ACKNOWLEDGMENTS

It is a pleasure to acknowledge extensive and fruitful discussions with Bruce van Dover, Dave Bishop and Tom Palstra in the course of preparing this review.

REFERENCES

1. S. Nakahara, G. J. Fisanick, M. F. Yan, R. B. van Dover, and T. Boone, J. Cryst. Growth **85**, 639 (1987); S. Nakahara, G. J. Fisanick, M. F. Yan, R. B. van Dover, T. Boone, and R. Moore, Proc. of the Matl. Res. Soc. 99, 575 (1988).
2. S. Nakahara, G. J. Fisanick, M. F. Yan, R. B. van Dover, and T. Boone, Appl. Phys. Lett. **53**, 2105 (1988).
3. M. F. Yan, R. L. Barns, H. M. O'Bryan, Jr., P. K. Gallagher, R. C. Sherwood and S. Jin, Appl. Phys. Lett. **51**, 532 (1987); D. W. Johnson, E. M. Gyorgy, W. W. Rhodes, R. J. Cava, L. C. Feldman and R. B. van Dover, Adv. Ceram. Mat. **2**, 364 (1987).
4. T. H. Tiefel, S. Jin, R. C. Sherwood, R. A. Fastnacht, S. Nakahara, and G. J. Fisanick, Mat. Res. Soc. Symp. Proc., Vol. 99, 365 (1988).
5. A. E. White, K. T. Short, D. C. Jacobson, J. M. Poate, R. C. Dynes, P. M. Mankiewich, W. J. Skocpol, R. E. Howard, M. Anzlowar, K. W. Baldwin, A. F. J. Levi, J. R. Kwo, T. Hsieh, and M. Hong, Phys. Rev. B, **37**, 3755 (1988).
6. S. W. Tozer, A. W. Kleinsasser, T. Penney, D. Kaiser, and F. Holzberg, Phys. Rev. Lett., **59**, 1768 (1987).
7. M. R. Beasley, Proc. Yamada Conf. XVIII (Sendai, Japan, 1987).
8. J. W. Ekin, Adv. Ceram. Mat. **2**, 586 (1987). (1981); J. R. Clem, B. Bumble, S. I. Raider, W. J. Gallagher, and Y. C. Shih, Phys. Rev. **B35**, 6637 (1987).
9. P. Chaudhari, J. Mannhart, D. Dimos, C. C. Tsuei, J. Chi, M. M. Oprysko, and M. Scheuermann, Phys. Rev. Lett. **60**, 1653 (1988); D. Dimos, P. Chaudhari, J. Mannhart and F. K. LeGoues, Phys. Rev. Lett. **61**, 219 (1988).
10. R. F. Cook, T. M. Shaw and P. R. Duncombe, Adv. Ceram. Mat. **2**, 606 (1987).
11. G. J. Fisanick, P. Mankiewich, W. Skocpol, R. E. Howard, A. Dayem, R. M. Fleming, A. E. White, S. H. Liou and R. Moore, Mat. Res. Soc. Symp. Proc. Vol 99, 703 (1988).
12. P. Gallagher, private communication.
13. A. G. Schrott, S. L. Cohen, T. R. Dinger, F. J. Himpsel, J. A. Yarmoff, K. G. Frase, S. I. Park, R. Purtell, "Thin Film Processing and Characterization of High-Temperature Superconductors", A. I. P. Conf. Proc. **165**, 349 (1988).
14. G. J. Fisanick, S. Nakahara, S. Jin, T. F. Tiefel, R. C. Sherwood, M. Yan, R. Moore and R. B. van Dover, in "High Temperature Superconductors" eds. J. Heiras, R. A. Barrio, T. Akachi and J. Taguena, pg. 88, (World Scientific) 1988.
15. S. Jin, T. H. Tiefel, R. C. Sherwood, M. E. Davis, R. B. van Dover, G. W. Kamlott, R. A. Fastnacht and H. D. Keith, Appl. Phys. Lett. **52**, 2074 (1988).
16. R. B. van Dover, L. F. Schneemeyer, E. M. Gyorgy, J. V. Waszczak, Phys. Rev. B., accepted.
17. E. M. Gyorgy, R. B. van Dover, S. Jin, R. C. Sherwood, L. F. Schneemeyer, T. H. Tiefel and J. V. Waszczak, Appl. Phys. Lett., accepted.
18. P. L. Gammel, L. F. Schneemeyer, J. V. Waszczak and D. J. Bishop, Phys. Rev. Lett. **61**, 1666 (1988).
19. T. T. M. Palstra, B. Batlogg, L. F. Schneemeyer, J. V. Waszczak, Phys. Rev. Lett. **61**, 1662 (1988).
20. R. B. van Dover, L. F. Schneemeyer, E. M. Gyorgy and J. V. Waszczak, Phys. Rev. Lett. submitted.

LIMITATIONS ON CRITICAL CURRENTS IN HIGH TEMPERATURE SUPERCONDUCTORS

C.C. Tsuei, J. Mannhart, and D. Dimos

IBM Research Division, Thomas J. Watson Research Center,
P.O. Box 218
Yorktown Heights, NY 10598-0218

ABSTRACT

The critical current density, J_c, of high temperature superconducting oxides such as $YBa_2Cu_3O_{7-\delta}$ can vary by more than four orders of magnitude, depending on whether the sample is in the form of an epitaxial film, a single crystal or a polycrystal. As a function of increasing reduced magnetic field, J_c in high temperature superconductors generally decreases more readily than that of conventional superconductors such as Nb_3Sn. These experimental observations are considered in terms of various critical current limiting factors which include: 1) depairing, 2) flux creep, 3) weak link effects. In the limit of depairing, the zero field J_c, is expected to be in the range of $10^{8-9} A/cm^2$ at 4.2 K and in the range of $10^{7-8} A/cm^2$ at 77 K. The highest experimental value for J_c, which was observed in [001] single crystal epitaxial films is at least an order of magnitude lower than the depairing limit. In these films, J_c is found to be limited by flux creep. With the aid of the techniques of epitaxial film growth and laser micro-patterning, a series of experiments has been designed to study the critical current limiting effects at isolated grain boundaries. These experiments indicate that J_c in polycrystals is further limited by the grain boundaries, which form SNS-type Josephson weak links. The dependence of the critical current density of the weak links on magnetic fields offers a possible explanation for the magnetic field sensitivity of J_c of polycrystalline samples.

INTRODUCTION

The recent discovery of high temperature superconductivity[1,2] in several Perovskite-type copper oxides has generated much hope for many technical applications of superconductivity at liquid nitrogen temperatures. Almost all of these potential applications[3,4] require that the high-T_c superconductors are able to carry a reasonable amount of current in the superconducting state (i.e. $J_c \gtrsim 10^5 A/cm^2$ at T = 77 K). For most large-scale applications, a relatively weak magnetic field dependence of the critical current density is also essential. In this article, various factors that limit the critical current density in high-T_c oxides will be examined. The discussion of the critical current density will be focused on experimental data of the critical current densities of the $YBa_2Cu_3O_{7-\delta}$ (YBaCuO) system, mostly because the YBaCuO system is well characterized in general, and because the temperature and magnetic field dependences of J_c have been studied extensively in this system. The critical current data of the Bi(Tl)-Sr(Ba)-Ca-Cu-O systems, which contain no rare-earth elements and have no twin boundaries, will be discussed briefly.

© 1989 American Institute of Physics

LIMITATIONS OF CRITICAL CURRENT DENSITIES

The critical current density J_c of high temperature superconducting oxides[5,6] such as $YBa_2Cu_3O_{7\%}$ varies typically by more than four orders of magnitude, depending on whether the sample is in the form of an epitaxial film, a single crystal or a polycrystal. As a function of increasing reduced magnetic flux density $b = B/B_{c2}$, J_c in high T_c superconductors generally decreases more readily than the critical current density of conventional superconductors such as Nb_3Sn. This is particularly true for the J_c (b) characteristics of polycrystalline high T_c superconductors. In order to understand the reasons of such a wide range of J_c variations, various critical-current limiting factors, namely 1) depairing, 2) flux creep and 3) weak link effects will be discussed.

1) Depairing Limit

As a benchmark for the highest theoretical value for the critical current density a superconductor can carry, one can estimate the depairing critical current density[7] by using the Ginzburg-Landau theory, which predicts that:

$$J_c(T) = \frac{4}{3\sqrt{6}} \frac{B_c(T)}{\mu_0 \lambda(T)} = J_c(0)(1-t^2)(1-t^4)^{1/2}, \qquad (1)$$

where $t = T/T_c$, and $B_c(T)$ and $\lambda(T)$ are the thermodynamic critical field and the London penetration depth respectively. This estimate of the depairing limit is hampered by great uncertainties in determining the critical fields. The recent observation[8] of huge magnetic relaxation in YBaCuO single crystals suggests that previous measurements of the upper critical fields in high-T_c-materials were probably measurements of the irreversibility line for flux depinning and represent therefore only a lower bound for B_{c2}. The determination of the lower critical field B_{c1} is equally controversial due to various experimental complications. Recently, new techniques have been developed to measure B_{c1} in a more reliable way by either identifying the onset of magnetic relaxation[9] or by studying the temperature dependence of the zero field cooled magnetization[10] Both techniques look for small variations in magnetization that signal the penetration of magnetic flux into the sample. The values of B_{c1} (with B∥c-axis), as determined by these techniques, are $B_{c1}(0) = $ 900 pm 100 G sup 9 and $B_{c1}(0) = 530 \pm 50$ G,[10] respectively. The values for the London penetration depth $\lambda(0)$ appear to converge to $\lambda_{ab}(0) \simeq 1400$Å for YBaCuO single crystals[10-12] Based on these values for B_{c1}, $\lambda_{ab}(0)$ and the BCS coherence length $\xi_0 \simeq 15 - 30$Å[5,6] one can calculate the depairing critical current density $J_c(T)$ as a function of temperature by using the Ginzburg Landau formula, Eq. (1). In Fig. 1, a range of $J_c(T)$ predicted by Eq. (1) is plotted as a function of temperature (for T > 50 K) for the YBaCuO system to reflect the range of $B_c(0)$ values. Although the Ginzburg-Landau theory might work as well at temperatures far below T_c, the London theory was chosen to estimate the low temperature range of $J_c(T)$. The depairing limit is derived in the London theory with by requiring that at J_c the density of the kinetic energy of the superconducting carriers equals the density of the condensation energy $(B_c^2(0)/2\mu_o)$, which yields

$$J_c(T) = \frac{B_c(T)}{\mu_o \lambda(T)}. \tag{2}$$

The depairing limit at low temperatures can also be inferred from the BCS-theory, which determines microscopically the critical velocity of the Cooper pairs beyond which quasi-particles are excited above the ground state[7]

$$J_c = \frac{ne\hbar}{\pi m \zeta_o}, \tag{3}$$

where n stands for the density of the Cooper pairs and m for the quasi-particle mass. In Fig. 1, the theoretical estimates for the zero-field depairing current density based on Eqs. (1)-(3) are shown along with some published experimental data.[13] Within an order of magnitude the estimates for the depairing limit agree with each other and suggest that the highest possible value for J_c at 4.2 K is about 10^{8-9} A/cm^2 and is reduced to about 10^{7-8} A/cm^2 at 77 K. The highest experimental values of J_c, as represented by those of epitaxial single crystalline ab-oriented films, are more than

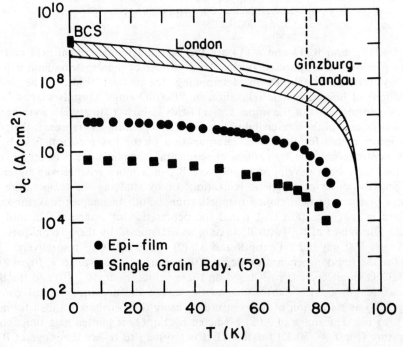

Fig. 1. Zero field critical current density J_c as a function of temperature as predicted by Eqs. (1)-(3) in the depairing limit. Also shown are experimental data for a single crystalline epitaxial film of YBaCuO (●), and of a bicrystalline YBaCuO film with a 5° tilt grain boundary (■).

an order of magnitude lower than the depairing limit. The critical current density of other types of YBaCuO samples can even be several orders of magnitude lower than the J_c of the single crystalline films, depending on the polycrystallinity (microstructure...) of the samples. It should be emphasized that, strictly speaking, J_c is not an intrinsic property of a superconductor (like e.g. the coherence length), because it can depend sensitively on the voltage criterion which is used for measuring J_c, on the microstructure and on the configuration of the sample. It is especially important to keep this point in mind, when experimental J_c data are compared with theoretical predictions or with other experimental results.

2) Flux Creep

As shown in Fig. 1, the highest experimental value for the critical current density in YBaCuO, which is found in single crystalline [001] films, is at least an order of magnitude lower than the depairing limit. It is therefore important to find out what mechanism actually limits J_c in these materials. In this case, it turns out that J_c is limited by creep of magnetic flux lines.[13] In a broad sense, the phenomena associated with flux creep can be described in terms of thermally activated motion of flux lines[14] which overcome pinning energies U_o that are created by structural defects in the superconductor (such as point defects, precipitates, grain boundaries etc.). The magnitude of the flux creep is characterized by the thermally assisted hopping rate for the flux lines:

$$\text{Hopping rate} \propto \exp(-U_o(B,T)/kT) \ . \qquad (4)$$

In principle, flux creep is expected to exist in any type-II superconductor and it has been observed in many conventional low T_c superconductors. In high T_c superconductors however, flux creep is very large compared to that in low T_c superconductors,[13,16-18] as a result of relatively weak pinning energies U_o in the high-T_c superconductor and of the high temperatures at which the superconductors can be operated (the factor U_o/kT_c in is for high-T_c superconductors about 10 to 100 times smaller than for the low T_c counterparts). Convincing experimental evidence for the presence of giant flux creep in high-T_c oxides (in both the YBaCuO and the Bi-Sr-Ca-Cu-O systems) is provided by strong magnetic relaxation[8,18] and the anomalous magnetic broadening of the resistive transition from the superconducting to the normal state.[19-21] The implications of large flux creep on the critical current density and on superconducting parameters like B_{c2} are enormous. At high temperatures (say above 50 K in YBaCuO) flux creep is quite strong and reduces J_c significantly. In other words, small transport currents will be enough to induce a considerable dissipative flux motion, by lowering the effective pinning energies U_o and thereby increasing the topping rate. Furthermore, strong flux creep implies that even if a superconductor with a T_c well above room temperature could be found, the critical current density of this superconductor would probably be relatively low at such a temperature, because of the considerable thermal activation involved.[16,17] A recent transport critical current measurement[13] provides evidence that J_c in the YBaCuO basal plane is indeed limited by flux creep over a wide temperature range.

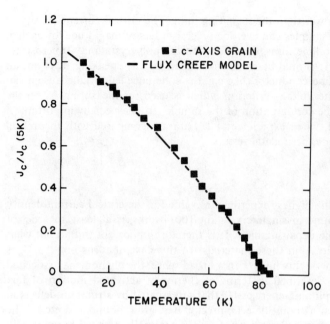

Fig. 2. Normalized critical current density for a [001] epitaxial single crystal film of YBa$_2$Cu$_3$O$_7$. The solid curve represents the temperature dependence of the critical current density as predicted by Eq. 5, with J$_c$ (0) = 6x10^{6}A/cm^2, (after Ref. 13).

As found in this experiment, the temperature dependence of the critical current density, J$_c$ (t) for [001] single crystalline films follows the prediction of a flux creep model[16,17] (see Fig.2):

$$J_c(B,t) = J_c(B,0)[1 - \alpha(B) t - \beta t^2] \quad (5)$$

where t = T/T$_c$ < < 1, B is the magnetic flux density and α (B) is given by

$$\alpha(B) = \frac{kT_c}{U(B,0)} \ln(aB\Omega/E_{min}) . \quad (6)$$

Here Ω is the attempt frequency with which the flux quanta try to escape from their pinning sites, the coefficient a stands for their average hopping distance, E$_{min}$ for the electric field criterion that is used to define J$_c$, and β reflects the temperature dependence of U$_o$. As shown in Fig.2, the experimentally found J$_c$ (0,t) characteristic is well described by Eq.(5) for α = 0.72 and β = 0.38. From the value of α, one can estimate the pinning energy U$_o$ (0,0) $\simeq$ 70 meV based on E$_{min}$ = 10^{-5} V/m and the assumptions that $\Omega \simeq 10^{10}$ Hz (typical phonon frequencies), B $\simeq$ 50 G (the self field of the transport current) and a $\simeq \Phi_o/\sqrt{B}$. The U$_o$(0,0) value determined from the transport measurement compares well with the value of 150 meV derived from magnetization data of an YBaCuO single crystal.[8] It is interesting to note that J$_c$ (t) does not strongly depend on the measuring criterion E$_{min}$, as it appears only logarithmically in the expression for α. The magnitude of the flux creep effect as manifested by the size of α, is mostly determined by the ratio kT/ U$_o$(B,0). From the YBaCuO system, the value of $\alpha(\simeq 0.7)$ as determined from the J$_c$ data is already almost an order of magnitude larger than that of a typical conventional superconductor ($\alpha \simeq 0.1$). For the case of the Bi-Sr-Ca-Cu-O system, a much larger current limiting effect by flux creep is suggested by an anomalously strong magnetic relaxation.[18]

3) Weak-links

It has been found experimentally that the critical current density of single crystals (bulk and films) is typically at least 10^3 times higher than that of polycrystalline samples. This finding strongly suggests that the disappointingly low critical current density in polycrystalline samples is caused by a combination of poor superconducting coupling across grain boundaries,[22,23] and of the anisotropy effects of J_c in the grains. To gain insight into this problem, we have carried out experiments designed to demonstrate the critical current limiting effects of the grain boundaries of YBaCuO by making use of the techniques of epitaxial film growth and laser micropatterning. Epitaxially grown [001] oriented films of YBaCuO on single crystalline SrTiO$_3$ substrates allow us to study directly the transport critical current in the basal plane. With epitaxial films grown on bicrystalline substrates (Fig. 3), one can

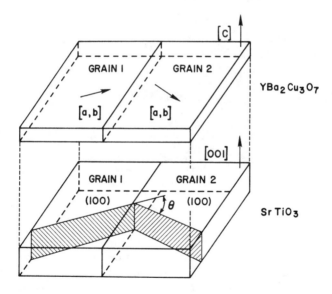

Fig. 3. Schematic diagram showing the epitaxial film growth of YBa$_2$Cu$_3$O$_7$ on a bicrystalline substrate of SrTiO$_3$. The acute angle, θ between the (100) planes in the two SrTiO$_3$ crystals defines the misorientation angle in the basal plane.

obtain further information on grain boundary critical current densities as a function of the misorientation of the grains. In order to make direct measurements of the superconducting properties of single grain boundaries and of their adjoining grains, we have used the techniques of laser micropatterning[24] of high T_c superconducting oxides. Laser ablation is a patterning method that offers direct experimental access to 1 μm wide lines of high T_c oxide films without degradation in T_c. In general, as shown in Fig.4, three lines were isolated by laser ablation. Two of these lines were within grains adjacent to each other, the third line straddled the grain boundary. Critical current densities were measured on lines typically 0.5μm thick, 5μm to 10μm wide and about 40μm long. The main results of these direct measurements on the critical current limiting effects of the grain boundaries in YBaCuO bicrystals are summarized as follows.

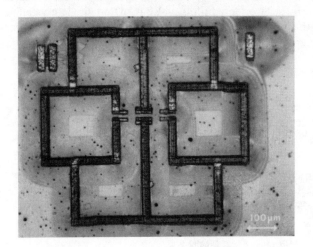

Fig 4. Laser ablated pattern which isolates three lines in an YBaCuO bicrystalline film: two lines are within the two adjacent grains, the middle line straddles the grain boundary.

(1) The transport critical current density of the grain boundary, as determined from I-V-measurements, was invariably found to be less than either of the two adjoining grains.[25] The reduction of J_c at the grain boundary in comparison with the grains varies from about a factor of 10 to more than 100. The J_c values of the boundaries at 5 K range from 750 A/cm^2 to 5x10^{4}A/cm^2, comparable to the J_c values reported in the literature on polycrystalline films and bulk samples. These results provide the first direct evidence for the conjecture that grain boundaries significantly limit the macroscopic critical currents of the polycrystalline high temperature superconductors.

(2) The grain boundaries in high T_c oxides behave as SNS-Josephson Junctions. The experimental evidence for this comes from the I-V characteristics of the grain

boundaries as a function of magnetic field and temperature.[13] Whereas the I-V characteristics of the grains are consistent with those expected from flux creep and flux flow, the I-V-characteristics of the grain boundaries are reminiscent of strongly coupled Josephson junctions. The magnetic field dependence of the zero voltage current, I_c shows great contrast for these samples; typical results are shown in Figs. 5a and 5b.

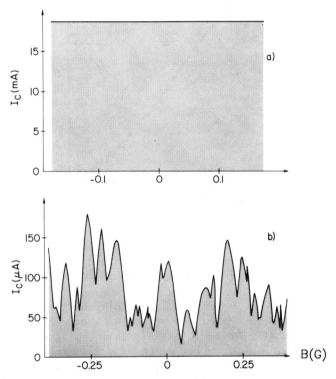

Fig. 5. Magnetic field dependence of the critical current for a) a line within a [001] grain of an YBaCuO bicrystalline film and b) a line across the grain boundary.

In this case the magnetic field was applied perpendicular to the film surface in the boundary plane. The ambient background field was less than 10 mG. As shown in Fig. 5a, the critical current in the grains is essentially independent of small magnetic fields. In sharp contrast, the values of J_c for the grain boundaries oscillates wildly with applied magnetic field, suggesting a composite interference pattern resulting from an spatially inhomogenous maximum Josephson current density. A recent detailed study of the temperature and magnetic field (B < 20 G) dependences of J_c across [001] tilt boundaries[13] shows that these grain boundaries are well described as SNS-type Josephson junctions. As an example, the critical current density of a

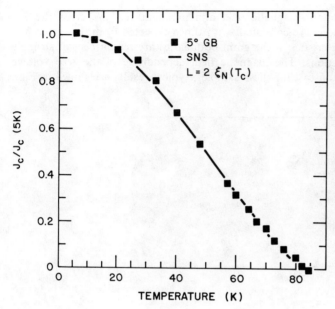

Fig. 6. Temperature dependence of the normalized critical current density for a line containing a 5° grain boundary. The solid curve reflects the trediction of the Likharev model (after Ref. 13).

grain boundary, normalized to its value at 5 K, is plotted as a function of temperature in Fig. 6. As shown in this figure, the grain boundary critical current density shows a different temperature dependence than the flux creep limited critical current density of the grains (Fig. 2). The solid curve in Fig. 6 represents the temperature dependence of J_c predicted by a model of Likharev[26] for an SNS-junction with the parameter $L/\xi_n(T_c) = 2$, where L is the thickness and ξ_n is the coherence length of the normal layer (i.e. of the grain boundary). This coherence length is estimated to be about 5Å. As shown in Fig. 6, the Likharev theory is in good agreement with the experimental data. Since the width of a typical grain boundary is 1-2 lattice constants (a≃3.9Å), the derived grain boundary thickness of ≃10Å is quite reasonable.

To further demonstrate the Josephson junction type behavior of the grain boundaries in YBaCuO-polycrystals we have fabricated an all-high-T_c DC-SQUID by laser patterning a superconducting loop which crosses the same grain boundary twice to create two Josephson junctions. The voltage response V(B) of the SQUID to an applied magnetic field B as a function of the bias current is shown in Fig. 7. The precise periodicity and the large modulation depth of V(B) clearly demonstrates the potential usefulness of the single grain boundary junctions.

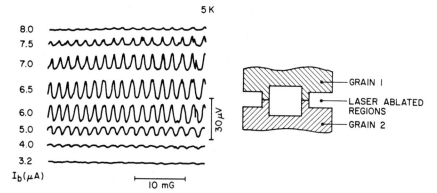

Fig. 7. An all-high-T_c DC-SQUID formed by laser patterning a superconducting loop that crosses a single grain boundary twice. Also shown is the voltage output of the SQUID as a function of the bias current I_b and applied magnetic field.

(3) The critical current density across the grain boundaries depends sensitively on the misorientation angle in the basal plane of the thin film bicrystals.[27] For small tilt angles θ ($< 15°$) the ratio of the grain boundary critical current density to the average value of the critical current density of the grains J_c^{gb} / J_c^g, decreases rapidly with increasing θ; for larger angles, this ratio levels off. The correlation between J_c and the grain misorientation underscores the importance of achieving nearly perfect grain-alignment in the basal plane for the optimization of J_c in polycrystalline YBaCuO materials. However, effects arising from percolation and from collective flux pinning at the grain boundaries should be considered also in developing high J_c polycrystalline superconducting materials.[28]

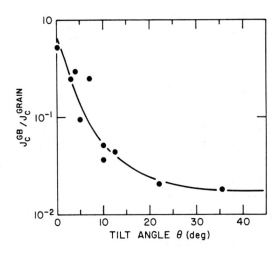

Fig. 8. Dependence of the critical current density of a grain boundary on the tilt angle θ of the boundary at 5 K (after Ref. 27). The solid line is supposed to be a guide for the eye.

SUMMARY AND CONCLUDING REMARKS

The wide range of critical current densities for high T_c superconductors such as $YBa_2Cu_3O_{7-\delta}$ has been considered in terms of various critical current limiting factors. In the limit of depairing, the zero field critical current density is expected to be in the range of $10^{8-9} A/cm^2$ at 4.2 K, and $10^{7-8} A/cm^2$ at 77 K. The highest experimental value for J_c, observed in [001] single crystal epitaxial films is at least an order of magnitude lower than this depairing limit. In single crystal films, the critical current density is found to be limited by flux creep which is a consequence of weak flux pinning and high operating temperatures. With the aid of the techniques of epitaxial film growth and laser micro-patterning, a series of experiments designed to study the critical current limiting effects at the grain boundary in $YBa_2Cu_3O_{7-\delta}$ bicrystals has revealed that:

1. In polycrystalline samples, J_c is further limited by weak superconducting coupling across grain boundaries.
2. The magnetic field dependence and the temperature dependence of J_c across grain boundaries strongly indicate that grain boundaries in YBaCuO act as SNS type Josephson junctions.
3. The weak-link magnetic field response of J_c explains why J_c of polycrystalline samples decreases rapidly with increasing field, especially in the low field regime. An all-high-T_c DC-SQUID using one grain boundary as a weak link has been demonstrated.
4. The correlation between J_c and the misorientation angle θ in the basal plane suggests that a propensity of low tilt angle grain boundaries is important for achieving high critical current densities in polycrystalline high-T_c materials.

In order to minimize the grain boundary limitation on the critical current density in polycrystalline materials, it is obvious that much work is needed to achieve clean, low tilt angle grain boundaries through proper fabrication processing and stochiometry control. The J_c limiting effects arising from microcracking and anisotropy should also be considered.

REFERENCES

1. J.G. Bednorz and K.A. Mueller, Z. Phys. **B64,** 189(1986).
2. M.K. Wu, J.R. Ashburn, C.J. Torng, P.H. Tor, R.L. Meng, L. Gao, Z.J. Huang, Y.Q. Wang and C.W. Chu, Phys. Rev. Lett. **58,** 908(1988).
3. A.P. Malozemoff, W.J. Gallagher and R.E. Schwall, in "Chemistry of High Temperature Superconductors", ACS Symposium Series 351, ed. D.L. Nelson, M.S. Wittingham and T.F. George (American Chemical Society, Washington DC, 1987), pp 208-306.
4. A.P. Malozemoff, Physica C, **153-155** 1049(1988).
5. Proceedings of the 18th Int'l Conference on Low Temp. Phys. (Kyoto 1987). Jpn J. Appl. Phys. **26,** Suppl. 26-3, (1988).

6. Proceedings of the Int'l Conference on "High Temperature Superconductors and Materials and Mechanisms of Superconductivity" ed. by J. Mueller and J.L. Olsen (North Holland, 1988). Also, Proceedings of the Int'l Conference on Critical Currents in High Temperature Superconductors (Snowmass, Colorado 1988), to be published in Cryogenics.
7. M. Tinkham, Introduction to Superconductivity, (McGraw-Hill, New York 1975) pp.116-120.
8. Y. Yeshurun and A.P. Malozemoff, Phys. Rev. Lett. **60**, 2202(1988).
9. Y. Yeshurun, A.P. Malozemoff, F. Holtzberg and T.R. Dinger (preprint).
10. L. Krusin-Elbaum, A.P. Malozemoff, Y. Yeshurun, D.C. Cronemeyer, and F. Holtzberg (preprint).
11. D.R. Harshman et al. Phys. Rev. B **36**, 2386(1987).
12. L. Krusin-Elbaum, A.P. Malozemoff, Y. Yeshurun, F. Holtzberg and R.L. Greene (preprint).
13. J. Mannhart, P. Chaudhari, D. Dimos, C.C. Tsuei and T.R. McGuire, "Critical Currents in [001] Grains and across their Tilt boundaries in YBaCuO Films", (to be published in Phys. Rev. Lett., Nov.1988).
14. P.W. Anderson, Phys. Rev. Lett. **9**, 309(1962); Y.B. Kim, Rev. Mod. Phys. **36** 39(1964).
15. M.R. Beasley, R. Labush and W.W. Webb, Phys. Rev. **181**, 682(1969).
16. M. Tinkham, Helv. Physica Acta, **61**, 443(1988).
17. D. Dew-Hughes (preprint).
18. Y. Yeshurun et al., "Magnetic Properties of YBaCuO and BiSrCaCuO Crystals- A Comparative Study of Flux Creep and Irreversibility", Proceedings of the Int'l Conference on Critical Currents in High Temperature Superconductors (Snowmass, Colorado 1988), to be published in Cryogenics.
19. M. Tinkham, Phys. Rev. Lett. **61**, 1658(1988).
20. Y. Iye, T. Tamegai, H. Takeya and H. Takei, in "Superconducting Materials", ed. S. Nakajima and H. Fukuyama, Japn. J. Appl. Phys. **Series 1**, 46, (1988).
21. T. Palstra, B. Batlogg, L.F. Schneemeyer, and J.V. Waszak, Phys. Rev. Lett. **61**, 1662(1988).
22. P. Chaudhari, F. LeGoues, and A. Segmuller, Science **238**, 342(1987); P. Chaudhari, Jpn J. Appl. Phys. **26**, 2023(1987).
23. J.W. Ekin et al. in "High Temperature Superconductors" Materials Research Society (MRS. Pittsburgh, PA, 1987), **Vol. EA-11**, 223.
24. J. Mannhart, M. Scheuermann, C.C. Tsuei, M.M. Oprysko, C.C. Chi, C.P. Umbach, R. Koch, and C. Miller, Appl. Phys. Lett. **52**, 1271(1988).
25. P. Chaudhari, J. Mannhart, D. Dimos, C.C. Tsuei, J. Chi, M.M. Oprysko, and M. Scheuermann, Phys. Rev. Lett. **60**, 1653(1988).
26. K.K. Likharev, Rev. Mod. Phys. **51**, 101(1979).
27. D. Dimos, P. Chaudhari, J. Mannhart, and F.K. LeGoues, Phys. Rev. Lett. **61**, 219(1988).
28. D.P. Hampshire, X. Cai, J. Seuntjens and D.C. Larbalestier, Supercond. Sci. Technol., **1**, 12(1988).

SPECTROSCOPIES

ARXPS-studies of $\hat{c}$-Axis Textured and Polycrystalline Superconducting $YBa_2Cu_3O_x$ 208

Novel Bonding Concepts for Superconductive Oxides: an XPS Study 216

An XPS Investigation of the Surface Layer Formed on '123' High T_c Superconducting Films Annealed in O_2 and CO_2 Atmospheres 232

XPS DOS Studies of Oxygen-plasma Treated $YBa_2Cu_3O_{7-\delta}$ Surfaces as a Function of Temperature 240

Temperature Effects in the Near-E_F Electronic Structure of $Bi_4Ca_3Sr_3Cu_4O_{16+x}$ 248

Photoemission Resonance Study of Sintered and Single-crystal $Bi_4Ca_3Sr_3Cu_4O_{16+x}$ 252

Electron Energy Loss Studies of the New High-T_c Superconductor $Bi_4Ca_3Sr_3Cu_4O_{16+x}$ 257

Oxygen and Copper Valencies in Oxygen-doped Superconducting $La_2CuO_{4.13}$ .. 262

AES and EELS Analysis of $TlBaCaCuO_x$ Thin Films at 300 K and at 100 K 269

Resonant Photoemission and Chemisorption Studies of Tl–Ba–Ca–Cu–O 276

Photoemission from Single-crystal $EuBa_2Cu_3O_{6+x}$ Cleaved below 20 K; Metallic-to-insulating Surface Transformation 283

Characterization of $Bi_2Sr_2Ca_1Cu_2O_8$ 289

Surface Analysis of H, C, O, Y, Ba and Cu on Pressed and Laser-evaporated YBCO 297

Photon-stimulated Desorption from High-T_c Superconductors 304

Surface and Electronic Structure of Bi–Ca–Sr–Cu–O Superconductors Studied by LEED, UPS, and XPS 312

Optical Properties of High-T_c Superconductors: Who Needed This Anyway 318

Photoemission Studies of High Temperature Superconductors 330

ARXPS-STUDIES OF ĉ-AXIS TEXTURED AND POLYCRISTALLINE SUPERCONDUCTING $YBa_2Cu_3O_x$

J. HALBRITTER*, P. WALK**, H.-J. MATHES**, W. AARNINK***,
I. APFELSTEDT*

* Kernforschungszentrum Karlsruhe GmbH., Postfach 3640, D-7500 Karlsruhe
** Universität Karlsruhe, IEKP, Postfach 3640, D-7500 Karlsruhe
***University Twente, P.O. Box 217, N-7500AE Enschede

ABSTRACT

XPS lines changing with photoelectron take-off angle allow the separation of different chemical shifts by a simultaneous fit. This yields the identification of chemical compounds, their amount and their spatial distribution. Such an ARXPS analysis is carried out for highly textured unscraped superconducting cuprate films and fractured sintered material, which are coated by reaction cinder.

The spatial separation of the cuprate and the thin (<4 nm) cinder yields the stoichiometry of the bulk cuprate, the cuprate "surface" being about a unit cell thick (< 1.2 nm) and the cinder. The XPS-signature of ĉ-axis textured $YBa_2Cu_3O_x$ surfaces show a specific "surface Ba(Ba-2)" on the Cu-oxide and Y-oxide layers yielding so an intrinsically insulating surface layer ($\approx$ 1.2 nm). Below this dead layer the "superconducting Ba(Ba-1)" is differing from Ba-2 and insulating Ba cuprates, whereas Y does not show such a difference. These XPS assignments are used to fit the ARXPS data of·fractured or polished $YBa_2Cu_3O_x$. The results show, that fracturing opens inter- and intragrain weak links. The amount of "schmutz", i.e. "unwanted" matter smutting the surface on intragrain fractures, is smaller (< 1 nm $BaCuO_2$?) than on textured, unscraped films, i.e. on grains and is by ARXPS clearly separable from reacton cinder (Y_2O_3, Y_2BaCu..., $=CO_3$, graphite) between the grains. These results show the difficulty to measure bulk cuprate properties through the cinder and through the surface unit cell, but explain the observed weak links.

INTRODUCTION

Perovskite superconductors are a class of materials, which were studied often in 1987 and 1988 by XPS - see Refs. 1 and 2 and references therein. To treat these XPS measurements properly the signatures of a clean "perovskite surface", of bulk cuprate a unit cell (> 1 mm) below the surface and of coatings due to the processes have to be known. Especially the proper signature of the cinder caused by O_2 annealing is needed because "clean" can be defined only if the signature of the "schmutz", i.e. "unwanted" matter smutting the surface, is known[2]. ARXPS (angle resolved x-ray photo electron spectroscopy) is a method[2] well suited to analyze this cinder and the cuprate because they are spatially separated for textured films.

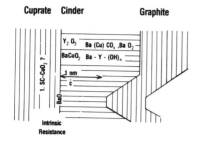

Fig. 1. Sketch of ĉ-axis oriented, highly textured $YBa_2Cu_3O_x$ surface with reaction cinder after O2 anneal. The initial cinder, being roughly 4 nm thick, is quite thin compared to epitaxial film prepared differently. This cinder grows with annealing time as $YBa_2Cu_3O_x$ partly decomposes.

Then by ARXPS minor amounts (≈ 0.3 nm) of interface compounds and their spatial distribution are identified in a depth range of 5 to 10 nm with negligible radiation damage. The latter two properties are crucial, because perovskite grains are known to be coated by a O2-reaction cinder of about 5 nm mean thickness even after scraping, and because these perovskites are extremely radiation sensitive[2].

As a first step the chemical compounds existing in the cinder of our highly textured films have been identified by ARXPS[2], showing graphite (see Fig. 1) as autermost layer ($\sim$ 2 nm) coating CO_x-compounds ($\sim$ 1 nm) bonded mainly to Ba. Underneath the carbon cinder Cu^{1+}, BaO_2 and Y_2O_3 ($\sim$ 1 nm) have been identified ($BaCuO_2$, Y_2BaCuO_5 ?) which coat the cuprate[2]. As signature of the superconducting cuprate two Barium ($\hat{=}$Ba-1 and -2) and one Yttrium ($\hat{=}$Y-1) line have been identified (YBa_2...) where Ba-2 and Y-1 are signatures of insulating cuprates $YBa_2Cu_3O_x$ (x < 6.5)[1] coating conducting bulk $YBa_2Cu_3O_7$ ($\hat{=}$Ba-1). With this assignment spectra of fractured crystals have been fitted showing similar reaction cinder as for the textured films. The findings, especially of two Ba shifts, whereas the bulk cuprate has only one Ba-site, ask for a rediscussion of cuprates and their surfaces taking the large unit cell into account.

For metals with atomic unit cells, the "surface" is about one atomic layer. In $YBa_2Cu_3O_x$ cuprates the situation is much more complicated:
- The cuprate is highly anisotropic where Y, Ba and Cu are confined to planes perpendicular to the ĉ-axis.
- The directional d-bonding yield nonlocal stoichiometry, i.e. the stoichiometry is achieved nonlocally over a unit cell, $d_u < 1.2$ nm.
- The "ĉ-axis surface" has different bond energies if formed by Y, Ba or Cu oxides and achieves bulk properties for distances larger d_u only, i.e.
- the ĉ-axis "surface" is at least $d_u \sim 1.2$ nm thick.
- The â-b̂-axis surfaces are drastically different compared to ĉ-axis surfaces and unstable[3].

This is discussed below where it is shown, that Ba-2 forms the stable ĉ-axis cuprate surface yielding a d_u thick dead, insulating layer underneath see Fig. 1. This dead layer is a reason for percolative path lengthening in the resistance and for a reduction of the critical current density[4]. Thus only fracturing perpendicular

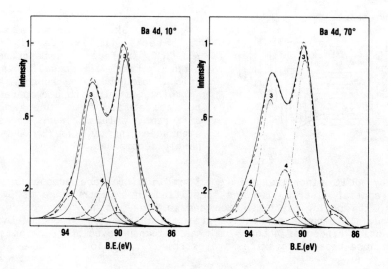

Fig. 2. Stable ARXPS fit for 10° and 70° for a Ba 4d photoelectron spectrum produced by $Al_{K\alpha}$ x-radiation onto the textured film.

to the ĉ-axis will open - for a short time - a surface reflecting bulk, cuprate properties being not shielded by 1 nm intrinsic - unavoidable - and > 2 nm extrinsic - avoidable - coatings.

EXPERIMENTAL

The photoelectron spectra are taken with a AEI ES 200 ESCA spectrometer[2]. The samples were mounted on a rotatable sample holder enabling the collection of photoelectrons (aperture < 10°) emitted at take-off angles θ between 10° and 70°. Before fitting the data with a Zenith Z-100 PC the background caused by inelastic processes and satellites is subtracted. These data are then fitted by a model function which is the sum of a Gaussian and a Lorentzian, where the Gaussian contribution increases with FWHM (full width at half maximum). The conducting compounds are fitted by an asymmetric model function, where to a symmetric Gaussian-Lorentzian an asymmetric part is added. As shown in Fig. 2 for our cuprate films the lines from different angles θ differ, because the stoichiometry depends on depth. This angle dependent line shape fitted simultaneously yields an enhanced sensitivity in stoichiometries and depths and makes ARXPS a sensitive, non destructive method to analyze interfaces up to depths of about 5-10 nm. The stability of the fits is checked by differently prepared surfaces. To evaluate the stoichiometry and the distribution quantitatively the fitted peak areas $F \propto n\sigma\lambda$ in their dependence on θ are used with

n: atomic concentration of the element producing the peak,
σ: photoelectron cross section for the atomic orbital, and

λ: mean free path of photoelectrons.
Interfaces show a sequence of stoichiometries corresponding to different depths and distributions[2] (TableI). Because buried[2] layers are shielded by coatings the cuprate shows the strongest decreases $F_i(\theta) \propto T_{i+1}(\theta)T_{i+2}(\theta)...$ with $T_i(\theta) = \exp(-d_i/\lambda_i \cos\theta)$ as transmission coefficients. The outermost layers show a weaker decrease with θ and for the top layer $F(\theta) \sim d/\cos\theta$ even increases. This qualitative concept has been modelled quantitatively by elaborate computer studies where the nonlinear dependencies on d_i and θ yield stable fits to the average distributions, shown, e.g., in Fig. 1.

The <u>preparation of the films</u> is summarized in Refs. 2 and 6, where also evidence for the excellent quality in structure, chemistry and superconductivity is presented. The reasons[2] for this excellent quality and the thin "schmutz" layer seem the highly oriented growth from an O rich, amorphous YBaCu oxide being dominated by metallic nuclei at the Al_2O_3- or MgO-interface. These excellent films have been measured as received, where the thin "schmutz" layer (< 4 nm) allows to identify various compounds in the reaction cinder and in the transition from the cinder to the bulk cuprate.

Cuprate	text.film unscraped on Al_2O_3		epit.film unscraped on MgO		sintered direct polished		sintered indirect fractured	
Angle	10°	70°	10°	40°	10°	50°	10°	50°
Ba-1	1.1	0.1	1.3	0.8	1.1	0.3	1.2	0.7
Ba-2	0.7	0.2	0.9	0.6	0.7	0.4	0.6	0.4
Ba-3	1.9	1.6	1.3	1.3	1.1	0.7	1.4	1.3
Ba-4	0.2	0.1	—	—	0.2	0.1	0.5	0.4
Y-1	1.0	0.2	1.0	0.8	1.0	0.5	1.0	0.5
Y-2	0.8	0.2	1.2	1.0	0.4	0.3	0.4	0.2
Y-3	0.2	0.1	0.4	0.3	0.2	0.0	0.2	0.2
C-5	1.8	0.7	—	—	1.0	0.5	1.5	0.7
Σ Ci	23.6	22.4	—	—	17.0	18.2	9.4	4.9

Table I: Fitted Ba 4d, C 1s and Y 3d peak areas for different angles θ for two films and two cuprate samples. The areas are divided by the photoelectron cross sections $\sigma_{Ba} = 5.86 \cdot \sigma_{C1s}$ and $\sigma = 5.98 \cdot \sigma_{C1s}$ and normalized by the intensity of the line Y-1 at 10°, being a measure of the cuprate $YBa_2Cu_3O_x$. For a single crystal fractured along the â-b̂ plane the photon impact (135 eV) gave ratios Ba-1/Ba-2 = 0.6 (7°) and 0.12 (73°) with traces of Ba-3 at 73°.

B.E./eV		Assignment	B.E./eV		Assignment [eV]
87.2	Ba-1	Superc. Cuprate	156.2	Y-1	Superc. Cuprate
88.2	Ba-2	Cuprate Surface	157.3	Y-2	$Y^{3+}:Y_2O_3$, ...
89.6	Ba-3	Ba^{2+}	158.6	Y-3	$Y(OH)_3$, ...
91.0	Ba-4	$Ba(OH)_2$			
284.0	C-1	Graphite,	528.0	O-0	Cupr. Cu-Ba-Chains
			529.1	O-1	Cupr. Cu-Plains
285.0	C-2	Graphite oxided	530.0	O-2	CO_x
286.1	C-3	Alcohole	531.4	O-3	Peroxide, Cupr.
288.1	C-4	Carboxyle	532.4	O-4	Co_y
289.0	C-5	Carbonate	534.0	O-5	Hydroxide, H_2O, metallic O

Table II: Binding Energy and assignment of fitted $Ba4d_{5/2}$-, $Y3d_{3/2}$-, C_{1s}- and O_{1s}- components.

The <u>preparation of the sintered samples</u> is described in Ref. 5 showing standard results, i.e. density ~ 93%, ρ_{100} > 200 μΩcm, T_c ~ 90 K and j_c ~ 500 A/cm^2. The sample "indirect" is prepared by various grinding cycles of sintered $YBa_2Cu_3O_x$ whereas "direct" stands for carefully grinding one time Y_2O_3, $BaCO_3$ and CuO and then pressing (200 MPa) and calcinating[5]. A direct sample is fractured, polished by SiC in CH_3CH_2OH and then analyzed by ARXPS. An indirect sample has been fractured and rapidly (< 5 min) transferred into the XPS set-up and analyzed.

The <u>simultaneous fits</u> are calibrated against the Au $4f_{7/2}$ line at 84 eV[2]. The <u>C_{1s} spectra</u> are fitted by 5 lines summarized in Fig. 3 and Table II. As obvious from Fig. 3 the "graphite" is the dominant feature, aside from (Ba-) carboxyles and carbonates underneath. The fit of the <u>O_{1s} spectra</u> in 5-6 lines (Table II) is not shown because we were not able to separate all the different oxides. The fit to the <u>Y 3d spectra</u> by <u>3 doublets with 2 eV peak distance</u> is summarized[2] in Table I and II. The <u>Ba 4d XPS spectra</u> are fitted by <u>4 doublets with peak distances of 2.6 eV</u>, see Table I and II and Fig. 2[2]. The <u>Cu-spectra</u> have <u>not been fitted</u> because of the lack of knowledge of the signatures of the different compounds involved and of the charge transfer and plasmon satellites[7]. The amount of Cu is obtained via the Cu 3s spectrum[2], which originates from the same depths range as Ba 4d and Y 3d.

DISCUSSION

We obtained the assignments[2] of the chemical compounds and their spatial distribution by the following procedure: Different components $F_i(\theta)$ of different elements have matching θ-dependencies

(Table I) if they belong to one chemical compound. This method yields for the highly textured $YBa_2Cu_3O_x$-films the assignments of Table II as elaborated in Ref.2. These assignments have been confirmed by differently treated films[2] and fractured surfaces. For the cuprate and its intrinsic surfaces we obtained the following result[2]: Ba-1, Ba-2 and Y-1 make up the cuprate $YBa_2Cu_3O_x$, where Ba-1 as the more metallic component makes up the bulk superconducting cuprate[1,2]. As shown by its weaker θ-decrease, the more ionic component Ba-2 forms the stable ĉ-axis surface for our as grown films[2]. Detailed fits reveal that Ba-2 forms a BaO_x double layer, which is confirmed by TEM measurements[8]. Our result of two Ba XPS lines Ba-1 and Ba-2 for ĉ-axis surfaces has recently been confirmed by UHV cleaved, cuprate single crystals[9]. There, the angle dependence (Table I) quantitatively fit to one Ba-2 layer hinting to reconstruction or to preferential cleavage along Ba-oxide stacking faults. The Ba-2 double layer of the textured film fits well to the growth dynamics because of the larger stability of BaO_x as compared, e.g., to Cu oxide. Y-1 does not show changes in the transition to superconducting cuprates spacewise[2] and 0-annealing[1] wise confirming its passive role in the superconducting interaction.

About an intrinsic â-b̂ surface, which is unstable below 450°C[3] our ARXPS-analysis gave no clue. The facturing of sintered samples yield a ratio r = Ba-1/Ba-2 ≈ 2 (Table I), which is in line with a randomly oriented, polycristalline sample. The smaller value r ≈ 0.5 for scraped samples[1] may be due to a lack of oxygen. For ĉ-axis surfaces r ~ 1.5 is obtained (Table I) indicating, that conduction or tunneling is more easily achieved via - fractured - â-b̂-surfaces with r > 2.

The oxidic cinder, occuring space-wise ontop of the cuprate, for the textured films consists of Ba-3,4 and Y-2,3, where $BaCO_3$, BaO_2 and Y_2O_3 play a dominant role. This sequence occurs for our fractured surfaces also, where the reduced amount of Y is remarkable when comparing with the textured films (Table 1). As shown by their strong θ-decreases as compared with the cuprate, Y-2 (Y_2O_3), Ba(Cu,Y)CO_3 and graphite are mainly between the cuprate grains. The large amount (≈ 3 layers) of Ba-3 on the fractures hint to $BaCuO_2$: This Cu-compound is more stable than CuO and Cu_2O and coats the cuprate epitaxially[10]. This stable, thin insulating layer coating epitaxially â-b̂- and ĉ-axis cuprate surfaces cannot be separated from the cuprate by TEM[8] but would explain STEM/EDAX[11]-results. Thus $BaCuO_2$ may be a reason for inter-or intra-grain weak links[10].

On top of the oxidic cinder, graphite is coating the highly textured films. As shown by Fig. 3, "graphite" occurs in sintered material and there between the grains accompanied by carbonates and Y3+. That this "graphite" is not an artifact from the ARXPS measurement is shown by this precipitation between the grains, i.e., as reaction cinder at grain surfaces[2], but and also by the "direct" sample. This sample shows an C-3 peak (ethanol) increase with θ = 10°

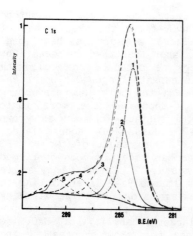 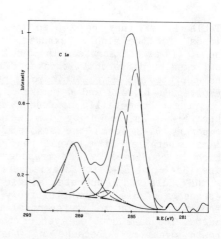

Fig. 3. C1s photoelectron spectrum emitted perpendicularly to the $YBa_2Cu_3O_x$ surface produced by Al_{K_α} x-radiation.
a) textured film (1 d air, 300 K)
b) sintered cuprate fractured.

to 50° by a factor 2, i.e. the ethanol from polishing is coating the graphite. The different amounts and positions of the graphite for the polished and unpolished fractured sample are due to smearing out the graphite along the as surface as has been reported before[12].

SUMMARY

The above ARXPS-analysis of the Ba 4d, Y 3d, C 1s and O1s spectra has revealed the stoichiometry of our reaction cinder, consisting of: carbonate, carboxyles and oxides of Cu, Ba and Y in a thickness below 2 nm on textured films and between grains, aside from graphite ($\approx$ 2 nm).
- The fractured surfaces are mainly coated by a Ba oxide (< 1 nm) - most likely $BaCuO_2$ - being a candidate for weak links.
- â-b̂-surfaces are unstable below 450°C and seem to fracture at $BaCuO_2$-precipitates.
- Free ĉ-axis surfaces show a specific "surface Ba" yielding a unit cell thick $d_u \sim 1.2$ nm insulating surface.
- Cleaving yields ĉ-axis surfaces showing also the "surface Ba", which may be the preferential path for cleaving or due to reconstruction.
- The bulk superconducting cuprate has a metallic Ba signature.

Because the above summarized thicknesses are mean values, holes in this coating may allow the measurement of bulk properties for, e.g., 10 or 20% of the surface area. This assignment "bulk properties" must be proven by quantitative ARXPS. For ĉ-axis surfaces with Ba-2 and the likelihood of reconstruction the separation of "bulk properties" by UPS and XPS will be difficult.

ACKNOWLEDGEMENT

The authors like to acknowledge helpful duscussions with
H. Küpfer, H. Rogalla, P. Steiner and Th. Wolf. The technical
support of the staff of IK II and ITP (KfK) is thankfully acknowledged.

REFERENCES

1. P. Steiner, S. Hüfner, K. Kinsinger, I. Sander, B. Siegwart, H. Schmitt, R. Schulz, S. Junk, G. Schwitgabel, A. Gold, C. Politis, H.P. Müller, R. Hoppe, S. Kemmler-Sack, C. Kunz, Z. Phys. B69, 449 (1988); W.K. Ford, C.T. Chen, J. Anderson, J. Kwo, S.H. Lion, M. Hong, G.V. Rubenacker, J.E. Drumheller, Phys. Rev. B 37, 7924 (1988)
2. J. Halbritter, P. Walk, H.-J. Mathes, B. Häuser, H. Rogalla, Physica C153, 127 (1988) Z. Physik B (1988)
3. H.W. Zandbergen, R. Gronsky, G. Thomas, phys. stat. sol. 105, 207 (1988)
4. J. Halbritter, H. Küpfer, B. Runtsch, H. Wühl, Z.Phys.B, 411 (1988)
5. T. Wolf, I. Apfelstedt, W. Goldacker, H. Küpfer, R. Flükiger, Physica C153, 351 (1988)
6. B. Häuser, E.G. Keim, H. Rogalla, A. van Silfhout, Appl. Phys. A46, 339 (1988); J. Halbritter, B. Häuser, E.G. Keim, J.-J. Mathes, P. Walk, H. Rogalla, IEEE Trans MAG 25, (1989)
7. J. Halbritter, H. Leiste, H.-J. Mathes, P. Walk, H. Winter, Solid State Comm. (1988)
8. H.W. Zandbergen, R. Gronsky, G. Thomas, Physica 153C, 1002 (1988); L.A. Bursill, Xu Dong Fan, phys. stat. sol. (a) 107, 503 (1988)
9. N.G. Stoffel, P.A. Morris, W.A. Bonner, D. La Graffe, M. Tang, Y. Chang, G. Margaritando, M. Ouellion, Phys. Rev. B 38, 312 (1988)
10. Y. Wadayama, K. Dudo, A. Nagata, K. Ikeda, Sh. Hanada, O. Izumi, Jap. Journ. Appl. Phys. 27, L 1221 (1988); Th. Wolf, private communication
11. S.E. Babcock, T.F. Kelly, P.J. Lee, J.M. Senntjencs, L.A. Lavanier, D.C. Larbalestier, Physica 152C 25(1988)
12. A.G. Schrott, S.L. Cohen, T.R. Dinger, F.J. Himpsel, J.A. Yarmoff, K.G. Frase, S.I. Park, R. Purtell, AVS series AIP conference proceedings 1988

NOVEL BONDING CONCEPTS FOR SUPERCONDUCTIVE OXIDES:

AN XPS STUDY

T. L. Barr*
C. R. Brundle**

A. Klumb,* Y. L. Liu,* L. M. Chen,* and M. P. Yin*

---

* Department of Materials and Laboratory for Surface Studies
  University of Wisconsin--Milwaukee, Milwaukee, Wisconsin 53201
** IBM Research, Almaden Research Center, San José, California 95120-6099

© 1989 American Institute of Physics

ABSTRACT

Detailed analyses of the x-ray photoelectron spectra (XPS), of the normal (above T_c) phase for the Ba-Y-Cu-oxide (1-2-3) superconductors, their precursors, and related systems have revealed a number of interrelated patterns, particularly that of the singular O(1s) peak (at 528.5 $\pm$ 0.3 eV) apparently produced by most, if not all, of these superconductive systems. This result (and its corresponding O(2s) value) has been shown to correspond closely to those oxygen binding energies produced by BaO and Y_2O_3, but not by any of the copper oxides. This suggests that the 1-2-3 systems are substantially influenced by the extreme ionic fields induced by the barium (and yttrium) ions, whereas the copper in these systems is, in large part, a structural cohabitant in the extremely ionic $Cu_3O_{7-x}^{-7}$ ions. The latter ions are shown to produce key electronic results that are quite different from those induced by the copper in the aforementioned copper oxides. The primary characteristics of that difference are suggested to be the substantial shifts and contractions experienced in the O(2p) part of the resulting valence band. In the case of BaO, its (O(2p) dominated) valence band is pushed quite close to its pseudo-Fermi (zero binding energy) edge. In fact, we predict a separation of only about 1 to 2 eV. Interjection of copper (into the 1-2-3 system) results in a substantial Cu(3d) band, that we suggest "double" perturbs the O(2p) band of the superconductors. Part of the latter band is pushed to higher binding energy (enhanced covalency), whereas the other side of the O(2p) band is shifted downfield to a point of contact with the Fermi edge. Thus, in our model, the origin of superconductivity is the mirror image of that proposed by most previous researchers.

A NOVEL BONDING CONCEPT FOR SUPERCONDUCTING OXIDES

The recent discovery of high T_c superconductivity in certain types of complex, mixed oxides has led to a series of x-ray photoelectron spectroscopy (XPS or ESCA) studies to assist in determining the important surface (and near surface) properties of these systems, and their related adducts. Because these XPS studies deal directly with the electronic structure of such materials, they also may provide some insight into the uncertain mechanism for this type of superconductivity.

Previous XPS studies in this area have generally concentrated on cleanliness,[1] the valence band region[2] and Cu(2p) core level spectra.[3]

It is our purpose in the following paper to point out that, although these analyses are all useful and the status of copper is very important, we have found in contradiction to the aforementioned generally accepted belief that the materials that form the T_c^N for these oxides superconductors are <u>not primarily copper oxides</u>, but rather <u>alkaline earth/rare earth oxides into which the copper is inserted and plays a moderate, but quite critical, role against a background of the dominating oxygen perturbation</u>. We will attempt herein to validate this obviously critical, unique view of these materials and its consequences.

It is certainly true that the electron density in the lattice structure of metallic oxides is substantially influenced by the contributions of the electron rich, oxide ions. In fact, as one of us has previously described, with the exception of those metals exhibiting valence region d and f states, almost all of the valence band electron density realized in oxides, (such as the precursors of the aforementioned T_c^N materials) originates in their O(2p) (valence) orbitals.[4] This implies that metallic oxides, such as those being considered herein, are primarily ionic, leaving as a principal question only the degree of ionicity, f_i.[5] Thus, it should be apparent that many important bonding features of the T_c^N materials may be uniquely revealed in XPS studies of the behavior of their oxygen core level and valence band states, particularly if these results may be correlated with the resulting ionicity of the materials.

In this regard, we have begun detailed analyses of the XPS results achieved by our group[1] and others[2,3,6] for the T_c^N materials, with particular emphasis on the so-called 1-2-3 systems, $YBa_2Cu_3O_{7-x}$.[7] Unfortunately, as implied in (1) above, all results achieved so far are compromised, to varying degrees, by the simultaneous presence of a variety of precursors and byproducts or impurities,[1] and thus the interpretation of the XPS data is, to some extent, a taxing exercise in deconvolution, and the discarding of unwanted parts. We, therefore, report herein only speculations regarding the various T_c^N materials analyzed. Final proof awaits further purification. As mentioned above, most of our XPS measurements have been achieved through examinations of the 1-2-3 T_c^N system or its precursors. This forces us to often resort to the use of analogies, particularly during attempts to extend these arguments to the other high T_c systems.[2,8]

CORE LEVEL RESULTS

In Figure 1, critical, core level binding energies for select precursor (and byproduct) systems are compared to those results realized by the various components in the (pre) superconductor system, particularly the T_c^N phase itself. In this regard, one should consider, in particular, the O(1s) binding energies produced by both the Cu(II) and Cu(I) oxides. These values prove to be typical of those achieved for most transition metal oxides. As has been seen in numerous XPS studies, there seems to be an often produced (still not entirely understood) small range of O(1s) values for <u>most</u> of these oxide systems, i.e., 530.0 ± 0.4 eV.[9] The copper oxides are no exception to this "rule," as they are found to be somewhat on the high side of this range, Figure 1. Cu(III) oxide has never been examined by XPS, but it is anticipated that it also should produce a binding energy in the aforementioned range.

In point of fact, however, close examination of the general results demonstrates that there are two major groups of exceptions to the aforementioned O(1s) = 530 eV oxide "rule." The first exceptional group is typified by those oxides that have substantial covalency in their bonding.[5] Obviously, these are best represented by the Group V, VI and VII oxides, but this type of covalency is also exhibited by such metalloid oxides as silica (O(1s) = 532.8 eV) and even alumina (O(1s) = 531.0 eV) (both for C(1s) ≃ 284.8 eV).[10] The second group of exceptions is composed of those oxides that are <u>extremely ionic</u>. Typically, the members of this group are represented by oxides formed with metals that initially (before oxide formation) have only s electrons, generally with relatively large quantum number, in their valence shell. Thus, some of the alkali and alkaline earth, and certain rare earth and transition metal oxides are found to be in this extremely ionic group, particularly from the bottom to the middle of these columns in the Periodic Table. These extreme ionic oxides are all typified by O(1s) values in the range 529.4 - 528.0 eV, with, for example, BaO O(1s) = 528.3 eV and Y_2O_3 O(1s) = 529.0 eV,[11] see Figure 1.

In this regard, it is very important to note that <u>the substantial drop in O(1s) binding energy of Ba, Sr, La, Y, and Tl oxides compared to the typical transition metal oxides, (e.g., copper) is primarily attributed to the enhanced ionicity of the former group over the latter</u>.[12] A detailed examination of this ionicity feature has recently been presented by some of the members of our group, including studies of particular oxides, and the general arguments concerning their ionic/covalent origins.[12] In those cases, we have discovered that enhanced ionicity in an oxide not only tends to diminish the O(1s) (and also, for that matter, the O(2s)) binding energies, but also produces certain characteristic alterations in the valence band structure.[12] In this regard, we have noted that, (1) the bands for extremely ionic oxides are essentially composed of only O(2p) density, (2) all of these oxides are also characterized by the presence of near core states separated from the previously described O(2p) density by less than 10 eV, see Figure 1. These near core states will, of course, act to perturb (push) the O(2p) band away from their position and closer to their zero (pseudo-Fermi) edge, and (3) as ionicity grows, the total width of the bands shrink to near core-like ($P_{3/2}$ + $P_{1/2}$) proportions.

With these features in mind, it is interesting to turn to a consideration of the XPS spectra of the T_c^N, high T_c superconducting precursors. In particular, one should note the series of O(1s) spectra realized during the cleaning of several 1-2-3 systems, as displayed in Figure 2. In this cleaning process most, but not all, of the precursor and byproduct materials are removed. (The various methods for removing these extraneous materials have been previously discussed in some detail by one of us,[1] and will not be repeated herein.) It is important to note that these processes appear to remove materials that are associated with larger O(1s) binding energies, particularly the carbonates, and indeed the O(1s) peaks above 531.0 eV (typical of metal carbonates) are among the last to disappear during cleaning (and preservation) of T_c^N and this disappearance seems to occur in conjunction with the corresponding loss of a carbonate-type C(1s) peak. <u>The important feature to note is that, along with the O(1s) peaks above 531.0 eV,</u>

all O(1s) peaks in or around 530.0 eV also disappear leaving only a rather singular photoelectron line, peaked at ~ 528.6 ± 0.3 eV, Figure 1! Since no system has yet been achieved without, at least, a modest O(1s) shoulder at higher binding energy, one cannot unequivocally state that the low binding energy peak is the <u>only</u> one produced by the T_c^N material,$^{(1-3,6)}$ but one can stipulate that the 1-2-3 T_c^N material produces <u>primarily</u> an O(1s) peak indicative of an extremely ionic, oxide system.

VALENCE BAND RESULTS AND IONICITY

Before considering farther the implications of the aforementioned copper perturbation and barium (and yttrium) ionicity, we need to review in a little more detail our results and conclusions regarding oxide valence bands. We noted above that the valence bands of all Group IA and IIA, and even most IIIA and IVA, oxides can be loosely construed as composed of only O(2p) orbital, i.e., the metal orbital contributions are never more than ~ 15% of the total (SiO_2), and usually substantially less.$^{(12)}$ In addition, as the oxide becomes more ionic, the band shrinks in size, and moves closer and closer to its zero binding energy point (quasi-Fermi edge).$^{(12)}$ As an example of these progressions, consider the results in Figure 1.

Unfortunately, no one has yet produced an absolutely clean valence band spectrum for BaO or Y_2O_3. Based upon the results achieved (for slightly contaminated systems) and extrapolating, we expect the leading edge peak for the valence band of BaO to be peaked somewhere between 3.0 and 2.5 eV upfield from its quasi-Fermi edge.$^{(12)}$ At the same time, the total (XPS measured) band width at 1/2 maximum is suggested at ~ 5.0 eV (as estimated from both our results,$^{(12)}$ and those of Van Dovern and Verhoeven$^{(13)}$ and extrapolated from the results achieved for other oxides). In addition, clean BaO should exhibit a leading edge onset* somewhere between 1.25 and 1.75 eV, thus producing one of the smallest (intrinsic) band gaps for any oxide, ~ 3.0 eV.

The valence band region for surface pure Y_2O_3 is as yet unmeasured, but is expected to produce results that are similar to those for BaO, except that the Y_2O_3 band (compared to that for BaO) should be slightly broader and somewhat shifted to higher binding energy, as reflected in the somewhat larger binding energy of the O(1s) and O(2s) in yttria.[11]

On the other hand, the valence band spectra for the copper oxides (e.g., CuO) are, of course, dominated by the relatively large cross-sectioned 3d band. This is peaked between 4 and 5 eV and is ~ 5 eV in width. This d band essentially "covers" part of the CuO O(2p) band that is peaked between 5.5 and 6.0 eV, and slightly broader than the Cu(3d) band.[14]

It should be apparent that from chemical considerations, it is as improper to view the 1-2-3 T_c^N system as copper (and yttrium) perturbed barium oxide, as it is to view it as barium and yttrium perturbed copper oxides. In fact, it is far more appropriate and realistic to consider the key system in question as the cuprate ion, $Cu_3O_{7-x}^{-7}$. This may seem to be a step backward, away from the arguments presented in the previous section where we placed great emphasis on the presence of barium, but it turns out that this is not the case. In a subsequent model we will, for convenience, consider how copper perturbs a barium oxide lattice, when, in point of fact, the copper must be intimately involved in the oxide lattice to start with. But, as we shall try to demonstrate, <u>our "cuprate" ion is quite unique, because of the near total ionicity induced in it by the combination of barium and yttrium ions (or for other high T_c-superconducting systems by strontium and lanthanum, or</u>

---

*Point where finite density begins for the valence band.

any other matched set that will introduce almost the same size field). A slightly weaker ionic field, such as that introduced by Na^+ or K^+, for example, would not produce the same electronic effect (as well as the resulting important size changes!). Thus, when the latter cations are employed, the superconductive properties are significantly altered. It is equally important to note that there is no similarity whatsoever in total field between our $Cu_3O_{7-x}^{-7}$ and the various copper oxides described and alluded to in many previous studies,[15] and this includes the interactions between the copper and the various oxygens that must be substantially perturbed from the interactions existing in a copper oxide. If the latter were not true, then one must ask why the O(1s) and O(2s) binding energies for the 1-2-3 system are so similar to those for BaO? The reason, (as we shall amplify below), is that it is more consistent to argue that there is in these systems an interjection of copper into a BaO (and Y_2O_3) lattice (to form the $Y^{+3} \cdot 2Ba^{+2} \cdot Cu_3O_{7-x}^{-7}$ system) than it is to argue the other way around. Further, this copper into BaO interjection seems to be accomplished through the use of a rather singular balance of the ionic and covalent forces, i.e., the inherent covalency of the ("additional") copper-oxygen bonds are roughly compensated by the strong inherent ionicity already "surrounding" the oxide ion site. Thus, only small perturbations of the O(1s) (and O(2s)) binding energies of BaO are anticipated in the 1-2-3 system and that is what is observed.[2] On the other hand, it is very important for superconductivity that this lack of alteration is not what we are predicting for the O(2p) valence band. The much more dramatic changes predicted for the latter are described in the next section.

It should be noted that we are not challenging those experiments and calculations that predict that the Cu(2p) and Cu(3d) states for CuO, and the 1-2-3- system, are similar. Indeed, they are, but we are suggesting that the

O(1s), O(2s) and (if detectable) O(2p) are, for the superconductor, quite different from CuO, and instead more closely related to copper perturbed BaO.

HIGH T_c^N SUPERCONDUCTING OXIDE MODEL BASED ON
LARGE IONIC FIELD AND COPPER PERTURBATION

We are now in a position to consider an alternative model for the oxide, high T_c^N systems. In this model (presently supported by evidence only for the 1-2-3 case), the basic "precursor" systems begin with the barium (and yttrium) oxides, with their <u>extreme ionicity</u> (and accompanying NaCl octahedral structure). In this vein, the key feature detected is the extremely low binding energy for the O(1s), Figure 1. The key accompanying feature (not well-demonstrated) is the corresponding shrinkage in width and down field shift of the relatively small (spherical) part of the near valence band, which is composed almost exclusively of O(2p) orbital density. It is important to realize that (with the possible exception of Cs_2O) BaO has its valence band shrunk in size and pushed closer to its pseudo-Fermi edge than any other simple oxide. This means that BaO is on the verge of being a relatively narrow, band gapped, intrinsic semiconductor, rather than an insulator. Thus, we have the peculiar situation of the creation of a Mott Transition (to a Mott Insulator), that essentially never occurs, i.e., as our consideration shifts from BeO to BaO down the Group IIA oxides the O(2p) bandwidth shrinks, and it also moves inexorably toward its quasi-Fermi edge (i.e., we end up with an intrinsic semiconductor rather than an insulator).

This movement and adjustment of the O(2p) band for BaO puts it in a unique position where it is, for example, much further downfield and narrower than the same band for CuO. In fact, the BaO, O(2p) band is located almost directly under the position where the (relatively) narrow and sizeable Cu(3d) band would occur in an oxide of copper. As a result, during 1-2-3, T_c^N formation, the Cu(3d) band might be depicted as now "sitting on" the BaO O(2p) band. In order to model this, it is helpful to note that rather detailed (but approximate) versions of the interactions of these two bands have already been considered for the high T_c superconductor systems based around the Anderson Hamiltonian and Resonance Valence Band Model.[16] Various forms of these interactions have been described[17] including the fairly detailed treatment of Shen et al.[18]

All of these treatments are characterized by the fact that the O(2p) density field, described with a receptive ligand hole state, L, and the latters hybridization interaction, T, with the Cu(3d) density, $T \simeq 2.4$ eV[18] are introduced as a perturbation that promotes a ligand-copper charge transfer path, Δ, (where $\Delta \simeq 0.4$ eV) to accompany the anticipated polar charge flow, u, in the Cu(3d) band, $d_i^n d_j^n \rightarrow d_i^{n-1} d_j^{n+1}$, where i and j denote two adjacent Cu sites, and n is the precharge flow 3d occupancy.

In our model, on the other hand, we invert the emphasis and assume that the principal loci for carriage of the current flow (in the <u>anticipated</u> superconductor) is through the O(2p) band, rather than the Cu(3d). The stage is set for this flow in the 1-2-3 system by the extreme ionicity of the barium (and yttrium) oxides, and the aforementioned unique features in the O(2p) band. The final critical impetus for flow, however, is established through the interjection of <u>appropriately oxidized</u> copper ions that modify the structure of the (previously octahedral) lattice to the distorted, defect filled triple layered, perovskite structure exhibited by the 1-2-3 system.[15] <u>More importantly, the aforementioned Cu(3d) band can be viewed as a perturbation of the O(2p)</u>. The mechanism for this perturbation is suggested to be just the reverse of that postulated previously for the perturbation of the Cu(3d).[18] This suggests that we may utilize the same scope and numbers realized by Shen et al.,[18] but add them as perturbative alterations to the O(2p) band, rather than to the Cu(3d). In this manner, we find a hybridization shift, T, of ~ 2.0 eV. This should be sufficient to push the leading edge of the O(2p) band to a point of ready coupling with the Fermi edge. The size of the charge transfer may also correspond to the aforementioned Δ.

Thus, our model also predicts that the mechanisms for superconductivity is through the resonance valence band path,[17,18] made possible through the hybridization of O(2p) and Cu(3d) bands, but the critical current carrying units are the O(2p) bands. Near conductive versions of the latter are "in place" as a result of the extreme ionicity, but they need the critical coupling of the "interjected" copper ions to complete the path for (2 dimensional) current flow, and provide the induced coupling force boost needed to implement the process.

In effect, one finds that the Cu(3d) band should perturb (through 2nd order) the O(2p) in two directions: (1) the former pushes the leading edge of the latter outward toward the Fermi edge. Interestingly, this part of the O(2p) band is dominated by nonbonding electrons, in π-like orbitals, that are essentially free for current flow. (2) It should also be noted that the same Cu(3d) perturbation should affect the trailing edge side of the O(2p) band, also pushing it outward (downfield) toward (in this case) higher binding energy. The latter movement of the O(2p) band has been shown by us[12] to signify an enhancement of covalency in the oxide system. This is, of course, consistent with the already realized[15] suggestion that the 1-2-3 (and related) high T_c^N systems exhibit reduced Cu-O bond distances, with enhanced covalency. The combination of these effects are quite unique with the band stretching in both directions, typical of the simultaneous enhancement of covalency (some electron cloistering) and (some) electron release. Interestingly, it is our supposition that the copper perturbs the O(2p) band in both the up and down field direction that (we suggest) explains why the binding energy of the O(1s) peak for the 1-2-3 system is so little shifted from that for BaO.

If these suppositions are true, then there should be some alteration of the densities in the valence band region that reflects this. Unfortunately, most previous investigators have studied in detail the intense top of the total band where the Cu(3d) part predominates. This part of the total band is indeed perturbed by the hybridization, but, in our estimation, it is only slightly altered. The O(2p) band, on the other hand, should be significantly shifted. Unfortunately, the latter is relatively small, i.e., $([Cu(3d)]/[O(2p)]_{XPS} > 5/1)$, and substantially covered by the Cu(3d), and therefore these shifts are not expected to be readily detectable. Through the use of careful XPS study, however, we have found that the (total) band structure of the 1-2-3 system does tend to protrude at its base in both the up and down field direction. These bulges, we suggest, are created by the aforementioned shifts induced into the O(2p) band through the perturbation (hybridization) of the Cu(3d).

REFERENCES

1. D. C. Miller, D. E. Fowler, C. R. Brundle, and W. Y. Lee, AVS Special Conference on High T_c Superconductivity, Anaheim, CA (AIP Conf. Proc., Vol. xx, Feb. 1988).

2. H. M. Meyer, D. M. Hill, T. J. Wagener Y. Gao, J. H. Weaver, D. W. Capone II and K. C. Goretta, Phys. Rev. B., to be published.

3. See for example, D. E. Ramaker, in Chemistry of High-Temperature Superconductor II, Ed. by D. L. Nelson and T. F. George, ACS Symp. Ser. 377, Washington, D.C., 1988, Ch. 8.

4. T. L. Barr, M. Mohsenian and L. M. Chen, J. Amer. Chem. Soc., to be published.

5. L. Pauling, The Nature of the Chemical Bond, Cornell University Press, Ithaca, 3rd Ed., 1960.

6. See also, P. Steiner et al., Z. Phys. B--Condensed Matter, 69, 449 (1988).

7. C. W. Chu, P. H. Hor, R. L. Meng, L. Gao and Z. J. Huang, Science, 235, 567 (1987).

8. C. R. Brundle, presented at ACS Symposium on High T_c Superconductivity, Los Angeles, 1988.

9. C. R. Brundle, Surface Sci., 48, 48 (1975) and T. L. Barr, J. Phys. Chem., 82, 801 (1978).

10. T. L. Barr, Appl. Surface Sci., 15, 1 (1983) and T. L. Barr, L. M. Chen, and M. Mohsenian, J. Phys. Chem., to be published.

11. T. L. Barr, in Quantitative Surface Analysis, Ed. N. S. McIntyre, ASTM, Philadelphia, 1978, pp. 83-104.

12. T. L. Barr and P. S. Bagus, to be published.

13. H. Van Dovern and J. A. Th. Verhoeven, J. Electron Spectrosc. and Related Phenomena, 21, 265 (1980).

14. G. K. Wertheim and S. Hufner, Phys. Rev. Lett., 28, 1028 (1972).

15. See for example, D. L. Nelson, M. S. Whittingham and T. F. George, Eds, Chemistry of High-Temperature Superconductors, ACS Symposium Series, 351, Washington, D.C., 1987.

16. See for example, P. W. Anderson, Science, 235, 1196 (1987). P. W. Anderson, G. Baskaran, Z. Zou, and T. Hsu, Phys. Rev. Lett., 58, 2790 (1987).

17. See for example, S. A. Kivelson, D. S. Rokhsar and J. P. Sethna, Phys. Rev. B, 35, 857 (1987).

18. Z.-X Shen, J. W. Allen et al., Phys. Rev. B., 36, 8414 (1987).

230

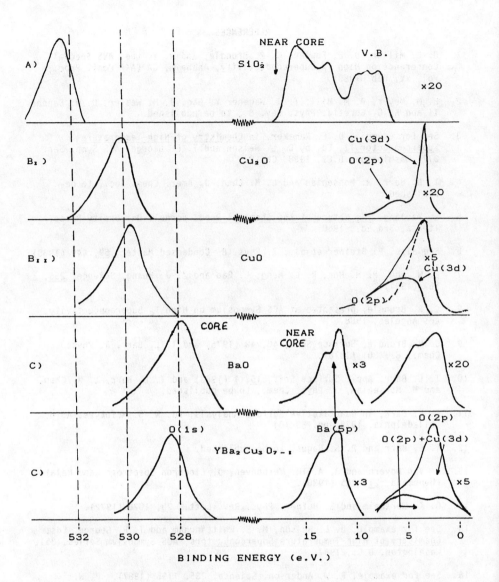

A = Covalent
B = Metallic
C = Very Ionic

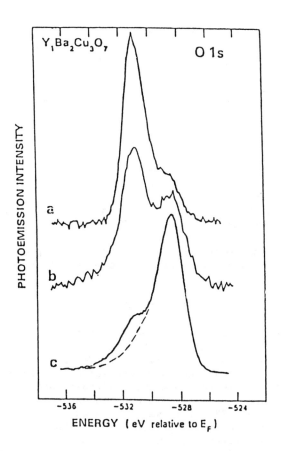

Evolution of O(1s) For 1-2-3 Bulk System During Drastic (But Necessary) Cleaning; a - b - c - d, Where d Is Anticipated Result For Total Cleaned System[2]

AN XPS INVESTIGATION OF THE SURFACE LAYER FORMED
ON '123' HIGH T_c SUPERCONDUCTING FILMS
ANNEALED IN O_2 AND CO_2 ATMOSPHERES

R. Caracciolo, M. S. Hegde, J.B. Wachtman,Jr., A. Inam
Rutgers University, Piscataway, NJ 08855

T. Venketesan
Bellcore, Red Bank, NJ 07701

ABSTRACT

$YBa_2Cu_3O_7$ thin films produced by laser deposition typically possess a layer consisting of several oxide forms. We attempt to identify this oxide layer using X-ray Photoelectron Spectroscopy (XPS). We alter the composition of the oxide layer by in-situ annealing in O_2 and CO_2 atmospheres. O_2 annealing predominantly produces a Ba_xCu_yO layer, whereas CO_2 predominantly produces a $BaCO_3$ layer.

INTRODUCTION

Pulsed laser deposition has been successfully used as one of the few techniques to produce smooth, high T_c, high J_c, superconducting $YBa_2Cu_3O_7$ thin films [1,2]. The interest in these films is in their applications in microelectronics, interconnects, Josephson junctions. Therefore it is desirable that the thin films be in the '123' superconducting phase up to the top monolayer of the surface. Although a '123' film, well oriented in the c-direction on $SrTiO_3$, have been obtained by this method, the top 10 - 20 A layer of the 'as-deposited' 90K film is found to be nonstoichiometric '123' [3]. The top nonsuperconducting layer is found to be barium rich. It was of interest to investigate the surface of this film, and thus we have employed x-ray photoelectron spectroscopy to study the surface composition and surface reactivity in a hope to produce a superconducting phase as near the surface as possible.

Carbon is a common impurity found on these materials. While hydrocarbon impurity does not affect the electronic structure of oxygen, a $CO_3^=$ on the surface does change the shape of the O1s signal. In order to isolate the contribution of oxygen from the surface carbonate, we first removed the surface carbon by in situ O_2 annealing, and then reintroduced carbonate by heating the '123' film in CO_2. Here we report the results of an XPS study where we examined a '123' film, as-deposited, O_2 annealed, and CO_2 annealed.

EXPERIMENTAL

This film was deposited by a laser deposition technique as described in reference 1. The films studied here are ~3000A thick on $SrTiO_3$ (100) substrates and showed zero resistance at 85K. The sample is mounted onto a resistively heated platinum foil with an

TRANSFER SYSTEM

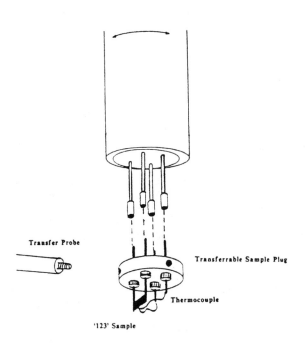

FIGURE 1. Schematic of the transferrable plug assembly with '123' film and thermocouple.

Alumel-Chromel thermocouple attached for temperature measurement. The YBaCuO film, the Pt foil, and the thermocouple, is mounted to a transferrable plug assembly with four electrical leads, two for resistive heating and two for the thermocouple, as illustrated in figure 1. This assembly can be transferred from the UHV analysis chamber to an attached reaction chamber, separated by a gate valve. The reaction chamber can be backfilled with gas, in this case O_2 and CO_2, to one atmosphere, evacuated, and the sample transferred back to the analysis chamber without breaking vacuum.

The surface analysis system is a KRATOS XSAM800 equipped with a dual anode (Mg/Al) X-ray source, and a single channel hemispherical electron energy analyzer. In this study, XPS spectra are obtained using Mg Kα radiation (1253.6 eV), with the analyzer operating in FAT mode (fixed analyzer transmission) and Hi magnification (2 x 2 mm^2 analysis area) due to the size of the substrate.

In all cases, high resolution spectra are obtained for the Cu2p, Ba3d, O1s, C1s, Y3d, and Ba4d signals. In the case of the O1s spectra, curve synthesis techniques are employed to determine the relative amount of the different chemical states of oxygen.

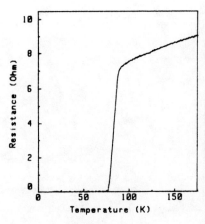

FIGURE 2. Resistance vs. T curve after CO_2 anneal.

The component peaks were combination Gaussian/Lorenzian peaks, with a FWHM of ~1.9 eV.

The YBaCuO thin film is analyzed under three conditions. It is first examined as prepared without any in-situ surface treatments. The sample is then transferred to the reaction chamber, backfilled with O_2 to one atmosphere in a continuous flow mode. The sample is heated to 400C and held there for 15 hrs. The sample is allowed to cool slowly to below 100C over a 3 hour period before the chamber is evacuated and the sample transferred back to the analysis position. The slow cooldown period is to allow for the tetragonal to orthorhombic phase transition. This procedure is repeated again and again until a carbon free surface is produced. This procedure is repeated once more except with CO_2 gas instead of O_2, to study the reactivity of '123' film with CO_2.

The sample was then removed from the vacuum system and its resistance measured. The resistance versus temperature curve is shown in figure 2. The film showed zero resistance below 77K.

RESULTS

Figure 3 compares the three Ba3d spectra. Clearly, the increasing intensity with O_2 and CO_2 annealing indicates that barium is segregating to form a Ba rich surface layer. Again it appears that CO_2 does not alter the Ba chemical state significantly. An important observation is the presence of a low BE state denoted 'B' in the 'as-deposited' film. This peak is attributed to Ba states in the bulk '123'[3]. The low KE of the Ba3d peaks results in a shallow escape depth (~22A), so that the signal from these states are attenuated by the Ba rich layer that is formed.

Figure 4 compares the Cu2p spectra of the '123' film for the three conditions described previously. There are some slight differences in the spectral shape between the 'as-deposited' case and the O_2 and CO_2 annealed case. The peak intensities do not vary significantly, indicating segregation or dissolution of copper does not occur. The O_2 annealing results in a decrease in the FWHM of the $Cu2p_{3/2}$ spectra suggesting the formation of a copper oxide species different than that found in the orthorhombic '123' lattice. Further, the relative intensity of the satellite peaks to the main peak increases indicating Cu^{2+} formation. However it appears that CO_2 annealing does not affect the chemical state of copper.

Figure 5 shows the C1s for the a) untreated, b) O_2 annealed, and c) CO_2 annealed YBaCuO film. In the case of the untreated surface, the C1s spectrum shows two forms of carbon. The large

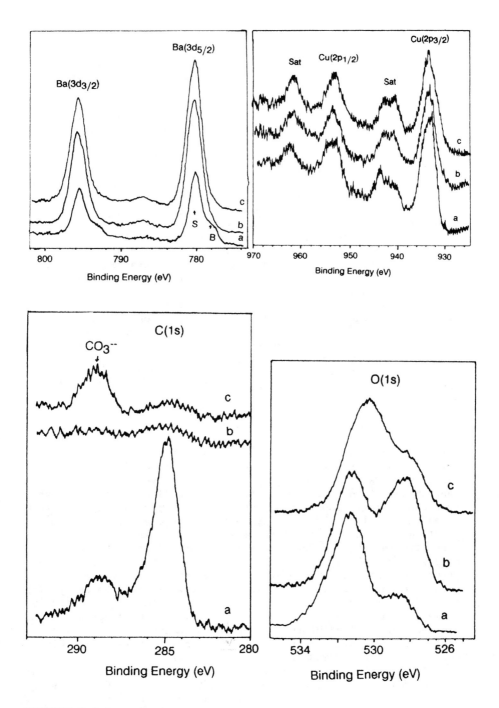

FIGURES 3, 4, 5, & 6. Ba3d, Cu2p, C1s, and O1s spectra respectively; a) as-deposited, b) O_2 annealed, and c) CO_2 annealed.

peak at 284.6 eV is the carbonaceous overlayer (graphite or pyrolitic carbon) typically found on most surfaces. Very often this carbon is removed by Ar^+ sputtering, but in this study, sputtering was purposely avoided so as not to alter the surface structure. This carbon is usually just an artefact and does not affect the chemistry of the surface. However the high BE peak (288 eV) is assigned to the presence of $CO_3^=$ on the surface. This form of carbon presents a problem since it is associated with the oxygen atoms in the carbonate state. These oxygens contribute significantly to the O1s signal making it difficult to interpret the O1s spectra accurately.

The O_2 anneal was effective in removing both forms of carbon as is evidenced by the C1s spectra (curve b). The corresponding O1s spectra is free of any oxygen signal originating from the carbonate compound. After CO_2 annealing (curve c), carbonate is reintroduced without reintroducing the carbonaceous overlayer. This reintroduces carbonate oxygen signal into the O1s spectrum. However by comparing the corresponding O1s spectra to that of the carbon free surface, it is possible to identify the carbonate oxygen contribution.

The most dramatic differences occur in the O1s spectra displayed in figure 6. Upon immediate inspection, it is obvious that there are two major chemical states, however upon closer inspection, more than two states can be observed. Curve fitting techniques are employed to determine the position and relative intensities of the various components. These are displayed in figure 7. A more in depth discussion is given in the following sections.

DISCUSSION

The '123' film was characterized by XPS for surface composition under three conditions, 'as-deposited' and after exposure to O_2 and to CO_2 atmospheres successively. Our initial hope in this investigation was to produce a '123' substrate with a superconducting phase as near the surface as possible by in-situ O_2 annealing. This would allow us to obtain accurate information about the electronic structure of $YBa_2Cu_3O_7$ by XPS, a surface sensitive technique. The most relevant and discernable information exists in the O1s spectrum. Theoretically, the orthorhombic '123' structure contains three distinctly different oxygen sites, populated in the ratio of 4:2:1, which we will refer to as type IV, II, and I oxygens respectively. However an oxide layer typically forms and this layer contributes oxygen signal which convolutes the O1s spectrum.. The information gathered in this investigation is useful in deconvoluting the O1s spectrum by determining the binding energy of oxides formed in the surface layer.

Initially the as-deposited film shows the presence of carbon in predominantly two chemical states (fig. 5a), a carbonaceous overlayer, on the order of 20A thick, and a second carbon form, presumably carbonate. The oxygen annealing is effective in removing this carbon, probably as the volatile product CO_2.

It is with the oxygen annealing that we were hopeful we would

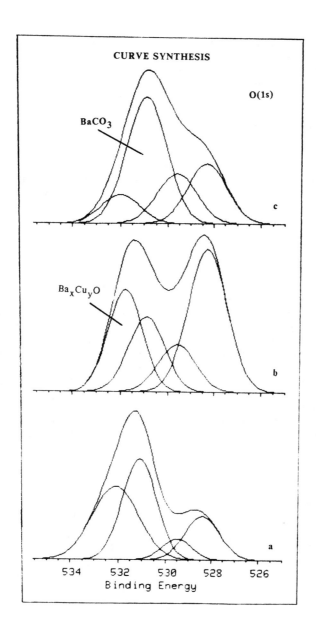

FIGURE 7. Curve fitting results of the O1s spectra displayed in figure 6. Curve b shows a large Ba_xCu_yO component, and curve c shows a large $BaCO_3$. The as-deposited O1s spectra (curve a) shows both Ba_xCu_yO and $BaCO_3$ components.

produce the superconducting phase in the near surface region. It turns out that the surface composition becomes enriched in barium and somewhat reduced in oxygen. Although RBS (Rutherford Backscattering Spectrometry) depth profiles show these films to be stoichiometric, it is possible that there is a slight excess in barium below the sensitivity of RBS. The extensive annealing allows segregation of excess barium to the surface forming a barium-copper-oxide layer. The change in the chemical state of copper is evidenced by the spectral changes shown in the Cu2p satellites. The bulk states denoted 'B' in figure 3a, are attenuated by this oxide layer. Subsequent annealing in CO_2 results in the formation of a $BaCO_3$ layer as is evidenced by the C1s peak at 288 eV and the increased oxygen signal. With this assumption, the O1s spectra were curve fitted and are displayed in figure 7.

The O1s spectra of the oxygen and carbon dioxide annealed samples were analyzed first so as to determine the binding energies of the oxide species in the overlayer. In the oxygen annealed sample, there are four components, the Ba_xCu_yO and three from the orthorhombic '123' phase. Referring to figure 7b, the high BE state at 532 eV is assigned to the Ba_xCu_yO layer. The three remaining peaks are assigned to the oxygen in the superconducting phase. When the surface is reduced in CO_2, CO_2 reacts with the BaO to form what is believed to be $BaCO_3$. Referring to figure 7c, this is evidenced by a reduction in intensity of the BaO component and increase in the component at 531 eV. It turns out that the type II oxygens have the same or nearly the same binding energy as oxygens of carbonate. Having identified some components of the nonstoichiometric layer, it is possible to curve fit the O1s spectrum of the as-deposited '123' surface (fig. 7a). As described

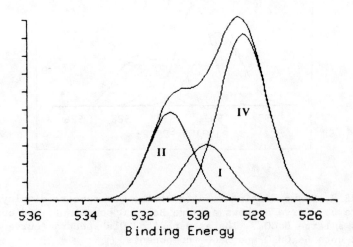

FIGURE 8. The resulting O1s spectra when the BaO component is subtracted from curve 7b.

before, there are five components, but only four peaks are used since the $BaCO_3$ oxygens fall in the same position as the type II oxygens.

To test how well our model describes the surface layer of the '123' film, we subtracted the BaO component from the O1s spectrum of the O_2 annealed sample. The resulting O1s spectrum (figure 8) approximates the theoretical O1s spectrum based on three different oxygen states in a ratio of 4:2:1. Similar spectra have been reported[4] of superconducting samples, prepared by pressing powder into a pellet and then mechanically scraping the surface in-situ until reproducible XPS spectra were obtained. Any discrepancies might be explained to some residual contamination in the chemically induced surface layer which is not easily identified. Such contamination might be native oxides of barium, copper, or yttrium, or even some carbonate. Curve 5b does show a trace of carbon residue. If this carbon is in the carbonate form, it is associated with three oxygen atoms. The total sum of these oxide contributions may be responsible for these differences.

SUMMARY

The surface of '123' films are characterized by XPS to determine the composition of a surface layer which is commonly present under typical processing conditions. The films are annealed in-situ under oxidizing and reducing conditions with oxygen and carbon dioxide respectively. These treatments alter the composition of the surface layer and these changes are identified by an in-depth analysis of the O1s spectrum. We are successful in identifying Ba_xCu_yO and $BaCO_3$, although it is very probable that native oxides of copper and yttrium do exist but are not easy to detect.

This work has been supported by the Center for Ceramics Research, Rutgers University, Piscataway, NJ.

REFERENCES

1. D. Dijkemp, T. Venketesan, X.D. Wu, S.A. Shaheen, N. Jiraswi, Y.H. Min-Lee, W.L. McLean, and M. Croft, Appl. Phys. Lett., 51, 619 (1987).

2. A. Inam, M.S. Hegde, X.D. Wu, T. Venketesan, P. England, P.F. Miceli, E.W. Chase, C.C. Chang, J.M. Tarasnen, and J.B. Wachtman, Appl. Phys. Lett., 53, 908 (1988).

3. X.D. Wu, A. Inam, M.S. Hegde, T. Venketesan, E.W. Chase, C.C. Chang, B. Wilkens, and J.M. Tarasnen, Phys. Rev. B (in press).

4. W. K. Foord, C. T. Chen, J. Anderson, J. Kwo, S. H. Liou, M. Hong, G. V. Rubenacker and J. E. Drumheller, Phys. Rev. B37, 7924(1988).

XPS DOS STUDIES OF OXYGEN-PLASMA TREATED $YBa_2Cu_3O_{7-\delta}$ SURFACES AS A FUNCTION OF TEMPERATURE.

T. Conard, J.M. Vohs, J.J. Pireaux and R. Caudano.
Facultés Universitaires Notre-Dame de la Paix,
Laboratoire LISE
rue de Bruxelles, 61, B-5000 NAMUR, Belgium.

ABSTRACT

Monochromatized AlK_α XPS has been used to monitor changes in the electronic structure of 123 sintered pellets as a function of temperature (90-650K). Since 123 surfaces are known to lose oxygen and react with water, a novel procedure was used to prepare the material surface that would be representative of the bulk material. In contrast with published results, we observed drastic DOS modification (i.e. the appearance of a new peak close to the Fermi level at low temperature).

INTRODUCTION

Since the discovery of high temperature superconductivity in the perovskites $La_{2-x}Sr_xCuO_4$ [1] and $YBa_2Cu_3O_{7-\delta}$ [2], there have been numerous experimental investigations of their physical properties, including a number of photoemission studies using ultraviolet, X-ray, or synchrotron radiation. This work is of great importance in the understanding of the temperature dependence in the electronic structure of these compounds.
It is now well known that the perovskite superconductors easily lose oxygen at their surfaces, even at room temperature. Several methods have been used to analyse compounds with a correct (i.e. $YBa_2Cu_3O_{7-\delta}$) oxygen composition such as annealing in oxygen [3] or scraping the superconductor at low temperature [4]. The present work used a novel procedure to prepare surfaces representative of the bulk oxygen stoichiometry by applying in situ an oxygen plasma discharge.
In this study, we have measured the density of states (DOS) of an YBaCuO sintered pellet in the valence band between the Fermi level and 50eV below it, over a wide temperature range (90 - 650 K). While doing this, we have also monitored the core levels of the oxygen, copper and barium atoms. In contrast with published results [4,5], we observed drastic changes in the DOS close to the Fermi level.

EXPERIMENTAL DETAILS

YBaCuO samples were powders prepared according to the original recipe [2]. For the photoemission experiments they were prepared as disks of approximately 10 mm in diameter and 1 mm thick. They were stored in air at room temperature.

The photoemission measurements were performed in a HP5950A spectrometer with hemispherical analyser. Monochromatic AlK_α X-rays (1486.6 eV) were used as exciting radiation and the base pressure in the analysis chamber fluctuated between 1.10^{-9} to 10^{-8} Torr depending on the sample temperature treatment. Indeed, the sample holder could be cooled with liquid nitrogen to a temperature of about 90 K and heated resistively to 650 K. The temperature was maintained in all this range by a thermoresistance temperature controller.

Initially, we recorded XPS spectra of the core levels without any sample treatment. Then the sample was brought out the ultravacuum chamber and severely scraped with a stainless steel blade. It was then rapidly reintroduced in the chamber and scraped again in vacuo with a diamond file. The aim of this surface treatment was to obtain a surface of the ceramic with a composition as close as possible to that of the bulk. The sample was then treated with an oxygen plasma discharge. Figure 1 shows the principle scheme of this discharge: A high purity silver ring was maintained at 500 V above the sample for twenty minutes in a presure of 10^{-2} mbar of oxygen. The discharge current was appproximately 0.7 mA : as a result, a blue glowing discharge bombarded the sample with ionized oxygen atoms. This procedure was used to restore the sample surface with the correct oxygen composition (i.e. $YBa_2Cu_3O_7$). After this surface treatment we began the temperature dependent analysis.

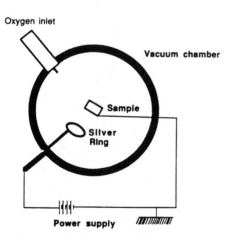

Fig. 1: Principle scheme of the oxygen plasma discharge

The core level spectra of all the constituent atoms exhibited more than one component. Curve fitting routines were used to distinguish them: a background proportional to the surface of the peaks, and gaussian-lorentzian mixed functions were used to reproduce the shape of the recorded spectra.

RESULTS AND DISCUSSION

CORE LEVEL SPECTRA

For all samples and stages of treatment, even after the severe scraping, significant contamination by carbon species was evident.There were two kinds of carbon species in the C1s spectra.One at approximately 289 eV which is at the same binding energy as the C1s peak for $BaCO_3$, one of the $YBa_2Cu_3O_7-\delta$ precursors.This peak could be due to unreacted precursor but is more probably due to reaction between the ceramic and some component of air like CO_2 [6,7]. The other component at 284.7 eV is attributed to hydrocarbon contamination and was used to calibrate all the spectra.In contrast with other results [8] we could not

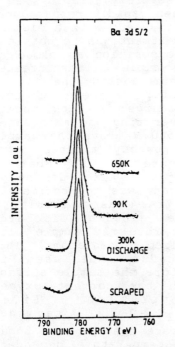

Fig. 2: Spectra of Ba 3d5/2 level: after scraping and after discharge in oxygen at room temperature, at 90K and 650K

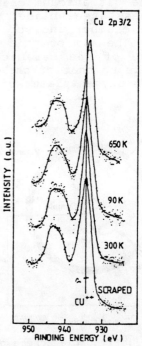

Fig. 3: Spectra of Cu 2p3/2 level: after scraping and after discharge in oxygen at room temperature, at 90K and 650K

eliminate this component after scraping: this peak is an indication of a large contamination in and around the micrograins in the ceramic.

The barium 3d5/2 spectra did not exhibit any change as a function of treatment procedure or sample temperature. Therefore, the barium peak was used as a reference in this work to calculate the atomic ratio of the other constituents. The Ba3d5/2 spectra are displayed in figure 2.

Figure 3 shows the spectra of the 3/2 component of the Cu2p doublet. These spectra exhibit a broad satellite structure near 940 eV typical of the Cu^{++} shake-up. No significant changes were observed in these spectra after scraping, oxygen plasma discharge and at low temperature. However, when the sample was heated, the Cu2p3/2 peak shifted to lower binding energy. The origin of this displacement can be attributed to reduction of copper from Cu^{++} to Cu^{+}. This could be clearly seen after computer curve resolution. This is also confirmed by the fact that the intensity ratio between the satellite peak and the main peak strongly decreased as the temperature was increased.

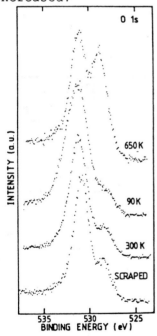

Fig. 4: Spectra of O 1s level: after scraping and after discharge in oxygen at room temperature, at 90K and 650K

Figure 4 displays the O1s spectra ; the calculated atomic ratios are listed in table I. After scraping, the amount of oxygen present at the surface decreased significantly, suggesting that these ceramics were highly contaminated in the surface region. After the O_2 discharge treatment there was no major modification in the room temperature spectra, however, the surface atomic concentrations were close to the bulk stoichiometry. At low temperature, the total amount of oxygen present at the surface increased significantly. This increase can be attributed to water contamination resulting from adsorption from the background gas. The component corresponding to O1s in water at about 533 eV increased and now represents a significant portion of the oxygen 1s signal. The high temperature spectra shows also significant modification that can be attributed to water desorption.

YBa$_2$Cu$_3$O$_{7-\delta}$ Sample treatment	Cu	O
untreated	4.3	14.1
scraped	4.2	6.5
discharge	2.7	7.0
low temp.	3.1	11.2
high temp.	1.8	4.4

Table I: Sample atomic ratios referred to barium. Note: the first two ratios for copper were obtained from a separate sample

VALENCE BAND SPECTRA

Figure 5 displays three valence band spectra recorded after the O$_2$ discharge at 300 K and subsequent cooling and heating to 90 and 650 K respectively. Table II lists the assignments of the main features of the room temperature spectra in accordance with recently published results [9]. In all these spectra, and even for other samples, we observed the presence of a W4f doublet at 36 and 38,5 eV which is likely due to bulk contamination resulting from the YBaCuO synthesis in a tungsten

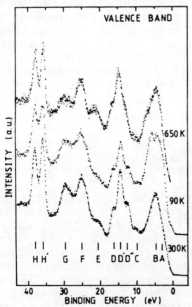

Fig. 5: Valence band spectra of oxygen plasma discharge pellet at room temperature, 90K and 650K

Structure	Origin
A	Cu3d-O2p
B	Hybrid state
C	O2p satellite
D	Ba3d1/2
D'	Ba3d1/2 Ba3d3/2
D''	Ba3d3/2
E	O2s
F	Y4p1/2 Y4p3/2
G	Ba5s
H	W4f5/2
H'	W4f7/2

Table II : Valence band structure attribution. (Adapted from ref 9)

crucible. No significant contribution of the W signal is expected at binding energies close to the Fermi level.

At low temperature (approximately 90 K), we observed drastic modification in the DOS (fig. 6). At 6 eV, a new strong feature (Y) is clearly apparent. This new peak can be explained if we recall that water adsorption is probably contaminating the sample surface at liquid nitrogen temperature. Another group [10] has reported similar DOS changes when they intentionally adsorbed H_2O on $YBa_2Cu_3O_7$. This interpretation is also consistent with changes we observed in the O1s spectra.

Another clearly visible change at low temperature is an increase in the DOS very near the Fermi level (structure X). This increase can not be explained by the presence of water/ice at the surface [10,11] as ice only contributes at higher binding energies and not at the Fermi level. This is the first time, to our knowledge, that such a temperature dependence in the DOS of YBaCuO has been observed by XPS. A similar modification in the DOS of the $Bi_4Ca_3Sr_3Cu_4O_{16+x}$ compound at low temperature has recently been reported [12] (fig.7) : it was measured by angle integrated ultraviolet photoemission and has been interpreted as the consequence of the opening of a superconductor gap. Another angle resolved UPS study [13] on $YBa_2Cu_3O_7$ also shows a modification in the DOS near the

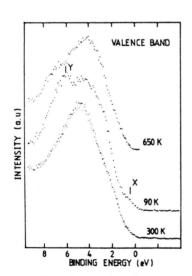

Fig. 6: Valence band spectra of oxygen plasma discharge pellets at room temperature, 90K, and 650K

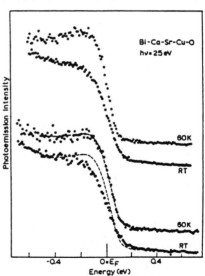

Fig. 7: Photoemission spectra taken at room temperature and at 60K for $Bi_4Ca_3Sr_3Cu_4O_{16+x}$ (ref 12)

Fermi level for certain direction of the Brillouin zone and for certain photon energy. In the present study, the fact that we are working in angle integrated mode allows us to exclude such an origin. This conclusion is reinforced by the fact that this modification is a temperature dependent structure since it does not appear in the room and high temperature spectra.

When the temperature of the superconductor was increased to 650 K, the structure at 6 eV is partially removed. This is interpreted as water desorption. However, there was still a strong shoulder at this energy, suggesting a reaction between this ceramic and the water previously adsorbed at the surface and formation of hydroxide species at the surface. This interpretation is consistent with the study of Kurtz et al.[7] which shows that another superconducting ceramic ($La_{2-x}Sr_xCuO_4$) is highly reactive with the different components of air (O_2, H_2O, CO_2, CO). The other modification at high temperature is the disappearance of stucture X at the Fermi edge.

CONCLUSION

$YBa_2Cu_3O_{7-\delta}$ samples whose surfaces has been treated in situ in an oxygen plasma discharge showed for the first time DOS modification close to the Fermi level, at liquid nitrogen temperature. This observation is consistent with the fact that the sample stoichiometry has been restored by the preparation process, and also probably with the fact that XPS, when induced by a monochromatized AlK_α source, is certainly less surface and grain boundary sensitive than previously published synchrotron photoemission results. As a consequence, we attribute the new DOS structure appearing at the Fermi level at 90K to an oxygen related band in the superconductor

REFERENCES

1. J.G. Bednorz, K.A. Müller, Z. Phys. B64, 189 (1986)
2. M.K. Wu, J.R. Ashburn, C.J. Torng, P.H. Hor, R.L. Meng, L. Gao, Z.J. Huang, Y.Q. Wang, C.W. Chu, Phys. Rev. Lett. 58, 908 (1987)
3. D. van der Marel, I. van Elst, G.A. Sawatzky, D.Heitman, to be published
4. A Samsavar, T.Miller, T.C.Chiang, B.G.Pazol, T.A.Friedmann, D.M. Ginsberg., Phys. Rev. B37, 5164 (1988)

5. P.D. Johnson, S.L. Qiu, L. Jiang, M.W. Ruckman, M. Strongin, S.L. Hulbert, R.F. Garrett, B. Sinkovic, N.V. Smith, R.J. Cava, C.S. Jee, D. Nichols, E. Kaczanowigz, R.E. Salomon, J.E. Crow. Phys. Rev. B35, 8811 (1987)
6. H. Fjellag, P. Karen, A. Kjehshur, J.K. Grepstad, Sol. Stat. Com, 64, 927 (1987)
7. R.L. Kurtz, R. Stockbauer, T.E. Madey, D. Mueller, A. Shih, L.T. Toth., Phys. Rev. B37, 7936 (1988)
8. D.C. Miller, D. Fowler, C.R. Brundle, W.Y. Lee, AIP Conference Proceedings n°165 "Thin Film Procesing and Characterization of High Tc Superconductors", Ed. J.M.E. Harper, R.J. Colton, and L.C. Feldman, page 336
9. P. Thiry, G. Rossi, Y. Petroff, A. Revcolevschi, J. Jegoudez., Europhys. lett. 5, 55 (1988)
10. S.L. Qiu, M.W. Runckman, N.B. Brookes, P.D. Johnson, J. Chen, C.L. Lin, M. Strongin, B. Sinkovic, J.E. Crow, C.S. Jee, Phys. Rev. B37, 3747 (1988)
11. N. Martenson, P.A. Malmquist, S. Svensson, E. Basilier, J.J. Pireaux, U. Gelius, K. Siegbahn., Nouveau Journal de Chimie, 1, 191 (1977)
12. Y. Chang, M. Tang, R. Zanoni, M. Onellion, R. Joynt, D.L. Huber, G. Margaritondo, P.A. Morris, W.A. Bonner, J.M. Tarascon, N.G. Stoffel, submitted to Phys. Rev. Lett
13. N.G. Stoffel., Y. Chang, M.K. Kelly, L. Dottl, M. Onellion, P.A. Morris, W.A. Bonner, G. Margaritondo, Phys. Rev. B37, 7952 (1988)

TEMPERATURE EFFECTS IN THE NEAR-E_F ELECTRONIC STRUCTURE OF $Bi_4Ca_3Sr_3Cu_4O_{16+x}$

Y. Chang,[a] N. G. Stoffel,[b] Ming Tang,[a] R. Zanoni,[a] M. Onellion,[a]
Robert Joynt,[a] D. L. Huber,[a] G. Margaritondo,[a] P. A. Morris,[b] W. A. Bonner[b]
and J. M. Tarascon,[b]
(a) University of Wisconsin-Madison, Stoughton, Wisconsin 53589-3097.
(b) Bellcore, Red Bank, New Jersey 07701

ABSTRACT

Since the discovery of superconductivity above liquid nitrogen temperature, photoemission spectroscopists have searched for changes in the electronic structure near the Fermi edge accompanying the superconducting transition. This extensive search was frustrated by the very small photoemission signal near E_F for materials in the $YBa_2Cu_3O_{7-x}$ family. We report the first successful observation of temperature effects in the near-edge photoemission spectra of a high-T_c superconductor, obtained for sintered and single-crystal specimens of $Bi_4Ca_3Sr_3Cu_4O_{16+x}$ ($T_c = 85$ K). The most important observed change is a decrease of the photoemission signal near E_F. The observed decrease is not explained by the normal temperature dependence of the Fermi-Dirac function, and it can be explained, in first approximation, by the opening of a gap 2Δ, with $\Delta = 20 - 40$ meV.

Since their discovery, high-temperature superconductors have been studied very extensively with photoemission techniques, which are excellent probes of the electronic structure.[1] Photoemission spectra yield excellent information on the chemical status of the elements in the material, and on many-body effects such as photoemission satellites and resonant phenomena. Furthermore, photoemission could directly contribute to the understanding of high-temperature superconductivity — since it should be possible to directly measure any changes in the electronic structure near the Fermi level, E_F, occurring during the superconducting transition. Such changes could be an excellent test of the different models of superconductivity, as it has been recently proposed, for example, by Tosatti and Rosei,[2] and by Huber.[3]

Until recently, two main obstacles appeared to block this line of research. First, one would expect the above changes to occur over an energy range whose magnitude is determined by the width of the superconductivity gap, 2Δ, which is a few meV for low-T_c superconductors, but should be tens of meV for higher T_c materials. In photoemission experiments, the resolution is typically worse than 100 meV, but can be improved at the expense of counting rate with high-resolution analyzers. This, however, conflicts with the need to collect data quickly to avoid problems with sample contamination and stoichiometry modifications. Second, no Fermi edge was visible in the spectra of $YBa_2Cu_3O_{7-x}$, casting serious doubts upon the very use of photoemission to study highly correlated systems such as the high-T_c materials.[1]

The second obstacle was removed early in 1988, when our group discovered that the photoemission spectra of the new high-T_c superconductor $Bi_4Ca_3Sr_3Cu_4O_{16+x}$ exhibited a clear Fermi edge.[4] This opened up the possibility of studying near-edge temperature effects. However, the reasons for the difference between $Bi_4Ca_3Sr_3Cu_4O_{16+x}$ and the older high-T_c compounds remain unclear. One can speculate that even high-quality cleaved surfaces of $YBa_2Cu_3O_{7-x}$ suffer from loss of oxygen, deviating from the stoichiometry required for high-T_c materials. Another possibility is that the relatively strong Fermi-edge signal for $Bi_4Ca_3Sr_3Cu_4O_{16+x}$ is related to the Bi-O planes.

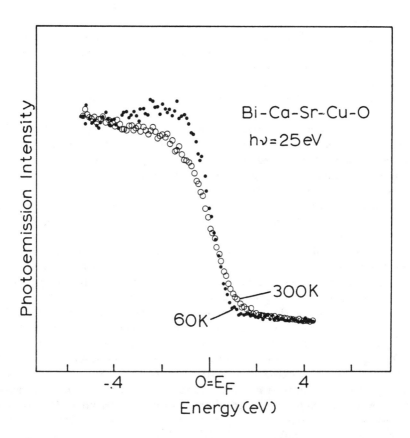

Fig. 1. Photoemission spectra taken at temperatures above and below T_c in the Fermi-edge region of the high-temperature superconductor $Bi_4Ca_3Sr_3Cu_4O_{16+x}$.

The problems associated with resolution are somewhat more subtle. First of all, it has been predicted[2,3] that different mechanisms of high-T_c superconductivity correspond to different features in the electronic structure over an energy range much more extended than 2Δ. Thus, 2Δ is not necessarily the right reference for the energy scale in all experiments. Even more important, however, is the fact that small *changes* in the spectra can be observed even with moderate resolution. This is the foundation of modulation spectroscopy, which can detect spectral changes occurring in an energy range smaller by orders of magnitude than the natural bandwidth of the corresponding feature.[5]

On the other hand, it is of course desirable to improve the resolution as much as possible, in order to extract the detailed spectral lineshape changes without excessive data processing. The improvements are limited by the quality of the instrumentation and, as already mentioned, by the need to avoid sample contamination. In these preliminary experiments, the resolution was not pushed to the limit, but it was sufficient to prove the existence of the temperature effects.

Figure 1 shows $Bi_4Ca_3Sr_3Cu_4O_{16+x}$ photoemission spectra taken at temperatures above and below T_c, on cleaved single-crystal specimens. The resolution (Gaussian FWHM) was 190 meV. The two spectra have been normalized to have equal intensity 0.5 eV below the Fermi level. This point was selected because it is much farther from the Fermi level than the expected gap width, and it corresponds to a relatively flat portion of the spectra, reflecting the plateau of the Fermi-Dirac distribution.

The normalization procedure is a delicate issue, since even minor modifications in the spectral intensity of the strongest valence band features can substantially affect the small temperature-related effects under investigation. We analyzed the effects of different normalization procedures, and found that they can indeed affect the apparent temperature induced changes in the region below the Fermi level. However, they do not qualitatively affect the most important effect, *i.e.*, the *decrease* of the signal at or above the Fermi level when the temperature decreases. Therefore, we conclude that this is a true temperature effect.

We compared the observed spectral changes when the temperature was lowered with the simple sharpening of the Fermi-Dirac function. After convolution with an instrumental Gaussian broadening function, we find that the sharpening cannot explain the observed changes. Specifically, the magnitude of the decrease on the right-hand side is smaller than experimentally observed, and it occurs at a higher energy. Thus, the signal decrease is not simply explained by the temperature dependence of the Fermi-Dirac distribution, and it suggests instead that the spectral edge shifts away from the Fermi edge as the temperature decreases. This is what one would expect from a simple BCS-like picture of the near-edge density of states. Can, however, a density-of-states picture describe photoemission from the superconducting state? This issue was explored by Huber, who developed a weak-coupling quasiparticle description of the process. Quite interestingly, with a constant density of states in the normal phase, the derived lineshape coincides with the BCS function, $|E|/(|E^2 - \Delta^2|)^{1/2}$, where E is the energy measured from the Fermi level, and a forbidden gap exists for $|E| < \Delta$. Thus, at least within the limits of this weak-coupling approach, it does not appear unreasonable to interpret photoemission spectral changes in terms of the opening of the superconducting gap, plus other distortions at lower energies.

A quantitative analysis of our data is made difficult by the limited resolution and by the problem of normalization. We analyzed the near-edge region using a Bayesian iteration to deconvolve the raw spectral data from the instrumental response. This approach is quite effective in deriving the *distance* in energy between the edges of two spectra. In the present case, extrapolation to the thermal-broadening limit indicates that Δ is in the 20-40 meV range. Values in this range also correspond the best description of the near-edge region at low temperature in terms of the BCS lineshape. A conservative estimate of the accuracy obtainable with the Bayesian deconvolution sets the limit values for Δ at 10 and 45 meV.

The observation of the above effects opens up exciting possibilities, but it also poses formidable challenges to theorists and experimentalists. Several theorists have suggested that a detailed study of these phenomena could discriminate, for example, between BCS-like and RVB models. Thus, photoemission could probe the very nature of high-temperature superconductivity.

On the other hand, several problems remain. For example, we must pinpoint the temperature at which the observed phenomena occur. We must explore ways to improve resolution, within the limits allowed by sample contamination. From a theoretical point of view, we must develop a description of the phenomena not relying on weak coupling, which appears inconsistent with high-temperature superconductivity and with the gap magnitude derived from our data. Note, for example, that our data suggest non-conservation of the area under the spectrum, inconsistent with weak coupling — although this effect is possibly affected by the normalization procedure. Because of this uncertainty, we cannot rule out conservation of the area. If the area is not conserved, then this could be qualitatively explained by an energy dependence of the parameter Δ, beyond the limits of a weak-coupling description. Ideally, we should be able to map where in k-space the electronic states are affected by the superconducting transition using angle-resolved photoemission.[6] Parallel experimental and theoretical studies are underway to attack these problems.

ACKNOWLEDGMENTS

This work was supported by the National Science Foundation, by the Office of Naval Research, and by the Wisconsin Alumni Research Foundation. We are grateful to Renzo Rosei and Erio Tosatti for the early disclosure of their results (Ref. 2). The experimental work was performed at the Wisconsin Synchrotron Radiation Center, a national facility supported by the National Science Foundation.

1. See, for example: *"Thin Film Processing and Characterization of High-Temperature Superconductors"*, J. A. Harper, R. J. Colton and L. C. Feldman Eds., AIP Conference Proc. **165** (1988).
2. E. Tosatti and R. Rosei, private communication.
3. D. L. Huber, submitted to Solid State Commun.
4. M. Onellion, Ming Tang, Y. Chang, G. Margaritondo, J. M. Tarascon, P. A. Morris, W. A. Bonner and N. G Stoffel, Phys. Rev. **B 38**, 881 (1988).
5. One of the many examples of this well-known class of experiments is described in U. M. Grassano, G. Margaritondo and R. Rosei, Phys. Rev. **B2**, 3319 (1970).
6. G. Margaritondo, C. M. Bertoni, J. H. Weaver, F. Levy, N. G. Stoffel and A. D. Katnani, Phys. Rev. **B23**, 3765 (1981); N. G. Stoffel, F. Levy, C. M. Bertoni and G. Margaritondo, Sol. State Commun. **41**, 53 (1982).

PHOTOEMISSION RESONANCE STUDY OF SINTERED AND SINGLE-CRYSTAL $Bi_4Ca_3Sr_3Cu_4O_{16+x}$

Ming Tang, Y. Chang, R. Zanoni, M. Onellion, Robert Joynt, D. L. Huber
and G. Margaritondo
University of Wisconsin-Madison, Stoughton, Wisconsin 53589-3097
and
P. A. Morris, W. A. Bonner, J. M. Tarascon and N. G. Stoffel
Bellcore, Red Bank, New Jersey 07701

ABSTRACT

We present soft-x-ray photoemission spectra that probe the valence and core electronic structure of the high-T_c superconductor $Bi_4Ca_3Sr_3Cu_4O_{16+x}$. The identification of spectral features was helped by the observation of the resonant behavior of a Cu-related satellite feature. The resonance occurs at photon energies near the Cu3p optical absorption edge, and affects a peak 12.5 eV below the Fermi edge. We identified this feature as a correlation satellite characteristic of Cu in the 2+ valence state. Other features observed in the spectra more than 7 eV below the Fermi edge are due to several different core levels. In particular, we observed a strong Bi5d doublet. Other core level peaks are due to the Sr4p and Ca3p orbitals, and to Bi, Sr and Ca s-orbitals. Within 7 eV of the Fermi edge, the spectra are dominated by valence states. The most important feature is the $Bi_4Ca_3Sr_3Cu_4O_{16+x}$ Fermi edge itself, which we observed for the first time on this, and whose existence was subsequently confirmed by several other groups. On the contrary, no edge was observed in the photoemission spectra of materials in the $YBa_2Cu_3O_{7-x}$ family. The observation of the Fermi edge has important implications for the theoretical interpretation of high-T_c superconductivity. Furthermore, it enabled us to see near-edge changes associated with the superconducting transition.

The electronic structure of the new high-temperature superconductor is a topic of very high interest in today's condensed matter science. Photoemission spectroscopy with synchrotron radiation provides an excellent tool for its study.[1] In particular, dramatic resonance effects in the dependence of photoemission cross sections on the photon energy can be used to extract crucial information on density of states features and on the chemical status of copper.[2] Recently, we extended this approach to the study of the new high-temperature superconductor $Bi_4Ca_3Sr_3Cu_4O_{16+x}$. The results revealed interesting similarities as well as differences with respect to other high-temperature superconductors such as $YBa_2Cu_3O_{7-x}$.[1,2]

Figure 1 illustrates the electronic structure of $Bi_4Ca_3Sr_3Cu_4O_{16+x}$, with the aid of a soft-x-ray photoemission spectrum. Before discussing in detail the different spectral features, we would like to call the reader's attention to the most relevant difference between the photoemission spectra of this material and those of $YBa_2Cu_3O_{7-x}$ and related compounds.[1,2] This is the presence, for the bismuth compound, of a clear Fermi edge.[3] On the contrary, no Fermi edge was observed for the Y compound.

The absence of a Fermi edge was attributed to a variety of possible causes, including, in particular, strong correlation effects.[1] This implied serious reservations about the use of photoemission to study high-temperature superconductors. The results of Fig. 1 show, reassuringly, that the absence of the Fermi edge is *not* an intrinsic feature of all high-temperature superconductors. Why, then, is the Fermi edge absent from the $YBa_2Cu_3O_{7-x}$ spectra while it is present in the Bi-compound spectra? This question is still the subject of a heated controversy.

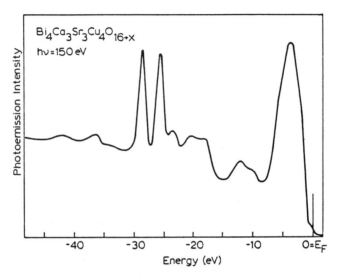

Fig. 1. Soft-x-ray synchrotron radiation photoemission spectrum of a sintered specimen of $Bi_4Ca_3Sr_3Cu_4O_{16+x}$, scraped in vacuum. The spectrum was taken with a photon energy of 150 eV. Note the Fermi edge.

We performed careful studies of the temperature dependence of the near-edge spectrum. These experiments demonstrated that a drop in temperature from room temperature to values below T_c does affect the near-edge lineshape. This phenomenon is discussed in detail in another paper of this conference.[4]

In order to identify Cu-related spectral features, we searched for photoemission resonances at the Cu3p threshold. As we can see in the single-crystal spectra of Fig. 2, resonant behavior is observed for, and only for, the peak 12.5 eV below E_F. This feature corresponds to the well-known Cu satellite, which has been observed for many Cu compounds including previous high-temperature superconductors. The nature of this peak and its resonance mechanism are explained in detail in Ref. 5. The important point is that the position in energy of this resonant feature is a fingerprint for the oxidation state of Cu.[2] In the present case, the position rules out monovalent copper, and is consistent with divalent copper.

Once established the position of the Cu satellite peak, we can proceed with the identification of the other features. The first four features below E_F clearly belong to the valence band. The next peak, 10.5-11.3 eV below E_F, is close to the expected[6] energy for Bi6s electrons, and therefore attributed to them. After the Cu satellite, the 17.7-17.9 eV feature remains unexplained. It was initially attributed to Sr4p electrons.[3] However, the 20.7 eV feature is a better Sr4p candidate because of its position in energy.[6] The 17.7-17.9 eV feature is attributed to O2s electrons.[6]

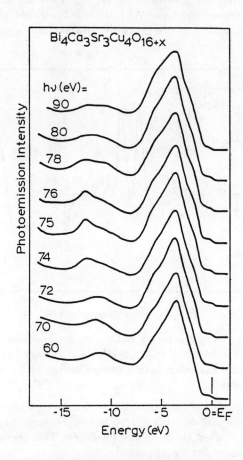

Fig. 2. Photoemission spectra taken at different photon energies on cleaved single-crystal $Bi_4Ca_3Sr_3Cu_4O_{16+x}$. The size of the cleaved surface was approximately 3×3 mm^2. Note the resonant behavior of the peak 12.5 eV below the Fermi energy, for photon energies close to the Cu3p optical absorption threshold.

Next, we see in the spectra three peaks which are due to the Ca3p and Bi5d spin-orbit doublets.[6] Specifically, the 25.7 and 28.8 eV peaks are primarily due to the Bi doublet. The 23.4 eV peak is due to Ca3p$_{3/2}$, and the Ca3p$_{1/2}$ contribution overlaps the Bi5d$_{5/2}$ peak. Note that, even after correcting for the weaker Ca3p contribution, the intensity ratio between the two Bi5d components is anomalous with respect to the ratio of the statistical factors. Effects of this kind are discussed, for example, in Ref. 6. The nature of the 34.6 eV peak is not clear. The peaks at 36.4 and 42 eV are attributed to Sr4s and Ca3s electrons. The identification of the spectral features is summarized in Table I.

In summary, the soft-x-ray photoemission spectra of $Bi_4Ca_3Sr_3Cu_4O_{16+x}$, and their analysis based on resonant cross-section effects, provide a detailed picture of the electronic structure in a wide energy region. The most relevant elements of the structure are the Fermi edge, and the resonant Cu satellite, which is consistent with a divalent state for this element. Strong temperature effects have been observed only in the Fermi-edge region.

Table I

Energy Referred to the Fermi Edge (eV)	Attribution
1.6, 3.6, 5.2	valence electrons
10.5-11.3	Bi6s
12.5	Cu satellite
17.7-17.9	O2s
20.7	Sr4p
23.4	Ca3p
25.7	Ca3p, Bi5d
28.8	Bi5d
36.4	Sr4s
42	Ca3s

ACKNOWLEDGMENTS

This work was supported by the National Science Foundation, by the Office of Naval Research, and by the Wisconsin Alumni Research Foundation. The photoemission spectra were taken at the Wisconsin Synchrotron Radiation Center, a national facility supported by the National Science Foundation.

1. See, for example: *"Thin Film Processing and Characterization of High-Temperature Superconductors"*, J. A. Harper, R. J. Colton and L. C. Feldman Eds., AIP Conference Proc. **165** (1988).
2. See, for example: M. Onellion, Y. Chang, D. W. Niles, Robert Joynt, G. Margaritondo, N. G. Stoffel and J. M. Tarascon, Phys. Rev. **B36**, 819 (1987), and: Ming Tang, N. G. Stoffel, Qi Biao Chen, David LaGraffe, P. A. Morris, W. A. Bonner, G. Margaritondo and M. Onellion, Phys. Rev. B **38**, 897 (1988).
3. M. Onellion, Ming Tang, Y. Chang, G. Margaritondo, J. M. Tarascon, P. A. Morris, W. A. Bonner and N. G Stoffel, Phys. Rev. B **38**, 881 (1988).
4. N. G. Stoffel, Y. Chang, Ming Tang, R. Zanoni, M. Onellion, Robert Joynt, D. L. Huber, G. Margaritondo, P. A. Morris, W. A. Bonner and J. M. Tarascon, these proceedings.
5. See: R. L. Kurtz, in Ref. 1.
6. G. Margaritondo, *"Introduction to Synchrotron Radiation"* (Oxford, New York,

1988). In particular, the identification of the 17.7-17.9 feature as due to O2s is supported by the fact that the signal near the Fermi edge exhibits a resonant enhancement at photon energies between 17 and 18 eV, and no resonance near 20.7 eV.

ELECTRON ENERGY LOSS STUDIES OF THE NEW HIGH-T_c SUPERCONDUCTOR $Bi_4Ca_3Sr_3Cu_4O_{16+x}$

Y. Chang, R. Zanoni, M. Onellion, G. Margaritondo
University of Wisconsin-Madison, Stoughton, Wisconsin 53589-3097
ans
P. A. Morris, W. A. Bonner, J. M. Tarascon and N. G. Stoffel
Bellcore, Red Bank, New Jersey 07701

ABSTRACT

Electron energy loss spectroscopy has been extensively used to investigate the electronic properties of high-T_c superconductors. In particular, we identified surface and bulk plasmon modes of 1-2-3 materials, and a number of peaks related to quasi-optical transitions. Together with photoemission spectra illustrating the initial states, these peaks provide information on the empty electronic states of these materials. We recently extended these studies to sintered and single crystal specimens of the new high-T_c superconductor $Bi_4Ca_3Sr_3Cu_4O_{16+x}$. There is an overall correlation between the energy loss spectral features and the photoemission features of this material. Specifically, we can explain almost all of the energy loss features by assuming that they are due to quasi-optical transitions to two groups of empty states, 0.5-1.1 eV and 1.5-2 eV above the Fermi edge. This analysis does not explain all the peaks, and in particular those related to collective electron modes. The identification of the bulk plasmon peak is complicated by the interference with features related to quasi-optical transitions involving the Bi5d doublet. We tentatively identify as bulk plasmon a peak at 27.3 eV, an energy similar to that of the bulk plasmon mode in 1-2-3 compounds. The corresponding best candidate for surface plasmon is at 15 eV, and it exhibits a dependence on the primary electron beam energy consistent with this identification.

Since the discovery of high-temperature superconductivity, many experiments have explored the occupied electronic states of the new material with photoemission spectroscopy.[1] Comparatively, only a very few experiments have explored the unoccuppied electronic states, with techniques such as inverse photoemission (or bremsstrahlung isochromat spectroscopy).[2] In the case of $YBa_2Cu_3O_{7-x}$, we successfully used low-resolution electron energy loss spectroscopy (EELS) as a simple and versatile probe to gain information not only on the unoccuppied electronic states, but also the collective electronic excitations.[3] These are very interesting topics, since they have been proposed as essential factors in some of the models of high-temperature superconductivity.

In this article, we extend our EELS studies to the new material $Bi_4Ca_3Sr_3Cu_4O_{16+x}$. The sample growth procedure and the technique to cleave single crystals under ultrahigh vacuum conditions has been described in Ref. 4. The base pressure during these experiments was in the low-10^{-10} Torr range, and the sample was kept at room temperature. The primary beam was provided by an electron gun collinear to the double-pass cylindrical mirror electron energy analyzer, used to study the energy distribution of the inelastically reflected electrons. We used primary energies in the range 100-400 eV. Table I presents a summary of the photoemission results from Ref. 5.

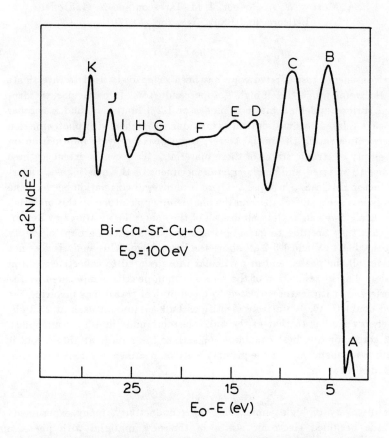

Fig. 1. Electron energy loss spectrum shown as a negative second derivative plot. The spectrum was taken on a cleaved sample of $Bi_4Ca_3Sr_3Cu_4O_{16+x}$ at room temperature, with a primary energy $E_o = 100$ eV.

Figure 1 shows the overall spectrum produced by a cleaved surface with a primary beam energy of 100 eV. The spectrum is shown in the conventional negative-second-derivative plot, to emphasize the fine structure. We can observe in Fig. 1 eleven features, labeled with the letters A-K. The energy losses corresponding to these features are 3.0, 5.5, 9.0, 12.8, 15.0, 18.4, 22.3, 24.3, 26.1, 27.3 and 29.3 eV. The typical accuracy in determining these losses is ±0.2 eV. Figure 2 shows EELS curves taken on the same cleaved surface for different values of the primary beam energy. Spectra taken on sintered pellets scraped under vacuum, not shown here, exhibit the same features with different relative intensities.

Ordinarily, the starting point in the analysis of EELS data like those of Figs. 1 and 2 is the identification of the plasmon peaks. This procedure was indeed used in the case of $YBa_2Cu_3O_{7-x}$. In the present case, however, we encounter a serious difficulty. The expected position for the bulk plasmon peak is in the energy loss region of peaks I, J and K. This region is dominated by losses due to quasi-optical transitions starting from the Bi5d doublet. These losses are quite intense, and make it difficult to identify the bulk plasmon peak.

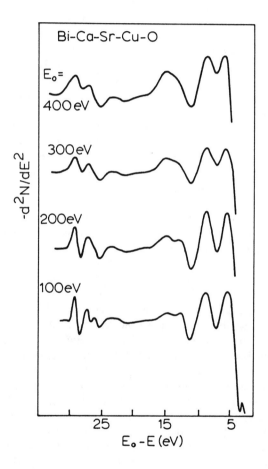

Fig. 2. $Bi_4Ca_3Sr_3Cu_4O_{16+x}$ electron energy loss spectra taken on cleaved specimens with different primary energies.

Based on the positions in energy of peaks I, J and K, and on the measured spin-orbit splitting of Bi5d from photoemission data, 3.2 eV,[5] we propose that peaks I and K are related to Bi5d transitions. In this scheme, peak J is not explained by such transitions, and it becomes a likely candidate as bulk plasmon. Note that the interference between the plasmon and the Bi5d transitions in the second-derivative curve makes it virtually impossible to estimate the intensity of each single feature. In particular, it is impossible to test the dependence of the plasmon peak on the primary beam energy, because this peak cannot be resolved from the nearest Bi5d feature at large primary energies.

The identification of peaks I and K in terms of Bi5d transitions implies that the final state for these transitions is 0.5-0.7 eV above the Fermi level. In fact, the measured photoemission binding energies for the Bi5d doublet are 25.6 and 28.8 eV.[5] Quite interestingly, the same approach can be used to explain almost all of the remaining EELS peaks, in terms of quasi-optical transitions, assuming only two final-state energies.

Specifically, a group of peaks can be explained using final-state energies equal or close to those of peaks I and J, 0.5-1.1 eV above the Fermi level — and initial-state energies derived from photoemission spectra.[5] Of this group, peak A appears related to a valence-band photoemission feature 1.9 eV below the Fermi level, E_F, and the final energy is approximately 1.1 eV. Peaks F and H are attributed to transitions from the photoemission peaks 17.9 and 23.4 eV below E_F, to energies 0.5 and 0.9 eV above E_F. These photoemission peaks are attributed O2s and Ca3p electrons (see discussion in Ref. 5).

A second group of EELS features can be explained in terms of quasi-optical transitions with final energies 1.5-2 eV above E_F. These are peaks B, D and G, related to the photoemissison peaks 3.6, 11.3 and 20.7 eV below E_F.[5] These photoemission peaks are attributed to valence, Bi6s and Sr4p electrons. The corresponding final energies are all in the range 1.5-2.0 eV above E_F.

This analysis leaves two features unexplained, peaks C and E. We propose that the latter is due to the excitation of surface plasmons. The hypothesis is based on the relative position of this peak and of the bulk plasmon loss. Although the ratio between these two losses is difficult to estimate *a priori*, we do not expect very large deviations from the $(2)^{1/2}$ value derived from simplistic boundary conditions for the dielectric function. In the present case, the ratio of the bulk and surface plasmon losses is 1.82, within the range of values found for metals and semiconductors. We observed that the intensity of peak E depends on the quality of the surface for sintered specimens, and this corroborates its attribution to the surface plasmon. Also note the dependence of this peak on the primary beam energy in Fig. 2. Specifically, both the bulk and surface plasmon features increase wiith primary beam energy, consistent with earlier results[3] on $YBa_2Cu_3O_{7-x}$.

The nature of the intense peak C is not clarified by our analysis. The position indicates that the peak is related to quasi-optical transitions involving valence states.[5] However, the final energy for the transition does not fall in the 0.5-1.1 eV and 1.5-2.0 eV ranges used for other peaks. We also note that the analysis in terms of these two final energy ranges does not guarantee that some of the peaks are not due to transitions to other final energies — it is merely an explanation based on the minimum possible number of hypotheses. However, the identification of the bulk plasmon peak appears quite reliable, based on the analysis of the possible contributions from Bi5d transitions.

Recently, we detected a significant temperature dependence of the near-edge photoemission lineshape.[6] Similar phenomena in the EELS curves would be quite interesting. A search for such phenomena is currently underway, using high-resolution EELS.

ACKNOWLEDGMENTS

This work was supported by the National Science Foundation, by the Office of Naval Research, and by the Wisconsin Alumni Research Foundation.

1. See, for example: *"Thin Film Processing and Characterization of High-Temperature Superconductors"*, J. A. Harper, R. J. Colton and L. C. Feldman Eds., AIP Conference Proc. **165** (1988).
2. H. M. Meyer III, Y. Gao, T. J. Wagener, D. M. Hill, J. H. Weaver, B. K. Flandermeyer and D. W. Capone II, in Ref. 1, p. 254.
3. Y. Chang, M. Onellion, D. W. Niles, Robert Joynt, G. Margaritondo, N. G. Stoffel and J. M. Tarascon, Solid-State Commun. **63**, 717 (1987).
4. M. Onellion, Ming Tang, Y. Chang, G. Margaritondo, J. M. Tarascon, P. A. Morris, W. A. Bonner and N. G Stoffel, Phys. Rev. **B 38**, 881 (1988).
5. Ming Tang, Y. Chang, R. Zanoni, M. Onellion, Robert Joynt, D. L. Huber, G. Margaritondo, P. A. Morris, W. A. Bonner, J. M. Tarascon and N. G. Stoffel, these proceedings.
6. Y. Chang, N. G. Stoffel, Ming Tang, R. Zanoni, M. Onellion, Robert Joynt, D. L. Huber, G. Margaritondo, P. A. Morris, W. A. Bonner and J. M. Tarascon, these proceedings.

OXYGEN AND COPPER VALENCIES IN OXYGEN-DOPED SUPERCONDUCTING $La_2CuO_{4.13}$

N. D. Shinn, J. W. Rogers, Jr., J. E. Schirber,
E. L. Venturini, D. S. Ginley and B. Morosin
Sandia National Laboratories, Albuquerque, NM 87185

ABSTRACT

Oxygen enrichment of La_2CuO_4 by a high-temperature and high-pressure (1-3 kbar at 875K) treatment produces a stable, bulk superconductor, $La_2CuO_{4.13}$ ($T_c \approx 30K$). Annealing in vacuum above 350K depletes the excess oxygen from $La_2CuO_{4.13}$ as $O_2(g)$, eliminates the superconductivity, and results in semiconducting La_2CuO_4. These annealed samples were subsequently recharged with oxygen, thereby reversibly restoring the bulk superconductivity. X-ray photoelectron spectroscopy has been used to identify the oxygen and copper valencies which correlate with the bulk superconductivity in this interstitially-oxygen-doped perovskite material. An XPS O(1s) binding energy of 532.1 eV is measured for the excess oxygen which, with additional analytical and quantitative magnetization data, indicates that the predominant excess oxygen species is a superoxide ion (O_2^-). The angular dependence of the O(1s) peak intensity indicates a near-surface concentration enrichment of the superoxide species. The majority copper valency is Cu(II) with a Cu(2p) binding energy of 933.0 ± 0.1 eV; however, evidence for a minority Cu(III) species with a Cu(2p) binding energy of 934.4 ± 0.2 eV is found only in the superconducting, oxygen-enriched material.

INTRODUCTION

The valency of oxygen and copper ions in the perovskite-based superconductors is one aspect of the continuing controversy regarding the electronic structure of these exciting new materials. This issue is compounded by the difficulty in obtaining unambiguous experimental data; multi-phase and carbonate contamination problems[1] and surface oxygen instabilities[2,3] can result in spectroscopic features which are erroneously attributed to the superconducting material. However, unlike the substitutionally-doped $La_{2-x}(Ba,Sr)_xCuO_4$ and 1-2-3 materials, the superconductivity in $La_2CuO_{4.13}$ is completely reversible and depends solely on the presence and extent of oxygen enrichment[4-7]. The control experiment on the reference material, namely La_2CuO_4, is performed in situ simply by thermally depleting the $La_2CuO_{4.13}$ of excess oxygen (T > 350K) to regenerate the antiferromagnetic, semiconducting La_2CuO_4 parent material[8]. Thus difference spectra can be used to identify features which are present only when the sample is superconducting.

Quantitative analysis[7] of the oxygen content in the enriched, superconducting material by weight gain and oxidation titer analysis lead to a discrepancy if the excess oxygen is assumed to be in the O^{2-} valency as are the lattice oxygen species. (The oxidation titer is the mole equivalent oxidized or reduced by a transfer of a unit quantity of charge, (e.g., $Cu^{2+} \rightarrow Cu^{3+} + e^-$ or 1/2 $O_2 + 2e^- \rightarrow O^{2-}$), and reflects the change in the concentration of more/less oxidizing species in the material.) Because of the complexity of the redox chemistry in the $La_2CuO_{4.13}$ ceramic, these two measurements of oxygen concentration cannot be compared directly to assign definitively a charge-to-mass ratio to the excess oxygen species. Magnetic susceptibility measurements[7,9] of $La_2CuO_{4.13}$ in its normal state (T > 30K) indicate the presence of isolated paramagnetic spins; assuming a spin of 1/2 and a g-factor of 2, a quantitative analysis yields 0.09 ± 0.04 spins/molecule of lanthanum cuprate. This implies a minimum of one spin per excess O_2 molecule and a maximum of one spin per excess O atom. These observations can be understood and the discrepancy among the quantitative measurements can be resolved by the proposition[7] that a large fraction of the excess oxygen in the superconductor is incorporated interstitially into the La_2CuO_4 lattice in a superoxide state (O_2^-). If all of the excess oxygen is in one chemical state, this is the only explanation that satisfactorily accounts for all of the available data.

EXPERIMENTAL

The sample preparation and characterization for these X-ray photoelectron spectroscopic (XPS) experiments has been described previously[4,7,8,10]. Both ceramic and single crystals of La_2CuO_4 were oxygen enriched by 1-3 kbar of oxygen at 875K for 12-48 hours followed by slow cooling (100K/hour to 300K). If the stoichiometry of the semiconducting reference material is defined to be $La_2CuO_{4.00}$, then the stoichiometry of the oxygen-enriched material is found to be $La_2CuO_{4.13 \pm 0.02}$. The definition of the exact stoichiometry for the unenriched material is not important since the photoelectron spectroscopic studies are concerned with identifying the chemically-inequivalent copper and oxygen species in the superconducting phase. All oxygen-enriched samples were checked with a SQUID magnetometer to verify bulk superconductivity (~30%).

The superconducting $La_2CuO_{4.13}$ and reference La_2CuO_4 samples were introduced into the VG ESCALAB 5 surface spectrometer through a load-lock system into an ultra-high vacuum (UHV) preparatory chamber (10^{-8} torr) followed by rapid transfer in UHV to the analysis chamber (2×10^{-10} torr). XPS data were obtained with non-monochromatic Mg K(α) X-ray excitation (12 kV, 20 mA) and the hemispherical electron energy analyzer was operated at a constant pass energy of 20 eV. The combined spectrometer resolution and photon bandwidth resulted in a working resolution of 0.9 eV. Binding energies were referenced to the Fermi level of gold [$Au(4f)_{7/2}$ = 83.8 eV] and were reproducible to ± 0.1 eV. Reference photoelectron spectra of each sample obtained immediately after introduction showed only La, Cu, O, and trace surface carbon contamination in the near-surface region probed by XPS (analysis depth[11] of 4-6 nm). Data were obtained at polar angles of 15° and 60° away from the sample normal to distinguish between intrinsic (bulk) and surface contamination contributions to the data.

RESULTS AND DISCUSSION

Photoemission spectra of the O(1s), Cu(2p) and La(3d) levels were recorded for all samples of La_2CuO_4 and $La_2CuO_{4.13}$ immediately after introduction into the analysis chamber, at several times thereafter during annealing to elevated temperatures, after cooling to 300K, and finally after subsequently dosing purified water onto the cool (300K) sample. The last step enabled the O(1s) XPS binding energies for adsorbed OH and H_2O on La_2CuO_4 to be measured directly, thereby identifying this contribution to the O(1s) lineshape for as-received samples.

Figure 1(a) shows representative O(1s) XPS data for the oxygen-enriched sample as received. In addition to the contributions from the intrinsic lattice oxygen (two inequivalent lattice sites[10]) and the excess oxygen in the enriched samples, the surface has adsorbed water and/or hydroxyl species which have O(1s) binding energies[12] in the range of 534-535 eV, resulting in the broad peak extending from 529 eV < E_B < 534 eV. Annealing to 400K is sufficient to desorb these surface contaminants, as indicated by the loss of the high binding energy shoulders and subsequent narrowing of the O(1s) peak shown in fig. 1(b). For the oxygen enriched samples, sporatic transient bursts of oxygen (p → 10^{-8} torr) were recorded by a UTI quadrupole mass spectrometer when the enriched $La_2CuO_{4.13}$ samples were heated to 350K, even before the surface contaminants were completely desorbed[8]. Above 400K, only mass 32 and 16 were observed. No pressure bursts were detected (p > 10^{-10} torr) for the blank La_2CuO_4 samples when heated to 400K and subsequently to ~700K. Thus the changes after annealing to 400K shown in fig. 1(b) are due to the removal of both surface contaminants and the onset of excess oxygen evolution from the superconductor.

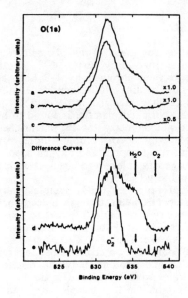

Fig. 1: XPS O(1s) spectra of oxygen-enriched $La_2CuO_{4.13}$

(a) as-received,
(b) after annealing to 400K,
(c) after 24 hours at 650K.

XPS difference curves showing

(d) the water and excess oxygen signal (a-c),
(e) only the excess oxygen signal (b-c)

Spectra of comparable peak shape to fig. 1(b) were recorded periodically during the first ten hours of an extended 675K annealing period. After 24 hours, the excess oxygen was completely depleted from the enriched samples as evidenced by the absence of further oxygen bursts. Fig. 1(c) shows an O(1s) spectrum after a 24 hour/675K annealing period and cooling to 300K. Whereas the XPS spectrum for the blank sample[10] (not shown) is unchanged with annealing once the surface water is desorbed, this is not true for the superconducting sample where both the peak shape and total integrated intensity vary during the 675K annealing.

The XPS data presented above can be used to identify the O(1s) XPS component that is directly correlated with the superconductivity. In addition to the qualitative identification of these species, the angular dependence of the XPS peak component intensities is used to derive quantitative estimates of their concentrations in the near-surface region (4-6 nm) sampled by XPS, subject to assumptions of XPS cross-section invariance. Figure 1(d) shows the difference curve obtained, after proper background subtraction and normalization[8,10], by subtracting the lattice oxygen signal, fig. 1(c), from the as-received superconductor O(1s) spectrum, fig. 1(a). Contributions from the excess oxygen and surface water are evident. The signal from only the excess oxygen in $La_2CuO_{4.13}$ is shown in fig. 1(e), obtained by subtracting fig. 1(c) from fig. 1(b).

A comparison of the binding energy measured for the excess oxygen peak (532.1 eV) and published binding energies for various oxygen species shows excellent agreement with a superoxide species (~532 eV)[13] and clearly rules out any mixture involving neutral O_2. The O(1s) binding energy for $H_2O(g)$ and $O_2(g)$ are 540.0 ± 0.2 eV and 544.0 ± 0.5 eV, respectively[14,15]. When adsorbed on $La_2CuO_{4.13}$, the measured binding energy for the H_2O transition is 535.4 ± 0.1 eV;

neutral O_2 would be expected to appear ~4 eV higher in binding energy. The data in fig. 1 show clearly that no intensity is observed in this binding energy region. In addition, neutral oxygen was not observed under any conditions throughout these experiments. Hence the presence of neutral O_2 can be completely excluded from further consideration.

The reported range[13,16] of O(1s) binding energies for a peroxide ion $(O_2)^{2-}$, 529-533 eV, suggests that if present in the $La_2CuO_{4.13}$ samples, the peroxide ion O(1s) intensity would overlap that of the O^{2-}. In order to test for this possibility, the low binding energy edges (below 529.5 eV) of all spectra - regardless of oxygen enrichment or annealing - were compared and found to be directly superimposible after only multiplicative scaling. This result demonstrates that (1) only the stable O^{2-} contributes to the O(1s) lineshape below 529.5 eV, (2) if the enriched samples contain peroxide ion, then its O(1s) binding energy must be greater than 530 eV, and (3) multiplicative scaling of fig. 1(c) to match any O(1s) lineshape below 529.5 eV properly accounts for the O^{2-} contribution. Previous work has shown that in order for the peroxide ion to be present in the superconductor in any appreciable amount, neutral di-oxygen must be present to account for the paramagnetism. The definitive absence of neutral O_2 by XPS completely rules out this alternative explanation.

A detailed analysis of all reasonable oxygen species other than the superoxide ion has been presented elsewhere[10]; the superoxide ion is most plausible assignment and, alone, accounts for all of the experimental observations. The relatively low temperatures (T > 350K) at which the excess oxygen rapidly evolves from the enriched samples strongly supports an interstitial, not substitutional, location for the excess oxygen; i.e., the excess oxygen could not be replacing oxygen vacancies in the lanthanum cuprate. A weak interaction with the lattice is consistent with both the extreme conditions required to incorporate the oxygen into the bulk and the ease with which it is liberated upon annealing. Although not known at present, the superoxide ions are postulated[7] to be centered at lattice positions equivalent to (1/4 1/4 1/4), between rectangles formed by two lattice O^{2-} ions and two La^{2+} ions located above or below the CuO planes. Four such positions are indicated in fig. 2 by ovals. These sites are large enough from ionic radii considerations and the observed c-axis expansion[4] to accommodate up to four superoxide molecular ions per La_2CuO_4 unit cell.

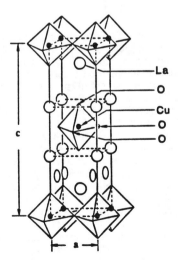

Fig. 2. Model for the location of the interstitial superoxide ions, indicated by ovals, in $La_2CuO_{4.13}$.

A quantitative estimate of the superoxide concentration in the outermost 4-6 nm of the sample may be obtained from the ratio of the O(1s) peak areas in figs. 1(e) and 1(c); i.e., the superoxide intensity compared to the lattice oxygen intensity. This indicates that approximately 40% of the total O(1s) signal is attributable to the superoxide chemical state. Assuming that the XPS cross-sections are comparable for lattice and superoxide oxygen, this implies that the near- surface region contains ~1.5 superoxide molecules per La_2CuO_4 unit. Recalling that the bulk averaged stoichiometry is 0.13 excess atoms or 0.065 excess molecules per La_2CuO_4, this result represents a considerable enrichment of superoxide ions near the surface of the samples. This suggests that a superoxide concentration gradient exists near the surface, which is not surprising considering the high oxygen pressures required to force the excess oxygen into the lanthanum cuprate. The majority of the superconducting phase has a much lower superoxide concentration of ~1/14 O_2^- per unit cell. We note that the superoxide-rich near-surface region (4-6 nm depth), although metallic[10], is probably not superconducting since the hole concentration greatly exceeds that of the superconducting phases[17] of the Sr and Ba doped La_2CuO_4.

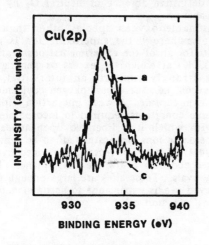

Fig. 3: XPS Cu(2p) spectra of oxygen-enriched $La_2Cu_{4.13}$

(a) as-received, and
(b) after annealing to 650K for 8 hours, (dashed curve)
(c) the XPS difference curve showing the 5-10% trivalent copper signal in the superconductor. (shaded region)

Figures 3(a) and 3(b) show typical Cu(2p) spectra for the superconducting (oxygen-enriched) and semiconducting (oxygen-depleted) ceramic samples, respectively. The signal-to-noise ratio of these spectra is lower than that of the previously shown O(1s) data as a result of shorter data acquisition times. This was done to prevent cumulative X-ray irradiation damage in these Cu(2p) data. In all cases using both ceramic and single-crystal $La_2CuO_{4.13}$ samples, the Cu(2p) spectra of the annealed sample was consistently narrower and showed reduced intensity only on the high binding energy side. Our data show that while the majority (>90%) of the near-surface copper ions are divalent, with the characteristic binding energy of 933.0 eV and the pronounced satellite structure[18], additional Cu(2p) signal is found reproducibly in the superconducting materials. After removing the Mg K(α) X-ray satellite intensity from these spectra and subtracting the conventional Shirley background, the difference curve shown in figure 3(c) is obtained. This additional intensity at 934.4 ± 0.2 eV is in good agreement with literature values for the Cu^{3+} ion in reference compounds[19] and other superconductors[18]. Because the gentle annealing of the oxygen-enriched lanthanum cuprate does not affect the copper or lanthanum concentrations, the minority Cu^{3+} concentration must be correlated with the presence of the interstitial superoxide ions. These XPS experiments are the first to provide definitive evidence that a higher oxidation state of copper is correlated with the superconducting state in a perovskite-based material, although its role in the superconductivity is not certain. This remains a controversial issue, particularly in other superconductors which can be plagued by contamination problems[20-23].

CONCLUSIONS

X-ray photoelectron spectroscopic and annealing studies of semiconducting La_2CuO_4 and superconducting $La_2CuO_{4.13}$ have shown that the excess oxygen incorporated into the bulk by high pressure treatment is chemically inequivalent to either of the intrinsic lattice oxygen species. The XPS O(1s) peak intensity due to to the excess oxygen is centered at E_B = 532 eV, in excellent agreement with published binding energies for a superoxide (O_2^-) species. At no time is neutral oxygen observed in any samples. This invalidates the hypothesis that a mixture of peroxide plus neutral di-oxygen accounts for the weight gain, paramagnetism and superconductivity. A quantitative estimate of the superoxide concentration using the angular dependence of the O(1s) intensity indicates a near-surface enrichment corresponding to ~1.5 superoxide ions, on average, per unit cell.

The Cu(2p) data for the blank La_2CuO_4 is in accord with copper ions in a Cu^{2+} valency. In addition, clear evidence for additional copper ions in a (formal) trivalent state, Cu^{3+}, is found in only the enriched samples and only concurrently with the superoxide species. Preliminary data on single crystal $La_2CuO_{4.13}$ samples confirms these observations.

ACKNOWLEDGEMENTS

This work was supported by the United States Department of Energy, Office of Basic Energy Sciences, under contract DE-AC04-76DP00789 with Sandia National Laboratories. The technical assistance of M. Mitchell, D. S. Blair, D. L. Overmyer and G. L. Fowler is appreciated.

REFERENCES

1. D. C. Miller, D. E. Fowler, C. R. Brundle and W. Y. Lee, in "Thin Film Processing and Characterization of High-Temperature Superconductors," American Vacuum Society Series 3, AIP Conf. Proc. 165, eds. J. M. E. Harper, R. J. Colton, L. C. Feldman (AIP, New York, 1988) p. 336.

2. R. S. List, A. J. Arko, Z. Fisk, S-W. Cheong, S. D. Conradson, J. D. Thompson, C. B. Pierce, D. E. Peterson, R. J. Bartlett, N. D. Shinn, J. E. Schirber, B. W. Veal, A. P. Paulikas and J. C. Campuzano, Phys. Rev. B (Rapid Communications), in press.

3. R. S. List, A. J. Arko, Z. Fisk, S-W. Cheong, J. D. Thompson, J. A. O'Rourke, C. G. Olson, A-B. Yang, T-W. Pi, J. E. Schirber and N. D. Shinn, Phys. Rev. Lett., submitted.

4. J. E. Schirber, E. L. Venturini, B. Morosin, J. F. Kwak, D. S. Ginley and R. J. Baughman, in "High-Temperature Superconductors", Mat. Res. Soc. Proc. Vol. 99, edited by M. B. Brodsky, R. C. Dynes, K. Kitazawa and H. L. Tuller (Materials Research Society, Pittsburgh, PA, 1988), p. 479.

5. J. Beille, B. Chevalier, G. Demazeau, F. Deslandes, J. Etourneau, O. Laborde, C. Michel, P. Lejay, J. Provost, B. Raveau, A. Sulpice, J. L. Tholence and R. Tournier, Physica 146B, 307 (1987).

6. J. E. Schirber, J. F. Kwak, E. L. Venturini, B. Morosin, D. S. Ginley, W. S. Fu and R. J. Baughman, Proc. 18th Inter. Conf. on Low Temp. Physics, Aug. 1987, in press.

7. J. E. Schirber, B. Morosin, R. M. Merrill, P. F. Hlava, E. L. Venturini, J. F. Kwak, P. J. Nigrey, R. J. Baughman and D. S. Ginley, Physica C152, 121 (1988).

8. J. W. Rogers, Jr., N. D. Shinn, J. E. Schirber, E. L. Venturini, D. S. Ginley and B. Morosin, Phys. Rev. B38, 5021 (1988).

9. E. L. Venturini, J. E. Schirber, B. Morosin, D. S. Ginley and J. F. Kwak, Mat. Res. Soc. Proc., to be published.

10. N. D. Shinn, J. W. Rogers, Jr., J. E. Schirber, Phys. Rev. B (submitted).

11. M. P. Seah and W. A. Dench, Surf. Interface Anal. 1, 2 (1979).

12. J. Fuggle, L. M. Watson, D. J. Fabian, and S. Affrossman, Surf. Sci. 49, 61 (1975).

13. C. T. Campbell, Surf. Sci. 173, L641 (1986).

14. D. W. Davis, J. M. Hollander, D. A. Shirley and T. D. Thomas, J. Chem. Phys. 52, 3295 (1970).

15. K. Seigbahn, C. Nordling, G. Johansson, J. Hedman, P. F. Heden, K. Hamrin, U. Gelius, T. Bergmark, L. O. Werme, R. Manne and Y. Baer, ESCA Applied to Free Molecules, (North-Holland, Amsterdam, 1969).

16. K. Prabhakaran and C. N. R. Rao, Surf. Sci. 186, L575 (1987).

17. J. B. Torrance, Y. Tokura, A. I. Nazzal, A. Bezinge, T. C. Huang and S. S. P. Parkin, Phys. Rev. Lett. 61, 1127 (1988).

18. P. Steiner, V. Kinsinger, I. Sander, B. Siegwart, S. Hufner, C. Politis, R. Hoppe and H. P. Muller, Z. Phys. B; Condensed Matter 67, 497 (1987).

19. F. Garcia-Alvarado, E. Moran, M. Vallet, J. M. Gonzalez-Calbet, M. A. Alario, M. T. Perez-Fr'as, J. L. Vincent, S. Ferrer, E. Garcia-Michel and M. C. Asensio, Solid State Commun. 63, 507 (1987).

20. P. Steiner, R. Courths, V. Kinsinger, I. Sander, B. Siegwart, S. Hufner and C. Politis, Appl. Phys. A44, 75 (1987).

21. P. Steiner, J. Albers, V. Kinsinger, I. Sander, B. Siegwart, S. Hufner and C. Politis, Z. Phys. B; Condensed Matter 66, 275 (1987).

22. N. Nucker, J. Fink, B. Renker, D. Ewert, C. Politis, P. J. W. Weijs and J. C. Fuggle, Z. Phys. B67, 9 (1987).

23. B. Viswanathan, S. Madhavan, C. S. Swamy, Phys. Status Solidi B133, 629 (1986).

AES AND EELS ANALYSIS OF TlBaCaCuO$_x$ THIN FILMS AT 300K AND AT 100K

A.J. Nelson, A. Swartzlander and L.L. Kazmerski
Solar Energy Research Institute, Golden, CO 80401

J.H. Kang, R.T. Kampwirth and K.E. Gray
Argonne National Laboratory, Argonne, Illinois 60439

ABSTRACT

Auger electron spectroscopy line-shape analysis of the Tl(NOO), Ba(MNN), Ca(LMM), Cu(LMM) and O(KLL) peaks has been performed in conjunction with electron energy loss spectroscopy (EELS) on magnetron sputter deposited TlBaCaCuO$_x$ thin films exhibiting a superconducting onset at 110K with zero resistance at 96K. AES and EELS analyses were performed at 300K and at 100K. Changes in the Auger line shapes and in the EELS spectra as the temperature is lowered below the critical point are related to changes in the electronic structure of states in the valence band (VB). Bulk and surface plasmon peaks are identified in the EELS spectra along with features due to core level transitions. Electron beam and ion beam induced effects are also addressed.

INTRODUCTION

The recent empirical discovery of superconductivity above 100K in the Tl-Ba-Ca-Cu-O system[1] has once again stimulated the synapses of the high-T$_c$ superconductor community. The fact that all of the recent high-T$_c$ materials research has been empirical in nature points to a clear need for experimental results which may help define the superconducting mechanism relevent to these new materials. Since Auger electron spectroscopy (AES) is sensitive to the variation of the local atomic charge density across the VB the technique is useful in characterizing states found near the VB maxima. Similarly, electron energy loss spectroscopy (EELS) stimulates transitions from core levels to empty states above or near the VB maxima.

In this paper, we report observed changes in the Tl(NOO), Ba(MNN), Ca(LMM), Cu(LMM) and O(KLL) Auger line shapes as well as observed changes in the EELS spectra for a TlBaCaCuO$_x$ film on yttrium-stabilized ZrO$_2$ after it was cooled to 100K. The observed changes are related to changes in the electronic structure of states in the VB as the material passes through its critical transition temperature.

EXPERIMENTAL

The TlBaCaCuO$_x$ films were prepared[2] by using a three-gun dc magnetron sputtering system equipped with a turbomolecular pump which provided a typical base pressure in the low 10^{-8} torr range. The three dc magnetron sputtering guns are aimed at a common point about 15 cm above the sources providing compositional uniformity to ±1% over a 2 cm^2 substrate area. Targets of Tl, Cu and a 1:1 BaCa mixture were simultaneously sputtered in a 20 mtorr argon atmosphere with an oxygen partial pressure of ≈0.1 mtorr being introduced directly adjacent to the substrate. A quartz crystal monitor is placed next to the substrate to determine the sputtering rates of each source prior to starting a deposition. The best films were deposited onto (100) oriented

© 1989 American Institute of Physics

single crystal or polycrystalline ZrO_2-9%Y_2O_3 substrates maintained at ambient temperature during deposition. Ex-situ post-annealing treatment was performed in a flowing oxygen atmosphere. In order to avoid the loss of the highly volatile Tl during the annealing process, the films were placed in a closed Au crucible, then placed in a flowing oxygen tube furnance and annealed at 850C for about 5 minutes.

Auger and EELS analysis were performed on a Perkin-Elmer/Physical Electronics Model 600 Scanning Auger Microprobe (SAM) system having a base pressure of 1×10^{-10} Torr. AES data was obtained with a primary electron beam energy of 5 keV and a current of 100 nA at an energy resolution of 0.2%. Primary electron beam energies of 300 eV and 600 eV were used for the EELS analysis (0.2% energy resolution) in order to distinguish between surface and bulk effects. Ion beam sputter etching was performed with a differentially pumped ion gun operating with a 3 kV Ar^+ ion beam (10^{-2} Pa Ar pressure) rastered over a 1.0×1.0 mm^2 area. Samples were cooled in vacuum to 100K using a LN_2 dewar equipped with a copper cold finger. The AES and EELS data were both recorded in N(E) (i.e. counts vs. energy) mode with the EELS data being displayed as $-d^2N/dE^2$. Quantitative compositional analysis was performed in a scanning electron microscope (SEM) using energy dispersive x-ray (EDX) analysis.

RESULTS AND DISCUSSION

Fig. 1 shows the variation of resistivity versus temperature, as measured by the standard four-probe technique, for a film on yttrium-stabilized ZrO_2. Result of the quantitative EDX compositional analysis indicated a film composition of $Tl_2Ba_2Ca_{1.3}Cu_2O_x$ which is in reasonable agreement with the 2212 structure. A single superconducting transition is observed which begins at about 110K and shows zero

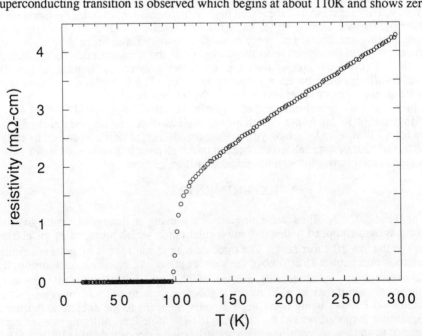

Fig. 1 Resistivity vs. temperature for $Tl_2Ba_2Ca_{1.3}Cu_2O_x$ thin film on ZrO_2:Y

271

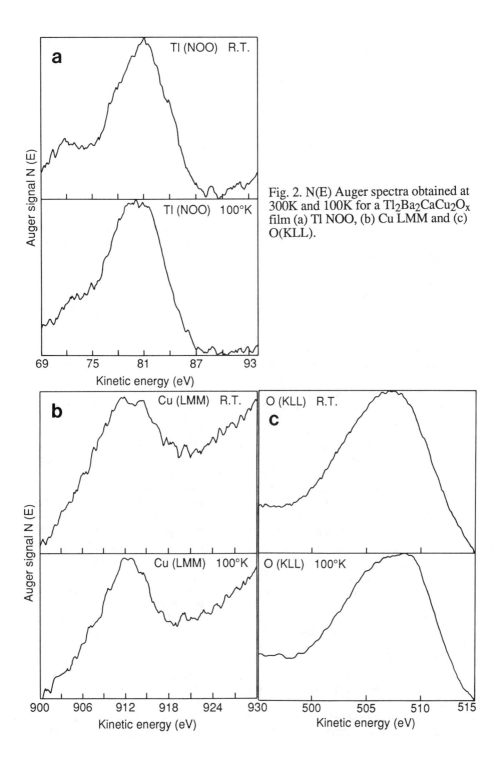

Fig. 2. N(E) Auger spectra obtained at 300K and 100K for a $Tl_2Ba_2CaCu_2O_x$ film (a) Tl NOO, (b) Cu LMM and (c) O(KLL).

resistance at $T_c(0)=96K$. The low temperature AES and EELS measurements were obtained at 100K (the limit of the apparatus) which is between the superconducting onset at 110K and $T_c(0)$ for this film.

Auger results for the Tl(NOO), Cu(LMM) and O(KLL) line shapes obtained at 300K and 100K are presented in Fig. 2a, b and c, respectively. The line shape and peak energy of these lines are strongly influenced by the chemical environment since the Auger electron emission involves valence electrons and the core level binding energy. Line shape analysis of the Ba(MNN) and Ca(LMM) Auger lines revealed no pertinent chemical or electron beam induced effects and thus are not included in this presentation. However, the change in the Tl(NOO) Auger line pictured in Fig. 2a as the film is cooled should be noted. Specifically, the 300K spectra centered at 81.0 eV with a small shoulder at 78.3 eV broadens and develops an additional peak at 79.3 eV as the temperature is lowered to 100K. This additional feature in the Tl(NOO) peak is possibly due to preferential electron beam assisted hydration/carbonation of Tl at the surface of the superconductor since the sample will act as a "cryogenic pump" for H_2O and CO at these lower temperatures or may be due to reordering of the Tl-O layer[3] as determined by neutron scattering. Additional evidence for one of these processes is seen in the O(KLL) spectra pictured in Fig. 2c. The 300K O $K_1L_{2,3}L_{2,3}$ spectra is a broad peak centered at 507.3 eV. The width of this peak indicates multiple states probably due to a continuum of holes in the K band and crystal-field effects. Upon cooling, the O $K_1L_{2,3}L_{2,3}$ peak broadens and develops two distinct features at 506.0 eV and at 508.6 eV. The additional feature at lower kinetic energy may again be due to the presence of an OH molecule on the surface or to the aforementioned reordering of the Tl-O layer which correlates with the probable causes of the change observed in the Tl spectra. Features representative of intrinsic physical effects (e.g., structural modifications of the square-planar CuO_4 clusters or of the Tl-O layers) as the material is cooled cannot presently be separated from extrinsic chemical effects and thus no definitive conclusions can be drawn from the Tl or O data concerning the occurance of this reordering phenomena.

The Cu $L_3M_{4,5}M_{4,5}$ measured at 300K and at 100K is presented in Fig. 2b. In the L_3VV Auger transition, a valence band (VB) electron fills a previously created core (L_3) hole. The excess energy causes ionization of a second valence band electron which is the measured Auger electron. The energy distribution of this Auger electron yields information about the VB density-of-states (DOS) spatially localized around the atom containing the core hole. The 300K spectra is rather broad and is composed of two main features evident at 911.4 eV and at 914.3 eV. The initial states of the Cu L_3VV Auger process are formed from the $3d^9$ and $3d^{10}\underline{L}$ states [$\underline{L}$ designates a ligand (O 2p) hole] of divalent copper with the final states formed by the $3d^7$ and the $3d^8\underline{L}$ multiplets with the $3d^8\underline{L}$ states dominate.[4-9] The width of the Cu L_3VV line is probably the result of the continuum of holes in the L band and crystal-field effects. As the sample is cooled to 100K, the Cu L_3VV line narrows and is centered at 912.2 eV. This change (narrowing) in the Auger line shape with decreasing temperature is distinct and opposite to the results for the Tl and O Auger lines and consequently is interpreted as possibly being due to a structural modification of the square-planar CuO_4 clusters[10] associated with a change in oxidation state (Cu^{+2}, Cu^{+1}) leading to a different hole-hole correlation energy.

Fig. 3a and b presents the EELS results obtained at 300K and 100K with primary beam energies of 300 eV and 600 eV, respectively. The 300 eV EELS spectra is more representative of the surface states while the 600 eV EELS spectra is more representative of bulk states. Also included in this figure is the EELS spectra at 100K after argon sputtering to remove adsorbed surface molecules due to the low tempera-

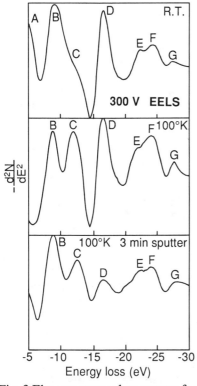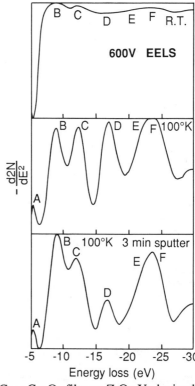

Fig. 3 Electron energy loss spectra for $Tl_2Ba_2Ca_{1.3}Cu_2O_x$ film on ZrO_2:Y obtained at 300K and 100K (a) 300 eV primary beam energy, (b) 600 eV primary beam energy.

Table I Electron energy loss features for $Tl_2Ba_2Ca_{1.3}Cu_2O_x$

E_p(eV)	T	\multicolumn{7}{c}{Loss Energies (eV)}
		A	B	C	D	E	F	G
300	300K	5.1	9.1	–	16.7	22.6	24.5	27.9
	100K	–	8.7	11.9	16.6	22.3	24.2	27.8
	100K (3 min. sputter)	–	8.6	12.5	16.5	22.5	24.1	28.0
600	300K	–	9.1	12.2	(18.1)	–	23.6	–
	100K	5.3	8.8	12.2	16.9	–	23.6	–
	100K (3 min. sputter)	5.3	8.9	11.9	16.8	–	23.6	–

tures. AES results on a sputtered sample of TlBaCaCuO$_x$ showed no significant change in the surface composition as long as the material was maintained at 100K. However, loss of Tl was evident if the sample was sputtered at 300K.

Table I summaries the measured loss energies, with respect to the elastic peak, of the seven observed features as labelled on the EELS spectra of Fig. 3. The low intensity of the 600 eV EELS spectra taken at 300K is due to instrumental effects and is not representative of any intrinsic physical phenomena. Interpretation of the peaks in the EELS spectra for $Tl_2Ba_2CaCu_2O_x$ is partially based on photoemission spectroscopy[4-9] and energy loss[11-13] results for $YBa_2Cu_3O_{7-x}$ superconductors. Utilizing these previous results, one may also infer the presence of two maxima in the density of states above the Fermi level (E_F) for the Tl-Ba-Ca-Cu-O system. The previously described empty-state level 2.3-2.5 eV above E_F, attributed to antibonding Cu 3d electrons, along with the other unoccupied state 4.3-4.6 eV above E_F determine the allowed transitions to be used for the interpretation of the observed spectral features. Based on these suppositions and the assignments found in the literature, feature A is interpreted as being due to a transition from the O $2p_{xy}$ eigenstate to the unoccupied Cu 3d antibonding state. This feature is not visible in the 300 eV spectra taken at 100K. Since the 300 eV spectra is more sensitive to surface overlayers, one concludes that the Cu 3d antibonding state is smeared out by the aforementioned surface contamination accumulated during cryogenic cooling. The fact that this feature is not observed in the 600 eV spectra taken at 300K can be accounted for by assuming that this unoccupied state for the bulk material is closer to the Fermi level at this temperature and/or has not fully developed. Therefore, the shift in energy and/or development of states in this band upon cooling could also be indicative of a structural modification of the square-planar CuO_4 clusters.

Further support of this interpretation may be evident from the energy shift upon cooling of feature B, identified as a transition from the occupied bonding Cu 3d level to an unoccupied state 4.3-4.6 eV above E_F. The peak intensity of this feature in the 600 eV spectra greatly increases as well upon cooling. Rearrangements in the steric configuration of the CuO_4 clusters would induce small observable energy shifts of states comprising the VB and thus would offer one explanation of the observed energy shifts in the EELS spectra.

Feature C is assigned to a transition between the O $2p_z$ eigenstate and the unoccupied Cu 3d antibonding state. The peak intensity of this feature in both the 300 eV and 600 eV spectra increases upon cooling, showing that more electrons are allowed to make this transition to the more developed Cu 3d antibonding band above the Fermi level. Feature D has previously been interpreted as a surface plasmon state, but is probably due to a transition from a Ba 5p level associated with $Ba(OH)_3$ since it is greatly diminished when the material is sputtered. Feature F is interpreted as a bulk plasmon state and exhibits no energy shift upon cooling. Features E and G are only resolvable in the 300 eV spectra with their intensities decreasing upon sputtering and thus are also believed to be associated with surface overlayers.

CONCLUSIONS

Auger line shape analysis and EELS analysis have been used to characterize the VB DOS of a $Tl_2Ba_2Ca_{1.3}Cu_2O_x$ thin film superconductor at 300K and at 100K. Changes in the Tl NOO Auger line shape as the film is cooled to 100K have been interpreted as being due to either hydration/carbonation of Tl at the surface or to reordering of the Tl-O layer. Changes in the Cu L_3VV Auger line shape as the film is cooled to 100K have been interpreted as being due to a structural modification of the square-planar CuO_4

clusters associated with a change in Cu oxidation state (Cu^{+2}, Cu^{+1}). EELS results give evidence which tends to support this interpretation.

Future work will include the construction of an apparatus to provide continuous cooling of a sample through its critical transition temperature and temperature cycling during AES and EELS analysis. Also, this study will be performed on Tl superconductor films exhibiting higher T_c's.

ACKNOWLEDGEMENTS

The work at SERI was supported by the US Department of Energy under Contract No. DE-AC02-83CH10093 and the work at Argonne was supported by the Department of Energy, BES-Materials Sciences, under Contract No. W-31-109-ENG-38.

REFERENCES

1. Z.Z. Sheng and A.M. Hermann, Nature **332**, 138 (1988).

2. J.H. Kang, R.T. Kampwirth and K.E. Gray, Phys. Lett. **A131**, 208 (1988).

3. W. Emowski, B.H. Toby, T. Egami, M.A. Subramanian, J. Gopalakrishnan and A.W. Sleight, submitted to Phys. Rev. B.

4. H. Ihara, M. Hirabayashi, N. Terada, Y. Kimura, K. Senzaki, M. Akimoto, K. Bushida, F. Kawashima and R. Uzuka, Japan J. Appl. Phys. **26**, L460 (1987)

5. Z. Iqbal, E. Leone, R. Chin, A.J. Signorelli, A. Bose and H. Eckhardt, J. Mater. Res. **2**, 768 (1987)

6. A. Balzarotti, M. De Crescenzi, C. Giovannella, R. Messi, N. Motta, F. Patella and A. Sgarlata, Phys. Rev. **B36**, 8285 (1987)

7. J.C. Fuggle, P.J.W. Weijs, R. Schoorl, G.A. Sawatzky, J. Fink, N. Nucker, P.J. Durham and W.M. Temmerman, Phys. Rev. **B37**, 123 (1988)

8. D.E. Ramaker, AIP Conference Proceedings No. 165, 284 (1988)

9. D. van der Marel, J. van Elp, G.A. Sawatzky and D. Heitmann, Phys. Rev. **B37**, 5136 (1988)

10. D.D. Sarma, Phys. Rev. **B37**, 7948 (1988)

11. Y. Chang, M. Onellion, D.W. Niles, R. Joynt, G. Margaritondo, N.G. Stoffel and J.M. Tarascon, Solid State Commun. **63**, 717 (1987)

12. M. Onellion, Y. Chang, M. Tang, R. Joynt, E.E. Hellstrom, M. Daeumling, J. Seuntjens, D. Hampshire, D.C. Larbalestier, G. Margaritondo, N.G. Stoffel and J.M. Tarascon, AIP Conference Proceedings No. 165, 240 (1988)

13. K. Jacobi, D.D. Sarma, P. Geng, C.T. Simmons and G. Kaindl, Phys. Rev. **B38**, 863 (1988)

RESONANT PHOTOEMISSION AND CHEMISORPTION STUDIES OF Tl-Ba-Ca-Cu-O

Roger L. Stockbauer, Steven W. Robey, and Richard L. Kurtz
Surface Science Division
National Institute of Standards and Technology*
Gaithersburg, MD 20899

D. Mueller, A. Shih, A.K. Singh, L. Toth and M. Osofsky
Naval Research Laboratory
Washington, DC 20375

ABSTRACT

Resonant photoemission has been used to study the electronic states and electron-electron interactions in a bulk sample of Tl-Ba-Ca-Cu-O high temperature superconductor. The electronic structure, i.e., broad peaks in the valence band at 3.2 and 5.5 eV and satellites at 10 and 13 eV binding energy, is similar to that observed for the 1-2-3 materials. This indicates that the electronic states and interactions are similar in these different classes of superconductor.

The surface reactivity has been probed with photoemission using controlled exposures of atmospheric gases. Both H_2O and CO_2 react with a sticking probability of near 0.2, forming hydroxide, and carbonate species, respectively. O_2 is non-reacting while CO reacts only slightly.

INTRODUCTION

Photoelectron spectroscopy has proved to be an invaluable tool in studying electronic interactions in high temperature superconductors.[1-2] Studies have typically concentrated on both the valence band and shallow core region[3-5] probed with photon energies between 20 and 100 eV or on the Cu and O core levels with x-rays.[6-7] Cross section measurements have shown that the CuO valence band is highly hybridized indicating a large amount of covalency in the bonding.[1] Resonant photoemission shows two satellite features which resonate at the Cu 3p excitation energy and a single satellite which resonates at the O 2s level.[8] From the binding energy and peak position of the resonant intensity of the Cu satellite one can infer that the Cu is primarily in a 2+ charge state, a fact verified by x-ray photoemission.[6] The resonant satellites also indicate a large hole-hole repulsion energy on the order of 6 eV for the Cu 3d holes[7] and XPS and Auger measurements indicate that it is even higher for the O 2p holes.[9]

Most of the early photoemission measurements on high temperature superconductors were made on scraped samples of bulk 1-2-3 material. Later measurements on single crystal samples have verified the earlier results. Recent results by List, et al.,[10] obtained by fracturing single crystals at low temperature (20 K), however, show major differences with these results. They observe a Fermi edge, a valence band sharply peaked to lower binding energy and much weaker or non-existent satellite peaks. Upon warming their samples to even

* Formerly the National Bureau of Standards

80 K, they observe an irreversible transformation to the spectra observed previously. This change, which they ascribe to O loss, though there is no direct evidence, would seem to indicate that photoemission measurements at room or even liquid nitrogen temperatures are indicative of oxygen deficient material. Our results from Tl-Ba-Ca-Cu-O samples scraped at room temperature are similar to those obtained on other high temperature superconductors. Such studies lead to an understanding of the electron-electron interactions in fully oxygenated as well as oxygen deficient material.[8] In particular the structure of the valence band peak and the existence of the Cu $3d$ and O $2p$ satellites indicates that the hybridization and the e-e correlations are the same in this material as in the 1-2-3 and Bi based high temperature superconductors.

Photoemission spectroscopy can also be utilized to quantitatively study the interaction of well prepared surfaces with gaseous molecules.[11] Photoemission can identify the reaction products remaining on the surface and using known exposures can determine the percentage of molecules which strike the surface and react. These studies give insight into the initial step of the degradation these materials suffer when exposed to atmosphere.

EXPERIMENT

The experiments were performed at the National Institute of Standards and Technology SURF-II Synchrotron Light Source. The monochromator and surface analyzer chamber have been described previously.[12-13] Briefly, photoemission spectra are taken between 25 and 105 eV photon energy and the emitted electrons are energy analyzed by a commercial double pass cylindrical mirror analyzer. A mask placed over the entrance of the analyzer allowed only electrons emitted normal to the sample surface to be detected. Clean surfaces of Tl-Ba-Ca-Cu-O were prepared by scraping the sample with a diamond file and stainless steel blade in the ultra high vacuum chamber whose base pressure was 1×10^{-10} Torr for these experiments. Care was taken not to exceed 100°C during

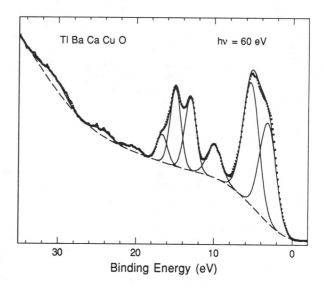

FIG. 1 Photoemission spectrum of Tl-Ba-Ca-Cu-O at 60 eV photon energy. The approximate secondary electron background (----), Gaussian peaks (——), and their sum (solid line thru data points) are shown.

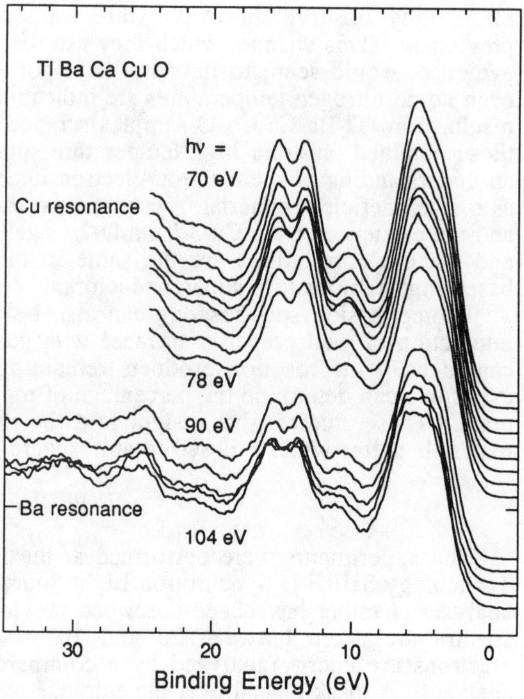

FIG. 2 Resonant photoemission spectra near the excitation onsets of the Cu $3p$ (hv=70-78 eV in 1 eV increments) and the Ba $4d$ (hv=90, 94, 98, 102, 104 eV).

the chamber bakeout to avoid reduction of the sample. Scraping has been shown to produce intra-granular fractures rather than inter-granular separation so that this procedure produces freshly exposed bulk material.[14]

Bulk superconducting material was prepared from mixtures of the metal oxides and carbonates in the appropriate ratios, calcined in air and sintered into pellets. A small amount of carbonate impurity was detected in x-ray diffraction. Zero resistance was obtained at 114 K.

The gas adsorption measurements were performed by back filling the chamber with a known pressure (measured with a hot filament ion gauge) of gas for a fixed period of time. Exposures, expressed in Langmuirs (1L = 10^{-6} Torr·sec), were not corrected for ion gauge sensitivity.

RESULTS

Fig. 1 shows the photoemission spectrum of Tl-Ba-Ca-Cu-O taken at 60 eV photon energy. The spectrum has been decomposed into a secondary electron background (dashed line) and a series of Gaussian curves (solid lines) fit to the various peaks. The sum of the secondary electron curve and Gaussian peaks is given as a solid line passing through the data points. The resulting peak positions and widths are shown in Table I along with the atomic assignments for the observed energy levels. Not apparent in any single photoemission spectra are the features that resonate with photon energy. Two series of spectra are shown in Fig. 2 for photon energies of 70-78 eV and 90-104 eV which correspond

TABLE I. Tl-Ba-Ca-Cu-O electronic states, binding energies referenced to E_F (±0.1 eV), and full widths at half maximum (FWHM ±0.2 eV).

Level	Energy	FWHM	Level	Energy	FWHM
Cu $3d$ / O $2p$	3.2	2.5	Tl $5d$ / Ba$5p$	16.7	1.6
Cu $3d$ / O $2p$	5.5	2.5	O $2s$	20.5	2.2
Cu sat. / O sat.	10.0	2.2	Ca $3p$	24.3	2.5
Cu sat. / Tl $5d$	13.1	1.7	Ba $5s$	30.6	3.0
Tl $5d$ / Ba$5p$	15.0	1.6			

to the onset of Cu $2p$ and Ba $4d$ excitations, respectively. The resonances can be used to identify the atomic origin of the peaks in the valence band; in particular the Cu features at 10.0 and 13.1 eV binding energy and the Ba $5p$ and $5s$ states at 15.0, 16.7 and 30.6 eV binding energy.

The two peaks which resonate at the Cu $3p$ core level near 74 eV are satellite features.[4] Their close similarity to those observed previously indicates that they have similar origins, i.e., Cu $3d$ satellite features split off from the main valence band by hole-hole repulsion energy in the final state configuration.

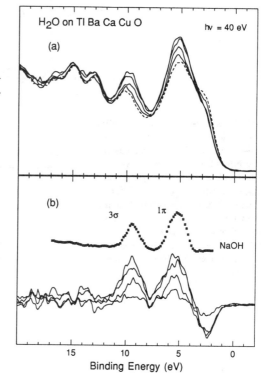

FIG. 3 (a) Photoemission spectra from a clean Tl-Ba-Ca-Cu-O surface (----) exposed to increasing amounts of H_2O. The curves for 0.3, 3, 30, and 300 L exposure are superimposed. (b) Difference spectra obtained by subtracting the spectrum from the clean surface from those exposed to H_2O. The dotted curve is a photoelectron spectra of NaOH (Ref. 15) shown for comparison.

The data for the adsorption of H_2O and CO_2 are shown in the upper panels of Figs. 3 and 4 respectively. The spectrum of clean Tl-Ba-Ca-Cu-O is shown as the dashed curve and increasing doses of adsorbed gas are shown superimposed as the solid curves. The curves are normalized to incident photon flux and therefore represent actual relative intensities. In order to obtain the spectrum from the adsorbate alone, the spectrum of the clean surface are subtracted from that of each exposure. These difference spectra are shown in the bottom panel of both figures. The negative dip near the Fermi energy in the difference curves is due to either attenuation of the electrons from the substrate by the overlayer or charge transfer from the substrate to the adsorbed layer.[11] Shown for comparison with the H_2O adsorption difference spectra in Fig. 3 is a photoemission spectrum of NaOH.[15] The close agreement of the peak separations and their relative intensities indicates that the species formed on an exposure of the surface to H_2O is an hydroxide. Similarly, the agreement between the photoelectron spectrum[16] of $CaCO_3$ with the difference spectra shown in Fig. 4 indicates that the species formed when the surface is exposed to CO_2 is a carbonate.

The results of two additional sets of measurements are not shown. These are the adsorption of O_2 and CO. These gases do not appear to react strongly with the surface. In exposures up to 300 L, no reaction was observed for O_2 and only a slight reaction was observed for CO. The product formed in the CO exposure appeared to be a carbonate species.

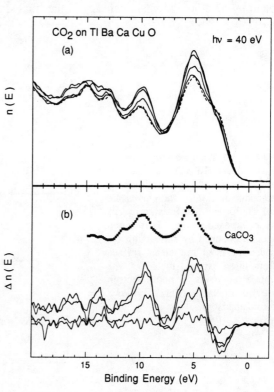

FIG. 4 (a) Photoemission spectra from a clean Tl-Ba-Ca-Cu-O surface (----) exposed to increasing amounts of CO_2. The curves for 0.3, 3, 30, and 300 L exposure are superimposed. (b) Difference spectra obtained by subtracting the spectrum from the clean surface from those exposed to CO_2. The dotted curve is a photoelectron spectra of $CaCO_3$ (Ref. 16) shown for comparison.

DISCUSSION

The valence band structure shown in Fig. 1 and the Cu $3d$ resonance satellites shown in Fig. 2 are similar to those obtained for the 1-2-3 compounds. This similarity indicates the Cu $3d$ - O $2p$ interactions are the same in both classes of materials.

From the measurement of the cross section of the valence band between 0 and 8 eV binding energy one can determine the individual contribution of Cu $3d$ and O $2p$ electrons. The relative intensity of the two components of this band should show a large change in relative intensity between 20 and 100 eV photon energy if the states were separable. The minor change observed indicates that the individual contribution of the Cu and O electrons are not separable, i.e., the valence band is highly hybridized.

The Cu $3d$ satellites at 10.0 and 13.1 eV resonate near a photon energy of 74 eV. These resonances are the result of coupling between final states reached by core level excitation and by direct excitation. The coupling results in a Fano interference between the two final state configurations and resonates as the Cu $3p$ core is excited to a quasi-bound state.

From the intensity of the hydroxide and carbonate peaks in the difference spectra in Figs. 3 and 4, and the measured exposures, we can estimate the reactivity of the Tl-Ba-Ca-Cu-O surface studied here. The surface saturates between 3 and 30 L for both molecules with the H_2O saturating at a lower dose than CO_2. Comparing the heights of the 3 L dose with that of the 30 L, we estimate that saturation coverage would be reached for ~ 5 L dose of H_2O and ~ 7 L dose of CO_2. This is less reactive by factors of 5 and 2, respectively, than the 1-2-3 materials studied previously.[11]

The chemistry of this multi-component surface is surely complex. Without knowing the surface geometry and composition, it is difficult to even speculate on the details of the reactions taking place. However, in more general terms, H_2O and CO_2 apparently react with O to reduce the surface while O_2 does not oxidize the surface. This does not, however, indicate that the surface studied here is fully oxygenated. The results of List, et al.,[10] suggest that the surfaces of the 1-2-3 superconductors lose O when scraped at room temperature. This loss, however, would involve only a small fraction of the O atoms on the surface with a large amount remaining to react with the adsorbates. That O_2 does not react to replace the lost oxygen could be due to the fact that the O lost from the surface is very weakly bound and there is not enough energy available to dissociate O_2.

That CO reacts only weakly to form a carbonate species, could be due to the fact that it must react with two surface O atoms, a process which is less probable than CO_2 reacting with a single surface O atom to form the same product.

SUMMARY

The similarity of the valence band and the satellite structure of the Tl-Ba-Ca-Cu-O and the Bi and 1-2-3 high temperature superconductors implies that the electron hybridization and electron-electron correlations are similar in all of these materials. The surface of the Tl based compound does not appear to be as reactive toward atmospheric gases as the 1-2-3 superconductors.

ACKNOWLEDGEMENTS

R.L.S., S.W.R., and R.L.K. wish to thank the United States Office Of Naval Research for support of this work. Assistance of the NIST SURF-II synchrotron light source staff is gratefully acknowledged.

REFERENCES

1. R.L. Kurtz, in *Proceedings of the American Vacuum Society Symposium on High-T_c Superconductors, Anaheim, California*, edited by J.M.E. Harper, R.J. Colton, and L.C. Feldman, AIP Conference Proceedings No. 165 (American Institute of Physics, New York, 1988), p. 222.
2. J.C. Fuggle, J. Fink, N. Nücker, Int. J. Mod. Phys. B (1988), in press.
3. P.D. Johnson, S.L. Qiu, L. Jiang, M.W. Ruckman, M. Strongin, S.L. Hulbert, R.F. Garrett, B. Sinkovic, N.V. Smith, R.J. Cava, C.S. Jee, D. Nichols, E. Kaczanowicz, R.E. Saloman, and J.E. Crow, Phys. Rev. B **35**, 8811 (1987).
4. R.L. Kurtz, R. Stockbauer, D. Mueller, A. Shih, L. Toth, M. Osofsky, and S. Wolf, Phys. Rev. B **35**, 8818 (1987).
5. M. Onellion, Y. Chang, D.W. Niles, R. Joynt, G. Margaritondo, N.G. Stoffel, and J.M. Tarascon, Phys. Rev. B **36**, 819 (1987).
6. P. Steiner, V. Kinsinger, I. Sander, B. Siegwart, S. Hüfner, C. Politis, R. Hoppe, and H. P. Müller, Z. Phys. B **67**, 497 (1987).
7. A. Fujimori, E. Takayama-Muromachi, Y. Uchida, and B. Okai, Phys. Rev. B **35**, 8814 (1987).
8. R.L. Kurtz, S.W. Robey, R. Stockbauer, D. Mueller, A. Shih, and L. Toth, Phys. Rev. B (1988), submitted.
9. D. Ramaker, Phys. Rev. B (1988), in press. D. Ramaker, N.H. Turner, and F.L. Hutson, Phys. Rev. B (1988), in press.
10. R.S. List, A.J. Arko, Z. Fisk, S.-W. Cheong, J.D. Thompson, J.A. O'Rourke, C.G. Olson, A.-B. Yang, T.-W. Pi, J.E. Schirber, and N.D. Shinn, Phys. Rev. Letts., submitted.
11. R.L. Kurtz, R. Stockbauer, T.E. Madey, D. Mueller, A. Shih, and L. Toth, Phys. Rev. B **37**, 7936 (1988).
12. R.L. Kurtz, D.L. Ederer, J. Barth, and R. Stockbauer, Nucl. Instrum. Methods **A266**, 425 (1988).
13. D.M. Hanson, R. Stockbauer, and T.E. Madey, Phys. Rev. B **24**, 5513 (1988).
14. J.E. Blendell, C.K. Chiang, D.C. Cranmer, S.W. Freiman, E.R. Fuller, Jr., E. Drescher-Kraisicka, W.L. Johnson, H.M. Ledbetter, L.H. Bennett, L.J. Swartzendruber, R.B. Marinenko, R.L. Myklebust, D.S. Bright, and D.E. Newbury, in *Chemistry of High-Temperature Superconductors*, D.L. Nelson, M. Stanley Whittingham, and T.F. George, eds., Amer. Chem. Soc., Washington, D. C., 1987, pg 240.
15. A. Conner, M. Considine, I.H. Hillier, and D. Briggs, J. Electron Spectrosc. Relat. Phenom. **12**, 143 (1977).
16. E. Tegeler, N. Kosuch, G. Wiech, and A. Faessler, J. Electron Spectrosc. Relat. Phenom. **18**, 23 (1980).

PHOTOEMISSION FROM SINGLE-CRYSTAL $EuBa_2Cu_3O_{6+x}$ CLEAVED BELOW 20K; METALLIC-TO-INSULATING SURFACE TRANSFORMATION

R. S. List, A. J. Arko, Z. Fisk, S-W. Cheong, S. D. Conradson, J. D. Thompson, C. B. Pierce, D. E. Peterson, R. J. Bartlett, and J. A. O'Rourke
Los Alamos National Laboratory, Los Alamos, NM 87545

N. D. Shinn and J. E. Schirber
Sandia Laboratories, Albuquerque, NM 87185

C. G. Olson, A-B. Yang and T-W. Pi
Ames National Laboratory
Iowa State University, Ames, IA 50011

B. W. Veal, A. P. Paulikas, and J. C. Campuzano
Argonne National Laboratory, Argonne, IL 60439

ABSTRACT

Valence band ultrviolet photoemission spectra (UPS) of single-crystal $EuBa_2Cu_3O_{6+x}$ ($x > 0.6$) samples cleaved in vacuum at 20 K demonstrate that the metallic superconducting phase undergoes an irreversible transformation via near-surface oxygen loss to an insulating state upon annealing above 50 K. Freshly cleaved surfaces at 20 K exhibit a density of states at the Fermi level comparable to that of copper, and have both O(2p) and Cu(3d) character at E_F based on the photon energy dependence of the intensity. Reasonably good agreement between band structure calculations and the present data would suggest theoretical models using the band state as a starting point.

INTRODUCTION

Photoelectron spectroscopy has been utilized extensively[1-6] to measure both the valence and core electronic structure of the perovskite-based high temperature superconductors. Surprisingly, all known valence band studies of the 1-2-3 type materials have found an extremely low density of states near the Fermi level.[2-6] Indeed, a Fermi edge is not clearly discerned. This evidence against a metallic structure both above and below the critical temperature was at least a contributing factor leading to theoretical models[7] based on a highly localized electronic structure. This paper reports the first studies which show unambiguously that a freshly cleaved surface of a representative 1-2-3 superconductor, $EuBa_2Cu_3O_{6+x}$ ($x > 0.6$), (1) is indeed a metal which exhibits a significant density of states with both O(2p) and Cu(3d) character at the Fermi level, and (2) undergoes an irreversible surface transformation to an insulating state most likely via oxygen loss at the surface upon annealing to 50 K or above.

© 1989 American Institute of Physics

Valence band UPS spectra from these annealed, oxygen-deficient surfaces strongly resemble the previously published[1-6] data on 1-2-3 materials, indicating that previous data must be regarded with caution. The data presented below establish new criteria for the evaluation of theoretical models and point to necessary experimental precautions for the 1-2-3 materials in the future. These results, together with the recent dispersive data[8] in the Bi-Sr-based system, as well as previous measurements[9] on surface transformations in $La_{2-x}Sr_xCuO_4$ with temperature, underscore the basic similarity of the electronic structure of the cuprate superconductors. We fully expect that it will be possible to observe band dispersion in the 1-2-3 compounds just as has been done in the Bi-Sr compounds. On the other hand, the weak Fermi edges observed in the other materials could perhaps be substantially improved if surface deterioration is avoided.

EXPERIMENTAL

The initial photoemission measurements were performed[10] on the the Los Alamos/Sandia U-3 Extended Range Grasshopper (ERG) beamline at the Natioanl Synchrotron Light Source (NSLS). Additional data were later obtained on the Iowa State/Montana State combined ERG/SEYA beamline at the Synchrotron Radiation Center (SRC) in Stoughton, WI. In both cases, a double pass cylindrical mirror electron energy analyzer was used. The combined spectrometer resolution and photon energy bandwidth was 500meV for the NSLS data and 200 meV or better for the SRC data. Relatively large single crystal samples (1.5x1x0.25mm) grown at Los Alamos from a PbO-CuO flux were oxygenated at Sandia[11] by high pressure oxygen treatment (3 kbar at 875 K for 3-5 days). The resultant oxygen stoichiometry (x>0.6) was inferred from the midpoint of T_c as obtained in diamagnetic susceptibility measurements (T_c>55K). The crystals were mounted on a liquid helium cryostat and electrically grounded with a film of graphite. Each sample was cleaved after the cryostat was fully cooled, and the chamber base pressure was below 5×10^{-11} torr. The behavior with temperature and reproducibility of the data show that these precautions were sufficient to eliminate sample contamination effects in our data.

RESULTS AND DISCUSSION

While evidence for the instability of the 1-2-3 vacuum interface above 50 K is derived from both valence band and core-level spectra, we will in this paper focus primarily on valence band spectra and their comparison to band calculations. Representative valence band spectra from a sample cleaved at 20 K and subsequently annealed at room temperature are shown in Fig. 1. Spectra (a) and (c) are from a fresh cleave at 20 K for hv=22eV and hv=70eV respectively, whereas spectra b and d show the same surface at the same photon energies after annealing to room tempeature for a period of about 2 hours. For comparison, spectrum (e) is a valence band spectrum of an as-grown (non-superconducting

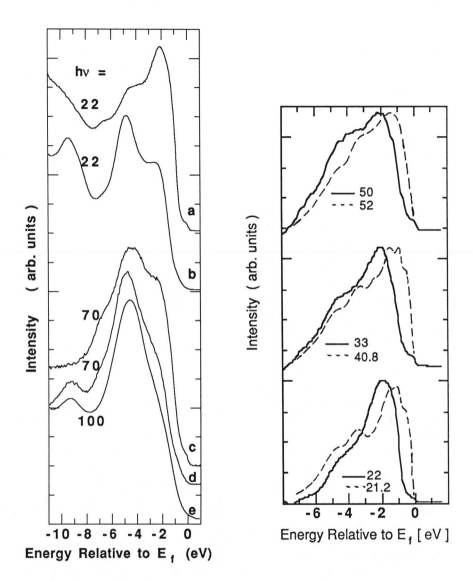

Fig.1. Valence band photoemission spectra for $EuBa_2Cu_3O_{6.7}$ (curves a through d), and unoxygenated $YBa_2Cu_3O_{6+x}$ (curve e), using indicated photon energies: a) Fresh cleave, T=20K; b) T=300K for 2 hrs. (note shift); c) Fresh cleave T=55K; d) T=300 K for 2 hrs (note 0.25 eV shift); e) T=20K, fresh cleave.

Fig. 2. Comparison of data (solid lines) to calculated photoemission spectra of Redinger et al. (dashed lines) at the photon energies shown in the figure. Above these energies tthe Eu 4f intensity insreases and results in discrepancies.

and hence oxyge-deficient) single crystal of $YBa_2Cu_3O_{6+x}$, also cleaved and measured at 20K, so that any differences are bulk rather than surface effects.

The freshly cleaved surface of the well-oxygenated sample (a and c) exhibits a clear Fermi edge at all photon energies with the intensity being as large as 10% of the valence band maximum for some photon energies. There is a strong intensity at -2 eV, and no appreciable intensity at -9.5 eV, a commonly observed satellite. Upon warming even up to 50 K, changes in the peak intensities at E_f, -2 eV, -4.5 eV, -6.5 eV and -9.5 eV are unmistakable. By heating the sample all the way to 300K one sees from Fig. 1 that (1) the states at E_F vanish, (2) the intensity reverses for the -2 eV and -4.5 eV peaks, (3) the shoulder at -6.5 eV vanishes, and (4) the -9.5 eV satellite peak grows to substantial intensity. A comparison of these spectra with that from the oxygen deficient Y-1-2-3 sample (spectrum e) shows that all the surface changes observed upon warming the oxygen-rich sample to room temperature are correlated with an oxygen deficient bulk 1-2-3 material. Further annealing at 300 K in vacuum eventually results in a rigid shift of the spectrum to higher binding energy (0.25 eV in Fig. 1), indicative of sample charging due to the formation of an insulating near-surface layer. It should be noted that all of the changes are irreversible as would be expected from oxygen loss.

These results demonstrate that the vacuum solid interface of the superconducting phase is unstable, with oxygen loss occuring in a facile process which is kinetically inhibited below 50 K. Additionally, these data show that previous photoemission data[1-6] obtained at or above 80K are unlikely to be representative of the correct superconducting phase. The most important observation to be made, however, is that there is a complete change of the band structure between the parent O_6 material (the Mott insulator) and the superconducting O_7 material. One cannot view the relationship between them as merely one of doping the insulator with impurity states near E_f to obtain the metallic superconductor. Large changes are observed throughout the entire width of the valence band in complete disproportion to the amount of oxygen lost.

Similar irreversible, temperature-dependent changes were measured for the Ba 4d level and O 1s level (not shown). Specifically, the low binding energy shoulder of the freshly-cleaved, 20K Ba 4d spectrum agrees well with the XPS data of Steiner et al.[12], reflecting a well-oxygenated surface phase. Upon annealing, the intensity ratios and energy positions of the two Ba 4d components undergo changes which appear to reflect the screening capability of the oxygen holes. There is a 3 eV shift in the O 1s core level between the two phases. A complete discussion of the core level spectra will form the basis of a separate paper.

Insight into the composition of the states near the Fermi level is obtained by photon energy-dependent measurements. High resolution data obtained at the O(2s 2p) and Cu(3p 3d) thresholds (22 eV and 74 eV respectively) show resonance structure near the Fermi energy, indicating that these states must include both copper and oxygen character. Admittedly, the Cu resonance at 74 eV is somewhat weak, rising just barely above the noise level. On the other hand the entire valence band resonance is very weak. However, the very existence of a substantial Fermi edge at 70 eV, and indeed at a photon energy as high as 139 eV, forces the conclusion that d-states are present at E_f.

A comparison of valence band data with the band structure calculations of Redinger et al.[13] (Fig. 2) shows remarkable qualitative agreement. We do not wish to imply with these comparisons the correctness of one calculation vs. others[14-16]. It is primarily that Redinger et al. have calculated UPS spectra. In Fig. 2 the secondary background has been subtracted from the data (solid lines) in order to better compare with the predicted spectra (dashed lines) obtained by convoluting the partial densities of states with the appropriate cross-sections and a 0.3 eV Gaussian broadening function. Both the overall shape of the spectra and the photon energy dependence seem to be in agreement with predictions. A major point of disagreement is a persistent 1 eV shift of the leading edge of the data to higher binding energy. This discrepancy could be due to several factors: 1) Assuming sufficient accuracy in the calculation, the difference may lie in the fact that we are measuring $O_{6.7}$ material while the calculation[17] is for O_7. We have actually measured a doubling of the Fermi edge between $T_c=55K$ and $T_c=70K$ material. 2) The existence of a Cu satellite at -12 eV clearly shows that we are dealing with correlated metals[18]. Creation of a hole in a highly correlated band could result in a shift of the data due to poor screening. 3) Perhaps even at 20K there is sufficient rapid oxygen loss to substantially decrease the Fermi edge. Additional work will clearly be necessary to pin down the cause of the discrepancy.

CONCLUSIONS

Photoemission spectra of single crystal $EuBa_2Cu_3O_{6+x}$ ($x > 0.6$) samples cleaved in vacuum at 20 K show that the superconducting phase is metallic with a density of states at the Fermi level comparable to that of copper. The data agree reasonably well with predicted spectra from a band structure calculation, with the exception of a 1 eV shift of the leading edge. Annealing to temperatures above 50 K results in dramatic spectral changes which correlate with spectra from non-superconducting 1-2-3 materials measured at 20 K. This clearly suggests a loss of oxygen in the near surface region, leading to an insulating layer. Measurements done at room temperature probably probe non-representative surface phases. Resonant photoemission indicates that a band model with copper/oxygen hybrid states at the Fermi level accounts for the essential aspects of our data.

ACKNOWLEDGEMENTS

This research was supported under the auspices of the U.S. Department of Energy. Sandia Laboratories wish to acknowledge contract #'s DE-AC04-76DP00789 and DE-AC02-76CH00016, while Ames Laboratory acknowledges contract #W/7405/ENG 82.

REFERENCES

1. See for example G. Wendin,*J. Phys. Colloq.* (Fr.) **48**, no. 9, 4 83-6 (1987), and references therein.
2. Z. X. Shen, J. W. Allen, J. J.Yeh, J.-S. Kang, W. P.Ellis, Spicer, I. Lindau, M. B. Maple, Y. D. Dalichaouch, M. S. Torikachlivi, J. Z. Sun,and T. H. Geballe,*Phys. Rev.* **B36**, 8414 (1987).
3. A. Samsavar, T. Miller, T. -C. Chiang, B. G. Pazol, T. A. Friedman, and D. M. Ginsberg, *Phys. Rev.* **B37**, 5164 (1988).
4. M. Onellion, Y. Chang, D. W. Niles, R. Joynt, G. Margaritondo, N. G. Stoffel, and J. M. Tarascon, *Phys. Rev.* **B36**, 819 (1987).
5. P. D. Johnson,S. L. Qui, L. Jiang, M. W. Ruckman, Myron Strongin, S. L. Hulbert, R. F. Garret, B. Sinkovic, N. V. Smith, R. J. Cava, C. S. Lee, D. Nichols, E. Kaczanovicz, R. E. Salomon, and J. E. Crow, *Phys. Rev.* **B35**, 8811 (1987).
6. M. Tang, N. G. Stoffel, Q. B. Chen, D. LaGraffe, P. A. Morris, W. A. Bonner, G. Margaritondo, M. Onellion, *Phys. Rev.* **B38**, 897 (1988).
7. For a condensed version of the various theoretical approaches, see T. M. Rice, *Z. Phys.-Condensed Matter* **67**, 141 (1987).
8. T. Takahashi, H. Matsuyama, H. Katayama-Yoshida, Y. Okabe, S. Hosoya, K. Seki, H. Fujimoto, M. Sato, and H. Inokuchi, *Nature* **334**, 691 (1988).
9. T. Takahashi, F. Maeda, H.Katayama-Yoshida ,Y. Okabe, T. Suzuki, A. Fujimori, S. Hosoya, S. Shamoto, and M. Sato, *Phys. Rev.* **B37**, 9788 (1988).
10. R. S. List, A. J. Arko, Z. Fisk, S.-W. Cheong, S. D. Conradson, J. D. Thompson, C. B. Pierce, D. E. Peterson, R. J. Bartlett, N. D. Shinn, J. E. Schirber, B. W. Veal, A. P. Paulikas, and J . C. Campuzano, *Phys. Rev. B Rapid Commun.*, (to be published).
11. J. E. Schirber, E. L. Venturini, B. Morosin, J. F. Kwak ,D. S.Ginley and R. J. Baughman, *Proc. 1987 Fall Meet. of Mater. Res. Soc.*, **99**, 4879 (1988).
12. P. Steiner, S. Hufner, V. Kinsinger, I. Sander, B. Siegwart, H. P. Muller, R. Hoppe, S. Kemmler-Sack and C. Kunz, *Z. Phys. B*, **69**, 449 (1988).
13. J. Redinger, A. J. Freeman, J. Yu, and S. Massida, *Phys. Lett.* **A124**, 469 (1987).
14. L. F. Mattheis, *Phys. Rev. Lett.* **58**, 1028 (1987).
15. W. M. Temmerman, G. M. Stocks, P. J. Durham, and P. A. Sterne, *Phys. F : Met. Phys.* **17**, L135 (1987) .
16. W. Y. Hsu, R. V. Kasowski, **Novel Superconductivity**, ed. by S. A. Wolf and V. Z. Kresin (Plenum Press, NY, 1987), p. 373.
17. J. Yu, A. J. Freeman, S. Massida, **Novel Superconductivity**, ed. by S. A. Wolf and V. Z. Kresin (Plenum Press, NY, 1987), p.367.
18. J. Zaanen, G. A. Sawatsky, and J. W. Allen, *Phys. Rev. Lett.* **55**, 418,(1985).

CHARACTERIZATION OF $Bi_2Sr_2Ca_1Cu_2O_8$

F. J. Himpsel, G. V. Chandrashekhar, A. Taleb-Ibrahimi,
A. B. McLean, and M. W. Shafer
IBM Research Division, T. J. Watson Research Center
Yorktown Heights, NY 10598

ABSTRACT

The electronic structure of the high temperature superconductor $Bi_2Sr_2Ca_1Cu_2O_{8+x}$ is determined from single crystal samples using photoelectron spectroscopy and polarization dependent near edge absorption spectroscopy. The valence band exhibits a O2p, Cu3d peak at 3.5eV below the Fermi level E_F. The photoemission intensity at E_F is higher than in the other high temperature superconductors. Also, the peak at E_F-9eV, previously associated with carbonate, is absent. A resonant $Cu3d^8$ satellite is seen at E_F-12.6eV, giving a hole-hole repulsion U = 5.6eV at the Cu site. The O1s edge exhibits a peak at threshold, $h\nu$ = 528.2eV, corresponding to O2p holes at E_F. This peak is excited only by the component of the electric field vector in the ab-plane. Using dipole selection rules for the O1s - O2p transition it is concluded that the oxygen holes have $p_{x,y}$ character, with x,y in the CuO planes.

INTRODUCTION

$Bi_2Sr_2Ca_1Cu_2O_8$ can serve as a model superconductor for surface-sensitive spectroscopies, because it cleaves like a layered compound, producing high-quality surfaces. The material also has a simple structure with all CuO_2 planes being equivalent. In $Y_1Ba_2Cu_3O_7$ there are copper - oxygen chains and planes, leading to a more complex electronic structure. An additional problem with $Y_1Ba_2Cu_3O_7$ is the presence of Ba, which makes surfaces and grain boundaries very reactive. This leads to $BaCO_3$ formation, which obscures surface-sensitive photoelectron spectra. On the other hand, $Bi_2Sr_2Ca_1Cu_2O_8$ exhibits less tendency to form surface carbonate. There exist several photoemission studies[1-5] of $Bi_2Sr_2Ca_1Cu_2O_8$.

© 1989 American Institute of Physics

SAMPLE PREPARATION

Single crystals of $Bi_2Sr_2Ca_1Cu_2O_{8+x}$, $x \approx 0.15$, were grown from polycrystalline material of composition $Bi_4Sr_3Ca_3Cu_4O_y$ maintained at about 875°C for 40-50 hours in air. The crystals were obtained as thin plates, which were up to about $3 \times 3 \times 0.5mm^3$ in size with the c-axis perpendicular to the plates. Typically, the onset of the superconducting transition was about 84-85K. An electron microprobe analysis indicated that the composition was near to $Bi_{2.1}Sr_{1.6}CaCu_2O$, with excess oxygen beyond the stoichiometric compound. Single crystal X-ray examination confirmed the incommensurately modulated $\sqrt{2} \times \sqrt{2}$ orthorhombic cell. Various methods for preparing clean surfaces were tried. Scraping produced clean surfaces, although the roughness increased upon prolonged scraping, and polarization effects became weaker. Surfaces cleaved in air gave a well-ordered surface with large polarization effects but exhibited an extra contamination peak in the O1s absorption edge at $h\nu = 533.5eV$ and in the C1s absorption edge. The latter is most probably due to the π^*-resonance in CO (see Ref. 6). This contamination could be removed by mild heating to a few 100°C, and the spectra became similar to those of samples that were scraped or cleaved in ultra high vacuum. Apparently this compound is rather stable against loss of oxygen upon heating in vacuum. The best method for preparing well-ordered and clean surfaces was cleaving in vacuum with a tab technique.

Carbonate contamination[7,8] showed up prominently in ceramic samples of $Y_1Ba_2Cu_3O_7$. It can also be found in $Bi_2Sr_2Ca_1Cu_2O_8$ at the natural growth surface (Fig. 1). Characteristic CO_3^{2-} orbitals are located at 5eV and 10eV below the Fermi level. They disappear upon scraping the surface in vacuum.

PHOTOEMISSION SPECTRA

In common with other CuO - based superconductors, the valence band of $Bi_2Sr_2Ca_1Cu_2O_8$ exhibits a dominant O2p, Cu3p peak 3.5eV below E_F with some fine structure (Fig. 2). Most strikingly, and as reported by other groups[1-3] there is clearly a metallic Fermi edge. For the other materials, the intensity drops off continuously towards the Fermi level, and a metallic Fermi edge has been obtained only occasionally. There are two possible explanations for this difference between $Bi_2Sr_2Ca_1Cu_2O_8$ and the other materials. The first is the better surface quality of $Bi_2Sr_2Ca_1Cu_2O_8$, evidenced by the absence of the carbonate peak at -9 to -10eV. As Fig. 1 shows, the intensity of the Fermi level becomes weaker relative to that of the main peak when the amount of

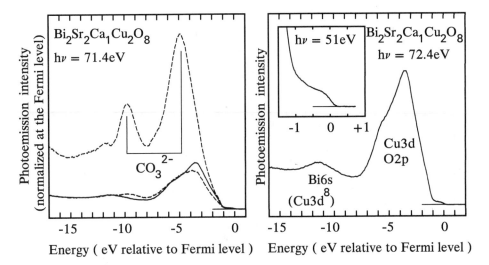

Fig. 1. Valence band spectra of $Bi_2Sr_2Ca_1Cu_2O_8$ before scraping in vacuum (top curve) and after scraping (bottom curves). Peaks at -4.8 and -9.8eV are likely due to carbonate surface contamination (compare Refs. 7,8).

Fig. 2. Valence band spectrum of $Bi_2Sr_2Ca_1Cu_2O_8$, scraped in vacuum. A clear metallic Fermi edge is observed (see inset) in addition to the O2p, Cu 3d valence band around -3.5eV. The peak at -11.3eV is mainly due to Bi6s states.

surface contamination increases. The second possibility is the existence of a slightly occupied Bi6p, O2p band, as predicted by one-electron band calculations[9-12]. In order to detect the Bi6p character of these states one would have to determine their photon-energy-dependent cross section. We have looked for their unoccupied counterpart at the Bi5d edge, but have found no obvious signal. An asymmetry in the Bi5d core levels (see Fig. 3) has been used as an argument[2] for metallic character of the Bi-oxide sheets. However, the Ca3p core level falls right between the Bi5d levels and may distort their line shape.

Below the valence band, one finds a broad peak at -11.3eV, which can be assigned to the Bi6s states according to band calculations[9-12]. There is a contribution from a resonant $Cu3d^8$ satellite (Fig. 4), that is enhanced at the Cu3p threshold. It is located at E_F -12.6eV and turns on at $h\nu$ = 73.6eV, coinciding with the threshold for a similar satellite in CuO and $Y_1Ba_2Cu_3O_7$ at $h\nu$ = 73.7eV (compare Ref. 8). This coincidence shows that Cu is mostly in the 2+ oxidation state. Compared with other materials the intensity of this satellite is weaker. This could be an effect of the short escape depth (about 5Å for the valence band spectra), which is less than the 31 Å long c-axis. Since $Bi_2Sr_2Ca_1Cu_2O_8$ cleaves (by symmetry) between the Bi-oxide layers, one has the CuO_2 layers buried rather deeply ($\approx$10Å) below the surface. This depth effect is less

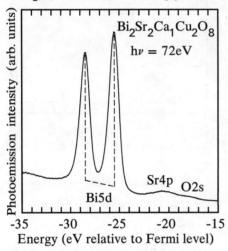

Fig. 3. Bi5d core level spectrum of $Bi_2Sr_2Ca_1Cu_2O_8$.

Fig. 4. Resonant $Cu3d^8$ satellite in $Bi_2Sr_2Ca_1Cu_2O_8$. A peak at E_F-12.6eV is resonantly enhanced above the Cu3p core level threshold ($h\nu$ = 73.6eV, see inset). The energy of the satellite allows one to determine the Coulomb repulsion of two holes on the Cu site, U_{dd} = 5.6eV.

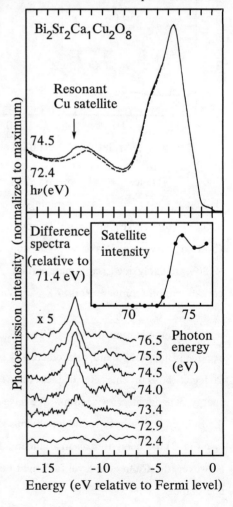

of a problem for the absorption spectra discussed in the following section, since photoelectrons with zero kinetic energy are chosen, which have more than 10 Å escape depth. From the energy of the Cu3d^8 satellite (-12.6eV) and the Cu3d^9 valence band (-3.5eV) one can obtain a hole-hole Coulomb repulsion U_{dd} = 5.6eV at the Cu site, assuming the first unoccupied state to be just above E_F. This is within the range of 4.6eV-6.0eV observed[13] for U_{dd} in various other Cu compounds and atomic Cu.

SOFT X-RAY ABSORPTION

Superconductivity in the CuO-based superconductors is correlated with the concentration of holes[14] in a nearly-filled Cu3d, O2p valence band. In the early models

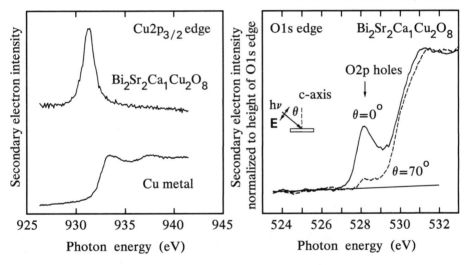

Fig. 5. The Cu2p absorption edge[17] of $Bi_2Sr_2Ca_1Cu_2O_8$ and of metallic Cu as a reference. The absorption is measured by detecting secondary electrons with zero kinetic energy.

Fig. 6. Polarization - dependence of the O1s absorption edge[17] for a $Bi_2Sr_2Ca_1Cu_2O_8$ single crystal cleaved in vacuum. The spike at hν = 528.2eV is excited by the component of the electric field vector E parallel to the a,b-plane, showing that there exist O2p holes of $p_{x,y}$ character near the Fermi level (Ref. 4).

of high-temperature superconductivity these holes are located on the Cu sublattice. They may spread to the oxygen sites via hybridization between the Cu3d and O2p valence states. Some models[15] assign a more dominant role to oxygen holes. If the hole-hole repulsion U_{dd} at the copper sites is large enough then any additional hole introduced by doping will prefer an oxygen site[16]. These hole states can be seen in near edge X-ray absorption fine structure (NEXAFS) as a peak at the threshold that corresponds to optical transitions from the core level to unoccupied states just above the Fermi level. At the Cu2p edge one finds a peak, as already expected from the existence of a Cu3d hole in stoichiometric Cu^{2+} (Fig. 5). No evidence for an extra Cu^{3+} feature is found. At the O1s edge we also find a spike at threshold, (Fig. 6) which indicates that some excess holes are located on the oxygen. In $Bi_2Sr_2Ca_1Cu_2O_{8+x}$, the doping with holes beyond the stoichiometric 3d hole in Cu^{2+} can be provided by two sources: For example, an additional amount x of oxygen may be present (our samples had $x \approx 0.15$). A second possibility is autodoping by having the Bi6p states incompletely ionized (see the band calculations in Refs. 9-12). The excess holes may convert some of the Cu^{2+} into Cu^{3+}, or create O^{1-} instead of O^{2-}. Our measurements provide evidence for O^{1-}, whereas no Cu^{3+} could be detected. A similar situation has already been found in $Y_1Ba_2Cu_3O_7$ (Ref. 8).

An interesting question is the orientation of the hole orbitals. This question can be answered by measuring the polarization dependence of the absorption edge and applying optical dipole selection rules (for a more detailed account see Ref. 4). With the electric field vector **E** parallel to the x-axis only transitions into p_x orbitals are allowed from an s-level, and likewise for the other directions. For single crystals of $Bi_2Sr_2Ca_1Cu_2O_8$ cleaved along the ab-plane, we find that the O2p holes are excited by **E** parallel to the surface (Fig. 6). Thus, they have $p_{x,y}$ character with x,y in the superconducting a,b-plane. Using a widely-adopted bonding concept for the CuO_2 planes the $p_{x,y}$ character of the O2p holes can be explained in a natural fashion. The ($Cu3d_{x^2-y^2}$) orbital has lobes pointing towards the in-plane oxygen that interact strongly with the $O2p_{x,y}$ orbitals. Their antibonding combination represents the highest occupied orbital, and will be the first one to be depleted when creating holes via doping. For $Bi_2Sr_2Ca_1Cu_2O_8$ there is another explanation for the $p_{x,y}$-character of the oxygen holes to be considered. A $Bi6p_{x,y}$, $O2p_{x,y}$ band is found to extend from 4 eV above the

Fermi level E_F down to 1 eV below E_F in band calculations[9-12]. This band is involved in an auto-doping process, as discussed above.

Acknowledgement: We acknowledge the assistance of A. Marx and J. Yurkas. Research was carried out in part at the NSLS, Brookhaven National Laboratory, which is supported by the U.S. Department of Energy, Division of Materials Sciences and Division of Chemical Sciences.

REFERENCES

1. M. Onellion, M. Tang, Y. Chang, G. Margaritondo, J. M. Tarascon, P. A. Morris, W. A. Bonner and N. G. Stoffel, Phys. Rev. B**38**, 881 (1988); Y. Chang, M. Tang, R. Zanoni, M. Onellion, R. Joynt, D. L. Huber, G. Margaritondo, P. A. Morris, W. A. Bonner, J. M. Tarascon and N. G. Stoffel, submitted to Phys Rev. Lett. (1988).

2. P. A. P. Lindberg, Z.-X Shen, I. Lindau, W. E. Spicer, C. B. Eom, and T. H. Geballe, Appl. Phys. Lett. **53**, 529 (1988); Z.-X Shen, P. A. P. Lindberg, I. Lindau, W. E. Spicer, C. B. Eom and T. H. Geballe, to be published.

3. T. Takahashi, H. Matsuyama, H. Katayama-Yoshida, Y. Okabe S. Hosoya, K. Seki, H. Fujimoto, M. Sato, and H. Inokuchi, Nature **334**, 691 (1988) and to be published.

4. F.J. Himpsel, G.V. Chandrashekhar, A.B. McLean, and M.W. Shafer, Phys. Rev. B, submitted.

5. E. G. Michel et al., Phys. Rev. B**38**, 5146 (1988).

6. Y. Jugnet, F. J. Himpsel, Ph. Avouris, and E. E. Koch, Phys. Rev. Lett. **53**, 198 (1984).

7. A.G. Schrott, S.L. Cohen, D.R. Dinger, F.J. Himpsel, J.A. Yarmoff, K.G. Frase, S.I. Park, R. Purtell, AIP Conf. Proceedings **165**, 349 (1988).

8. J. A. Yarmoff, D. R. Clarke, W. Drube, U. O. Karlsson, A. Taleb-Ibrahimi, and F. J. Himpsel, Phys. Rev. B**36**, 3967 (1987).

9. M. S. Hybertsen and L. F. Mattheiss, Phys. Rev. Lett. **60**, 1661 (1988); L. F. Mattheiss and D. R. Hamann, Phys. Rev. **B38**, 5012 (1988).

10. H. Krakauer, and W. E. Pickett, Phys. Rev. Lett. **60**, 1665 (1988).

11. F. Herman, R. V. Kasowski and W. Y. Hsu, Phys. Rev. **B38**, 204 (1988).

12. S. Massidda, J. Yu and A. J. Freeman, Physica C, to be published; P. Marksteiner et al. Phys. Rev. **B38**, 5098 (1988). B. A. Richert and Roland E. Allen, Phys. Rev. B, submitted.

13. J.A. Yarmoff, D.R. Clarke, W. Drube, U.O. Karlsson, A. Taleb-Ibrahimi, and F.J. Himpsel, AIP Conf. Proceedings **165**, 264 (1988).

14. M. W. Shafer, T. Penney, and b. L. Olson, Phys. Rev. **B36**, 4047 (1987); H.Takagi, H.Eisaki, S.Uchida, A.Maeda, S.Tajima, K.Uchinokura and S.Tanaka, Nature **332**, 236 (1988).

15. See e.g., V. J. Emery, Phys. Rev. Lett. **58**, 2794 (1987); F.C. Zhang and T. M. Rice, Phys. Rev. **B37**, 3759 (1988); D.M. Newns, M. Rasolt ,P. Pattnaik, and D.A. Papaconstantopoulos, Physica C **153-155**, 1287 (1988).

16. For the various energy parameters involved in this picture see J. Zaanen, G. A. Sawatzky, and J. W. Allen, Phys. Rev. Lett. **55**, 418 (1985).

17. The energy calibration of the O 1s and Cu 2p edges and the O 1s binding energies are only accurate to ± 1 eV. The O 1s binding energy (528.4eV) coincides with the spike in the absorption edge (528.2 eV) within our accuracy.

SURFACE ANALYSIS OF H,C,O,Y,Ba, AND Cu ON PRESSED AND LASER-EVAPORATED YBCO

J. Albert Schultz and Howard K. Schmidt
Ionwerks, 2215 Addison
Houston, TX 77030

P. Terrence Murray
Research Institute
University of Dayton
Dayton, OH 45469

Alex Ignatiev
Space Vacuum Epitaxy Center and Physics Department
University of Houston
Houston, TX 77204-5507

ABSTRACT

Direct Recoil Spectroscopy (DRS), Low Energy Ion Scattering (LEIS), SIMS, AES, and XPS have been compared from two YBCO high temperature superconductor samples. The increased surface sensitivity of DRS and LEIS in symbiosis with electron spectroscopy has shown copper depletion at the surfaces. This combination of spectroscopies should allow detailed studies of segregation and reactions, especially those involving H and O, at the surface of these materials.

INTRODUCTION

Direct Recoil Spectroscopy (DRS) by time-of-flight (TOF) analysis of forward recoiled neutral (and ionized) surface atoms during grazing incidence, pulsed keV argon ion bombardment of a surface was suggested[1] and proven[2] as an effective way of detecting surface hydrogen. In this technique, primary argon (potassium) ions from a collimated source desorb surface hydrogen atoms by direct binary collisions into a well defined forward scattering angle. The recoil energy of a surface atom can be calculated by the following expression

$$E_R = E_P \left[\frac{4 M_P M_R}{(M_P + M_R)2} \right] \cos^2 \phi$$

where E_R, M_R and E_P, M_P are the energy and mass of the recoil and primary and ϕ is the recoil angle (see insert on figure 1). The energy of the recoiled hydrogen is typically several hundred eV. At these energies, H is detected with near unity efficiency by a channel electron multiplier (CEM) irrespective of its charge state. The ability to detect both neutrals and ions removes the difficulty of unknown ion survival probability in making ion scattering and DRS quantitative. A TOF spectrum is obtained by pulsing the primary beam onto the sample. The directly recoiled hydrogen, H(DR), is faster than the multiply scattered primary ions and appears as a well resolved peak. The application of this technique for studies of metal-hydrogen surfaces has been reviewed[3]. A more general review of DRS including its application to the identification of surface elements other than hydrogen has been recently published[4]. Our purpose in this work is to extend the utility of DRS, performed simultaneously with low energy ion scattering (LEIS), to include analysis of the heavier surface elements found in superconductors.

Two samples of YBCO superconductor, one a laser evaporated film (LF) and one a pressed oxide (PO), were examined by five surface analytical techniques: X-ray photoelectron spectroscopy (XPS), Auger electron spectroscopy (AES), secondary ion mass spectrometry (SIMS), time-of-flight low energy ion scattering (LEIS), and time of flight direct recoil spectroscopy (DRS). In order to make a valid comparison between techniques, the intensity given by a particular technique for each elemental peak was ratioed between samples after each spectrum had been normalized to a barium intensity of two arbitrary units. Thus a comparison can be made between techniques without resort to sensitivity factors which at present are poorly known for DRS and LEIS and may be indeterminable for SIMS.

EXPERIMENTAL

The film sample was grown by pulsed laser evaporation (frequency doubled Nd:YAG) of a bulk YBCO target onto a $SrTiO_3$ substrate held at 400°C. The film was then annealed in 1 atm of oxygen at 900°C for 30 minutes followed by slow cooling to room temperature. The measured T_c for this film was 60°K. The pressed oxide sample was prepared by the published procedures[5] and had a T_c of 98°K.

XPS and AES were obtained with a cylindrical mirror analyzer using a Mg anode and an electron energy of 5 keV. The XPS and AES measurements were made after bombardment in a separate spectrometer by 700 eV Ar^+ (10^{16} ions/cm^2) during the LEIS/DRS/SIMS analysis and after subsequent transport in air for two days between spectrometers. The ion scattering results were obtained with a 10 keV, 10 nsec pulsed potassium beam impinging the surface at a 77.5° angle of incidence (see insert on fig. 1). The DRS line of sight detector was positioned in a forward scattering/recoil direction of 25°. The LEIS and SIMS were obtained through a Poschenrieder[6] time of flight sector situated at a scattering angle of 77.8°. The LEIS results were obtained by grounding the sector and viewing the sample through a hole in the outer sector half. SIMS spectra were then obtained by biasing the sample to +700 volts and timing the ions through the energized sector. The DRS, LEIS, and SIMS can in principle be performed simultaneously.

RESULTS AND DISCUSSION

Figure 1 (bottom two spectra) shows the comparison between DRS from LF (A) and PO (B) after 700eV Ar$^+$ sputtering (10^{14} ions/cm^2) which was just enough to remove a hydrocarbon overlayer from the "as is" sample. DRS from the "as is" LF and PO samples (spectra not shown) have equal intensities of C and O and intensities of Y and Ba comparable to those shown in the top of each panel for the recontaminated sputtered sample. The absence of Cu in the LF and only a trace of Cu

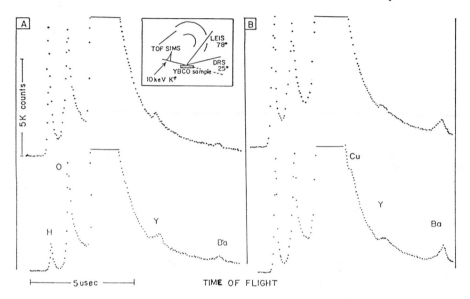

Figure 1. DRS from Laser Film (A) and Pressed Oxide (B). Bottom spectra are after sputtering. Top spectra were taken within one minute after sputtering was stopped.

visible in the PO, as well as the reversal of Y/Ba intensities between the two samples, are the most notable features in the sputtered samples. Some reduction in the metal intensities and an increase in H and O intensities is observed as the samples recontaminate after sputtering. Not all of the increase in H and O can be attributed to recontamination from the residual gases. The base pressures after sputtering the LF and PO were 5 and 2 x 10^{-9} Torr respectively, yet in the case of the LF (poorest vacuum) it was possible to sputter most of the H while in the case of the PO no decrease of surface H could be achieved with identical argon ion fluxes. This likely is caused by more grain boundary diffusion in the PO compared to LF. Even in the case of the LF, however, resaturation of the surface with H and O was complete within 30 seconds which is difficult to explain by recontamination from residual gases for a hydrogen surface coverage greater than 0.2 monolayer.

The metal to oxygen stoichiometry by DRS has not been determined and will require additional experimental and theoretical work. The metal DRS intensities should be very nearly equal for equal and unblocked numbers of metal atoms on the surface as will now be discussed.

An inverse square potential predicts that the recoil cross-section for Y and Ba should be 10 and 40% greater than the Cu cross-section. This is in agreement with results using the more sophisticated Kr-C potential predicting a factor of 2 increase in the DR cross-section for Ar bombardment of target masses between Ni and Au[4]. Because more energy is transferred to the lighter Cu than Y, the subsequent velocity will be higher by a factor of 1.25. Since the detection efficiency of a CEM is proportional to velocity it is probable that an 8500 eV Cu atom will be more easily detected than a 7700 eV Y atom. The same argument can be made for Y and Ba at 7700 and 6200 eV respectively, so that the increased detection efficiency will compensate the small decrease in recoil cross-section as the mass of the DR becomes smaller. If Cu were predominant in the surface, we would expect to see it easily.

The same trend is seen by the LEIS data in Fig. 2 where the metal atoms are detected by the energy loss of the potassium scattering into a laboratory angle of 77.8°. It is more difficult, because of overlap and a large multiple scattering background, to assign peak intensities here compared to DRS. A scattering/impurity surface recoil background has been estimated, as shown by a dashed line, and peak heights have been measured. As can be seen, this procedure underestimates the amount of Y because of its much wider peak shape. Thus, ratios of Y/Ba will be considered lower limits. Changing the TOF scale to an energy scale, applying deconvolution procedures, and careful calibration experiments to measure relative cross-sections/detector efficiency will clearly make LEIS useful in surface studies of these materials. Qualitative agreement between LEIS and DRS is demonstrated by comparing Figures 1 & 2.

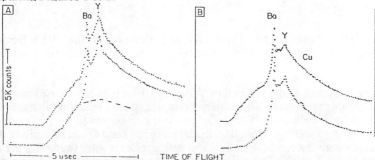

Figure 2.   **LEIS from YBCO samples.** (A) Laser Film sputtered (bottom) and one minute in residual (top). (B) Pressed oxide sputtered (bottom) and one minute in residual gasses (top).

Positive SIMS spectra (not shown) were obtained from the LF after Ar sputtering. No Cu is on the surface as shown by DRS and LEIS. The ratio of positive secondary ions Cu/Y/Ba is 1.0/3.5/7.0. One would expect the secondary ion yield of Ba to be larger than Cu and in fact over an order of magnitude difference has been reported[7]. This means that a substantial amount (> .14 Ba) of copper is being sputtered by the 10 keV potassium beam even though none is on the surface. The reversal of the Y/Ba SIMS ratio compared to that obtained by DRS and LEIS is likely explained by a difference in ionization probability and possibly a difference in depth of origin. It is clear that the combination of simultaneous LEIS, DRS, and SIMS will allow conclusions to be drawn on depth of origin and ionization probability of secondary ions which are not usually possible from SIMS measurements alone.

Table I gives a comparison between the spectroscopies used in this study. The Auger and XPS data have been corrected for peak height sensitivity factors where possible[8] and normalized to a barium stoichiometry of 2.0. The use in the XPS of low binding energy (BE) Ba and Cu lines (given parenthetically as b, c) was necessary to remove ambiguity caused by line shape changed in the high BE Cu transition. The LF comprises mostly a Cu^+ oxidation state with a narrow peak shape while the PO contains mostly the Cu^{+2} oxidation state with its broader XPS line. Peak height measurements thus tend to overestimate the amount of Cu in the LF and underestimate the amount in the PO. The 127 eV Y AES transition for which no sensitivity factor has been published was used to determine Y ratios between the LF and PO samples. As discussed above, the LEIS data is difficult to quantitate. The

Table I. Comparison of Elemental Ratios from Laser Film(LF) and Pressed Oxide(PO) YBCO Samples. XPS and AES Results are Corrected by Sensitivity Factors Except as noted Parenthetically.

	XPS		AES		LEIS[a]		DRS[a]	
	LF	PO	LF	PO	LF	PO	LFF	PO
Y	3.10	.61	$(.747)^d$	$(.181)^d$	2.6	1.38	3.1	0.7
Ba	2.00 $(2.59)^b$	2.00 $(1.31)^b$	2.00	2.00	2.0	2.0	2.0	2.0
Cu	1.73 $(.665)^c$	0.74 $(.755)^c$	0.81	1.35	—	0.2	—	0.2
O	14.0	7.1	3.32	2.03	—	—	28.3	10.1
C	14.6	4.6	1.88	1.48	—	—	—	—
H	—	—	—	—	—	—	35.7	19.0

a Ion scattering and direct recoil peak height are presented. See text for derivation of elemental concentration from these tabulated intensities.
b Ba 4D transition intensity.
c Cu 3p transition intensity.
d 127 eV Y Auger transition.

peak intensities are given with the Ba peak normalized to 2.00 arbitrary units. The Ba(DR) peak is likewise normalized to 2.0 arb. units, however, arguments have been made that the ratio of intensities should equal the ratio of elements visible at the surface. Therefore, a rough comparison between the XPS, AES, and DRS metal elemental composition measurements can be made for the PO sample. The agreement between XPS and DRS for relative Ba and Y is excellent not only for the PO but for the LF samples as well. Irrespective of which transition is considered for Cu in the XPS, the electron spectroscopy shows the surface to be significantly deficient in Cu (compared to the $Y_1Ba_2Cu_3O_x$ ideal stoichiometry) in rough agreement with the DRS. More work using samples richer in copper will be required.

Table II.   Ratio of elemental concentration between Laser Film(LF) and Pressed Oxide(PO) Samples.

	Y	Cu	O	C	H
XPS	5.1	2.3(0.5)[a]	2.0	2.1	—
AES	4.1	0.61	1.6	1.3	—
LEIS	(1.9)[b]	—	—	—	—
DRS	4.4	—	2.8	—	1.9

a   Low binding energy Ba and Cu transitions used (see text).
b   See text.

Table II gives a comparison between intensity ratios for each element from both samples as determined by each of the four techniques. By normalizing to the Ba intensity and taking the ratio, the dependence on sensitivity factors, some of which are unknown, disappears. As discussed previously, the low BE Cu line is preferred and gives a Cu ratio of 0.5 in agreement with the AES result of 0.61. The Y ratio is in excellent agreement between techniques except for the LEIS result which gives a lower limit as already discussed. The results for O are in agreement if one assumes that C occludes sites normally occupied by O. This has in fact been seen in the DRS during the initial spectral acquisition when C was of comparable intensity to O. Hydrogen coverage can be determined only by DRS and appears to be twice as prevalent in the LF.

CONCLUSIONS

A comparison between ion scattering techniques and electron spectroscopies indicates the usefulness of DRS and LEIS for quantitation of Y,Ba,Cu,O,C, and H. The surface layer of both YBCO samples is found to be depleted in Cu. The origin of copper sputtered by grazing incidence 10 keV potassium is at least one layer below the surface on the Y rich LF surface. Hydrogen is rapidly replenished at the surface of the PO during Ar ion bombardment probably as a result of grain boundary diffusion.

Sensitivity analysis assumes homogeneous elemental mixtures in a material. We have shown, however, by a symbiosis of spectroscopies that, at least for Cu, this is not true for these surfaces. Nevertheless, the agreement obtained between techniques is encouragement for a more extensive study combining angularly resolved electron spectroscopies with ion scattering on better characterized thin film samples.

ACKNOWLEDGEMENTS

We would like to thank Paul Chu for supplying the pressed oxide superconductor sample. Alex Ignatiev wishes to acknowledge support by NASA Grant NAGW-977 for a portion of this work.

REFERENCES

1   Y.S. Chen, G.L. Miller, D.A.H. Robinson, G.H. Wheatley and T.M. Buck, Surface Sci. **62** (1977) 133.

2   S.B. Luitjens, A.H. Algra, E.P.Th.M. Suurmeijer and A.L. Boer, Appl. Phys. **21** (1980) 205.

3   M.H. Mintz and J.A. Schultz, J. Less Common Metals **103** (1984) 349.

4   W. Eckstein, Nucl. Instr. and Meth. B**27** (1987) 78.

5   C.W. Chu, P.H. Hor, R.L. Meng, L. Gao, Z.J. Huang and Y.Q. Wang, Phys. Rev. Lett. **58** (1987) 405.

6   W.P. Poschenrieder, Int. J. Mass. Spectrom. Ion Physics., **9** (1972) 357.

7   H.A. Storms, K.F. Brown and J.D. Stein, Anal. Chem. **49** (1977) 2023.

8   Practical Surface Analysis by Auger and Photoelectron Spectroscopy, D. Briggs and M.P. Seah, John Wiley Ltd., 1983, Chichester.

PHOTON-STIMULATED DESORPTION FROM HIGH-T_c SUPERCONDUCTORS

R. A. Rosenberg and C.-R. Wen
Synchrotron Radiation Center and
University of Wisconsin-Madison, Stoughton, Wisconsin 53589-3097

ABSTRACT

We present results of photon-stimulated desorption (PSD) measurements from the high-T_c, 123 superconductors, $MBa_2Cu_3O_{7-x}$ where M = Y or Dy. The predominant desorbing species is O_2. Excitation spectra in the energy range 60-140 eV show structure due to excitation of Cu(3p) and Ba(4d) electrons. In the energy range 10-30 eV spectral features may be related to excitation of Ba(5p) and O(2s) electrons with possible contributions from bulk and surface plasmons. Irradiation with intense, unmonochromatized light results in a multi-component time decay of the oxygen PSD signal. Through analysis of the time-dependent data, we derive cross sections for O_2 PSD which may be as high as 3.5±3 Mb. Core level PSD spectra taken after white-light irradiation show a loss of the Ba-related feature and an energy shift in the Cu-related feature. These changes are related to modifications of the local oxygen environment caused by radiation damage.

INTRODUCTION

The properties of high-T_c superconductors such as $YBa_2Cu_3O_{7-x}$ (YBCO) and $DyBa_2Cu_3O_{7-x}$ (DBCO) are very sensitive to the oxygen content[1-3] (an orthorombic to tetragonal phase transition occurs at x = 0.5). Many recent papers have addressed the issue of oxygen loss in the high-T_c superconductors as a result of thermal[4] or radiation-induced processes.[5-8] The general conclusion is that these materials have a high propensity to lose oxygen.

Many applications of the new materials will necessitate exposure to radiation either during their processing or their implementation. When used in electronic devices, the materials may be exposed to radiation during plasma etching or X-ray or electron-beam lithography. These devices may then be used in radiation environments, such as accelerators or outer space. When used as materials for superconducting magnets for accelerators, they will be exposed to radiation produced by the accelerating or colliding particles. In order to investigate the mechanisms by which radiation-induced modifications of these materials occur, we have performed experiments to examine the process of photon-stimulated desorption (PSD) in the high-T_c superconductors.

Previous studies have examined the radiation-induced oxygen loss processes in YBCO indirectly, by examining changes in the electronic structure as the result of photon or electron irradiation using photoelectron spectroscopy[5,6] (PES), inverse photoelectron spectroscopy[6,7] (IPES), or electron-energy loss spectroscopy[8] (EELS). These studies found radiation-induced changes in the spectra in periods ranging from 10 min to several hours.

© 1989 American Institute of Physics

The experimental results reported here were obtained at the Synchrotron Radiation Center (SRC), University of Wisconsin. Experiments involving irradiation by intense, unmonochromatized synchrotron radiation (SR) were performed on the white-light beamline, while experiments utilizing monochromatized photons were performed on the 3m TGM beamline. The remaining experimental details have been reported previously.[9]

WHITE-LIGHT EXCITATION

In the first study to examine the radiation-induced desorption product directly, we reported on the time-dependence of the photodesorbed O_2 resulting from exposure of a freshly scraped, sintered YBCO surface to intense, unmonochromatized synchrotron radiation (SR).[9] Using a quadrupole mass spectrometer (QMS), we found O_2 to be the only detectable, **neutral** desorbing species. (The only positive ion observed was H^+; no attempt was made to detect negative ions.) Figure 1 shows the results of time-dependent O_2 PSD measurements resulting from the exposure of a freshly-scraped surface to white (unmonochromatized) light.

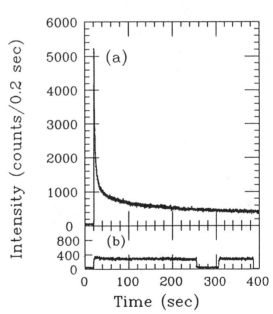

Fig. 1. Time dependence of the O_2 yield from YBCO upon irradiation by intense, white synchrotron radiation. At t=0, the photon shutter is closed. At t=20 sec the shutter is open. (a) Yield from a freshly-scraped, previously not irradiated sample. (b) Yield from the same, irradiated sample of (a) after closing the shutter briefly. The shutter is closed between t=255 and 308 sec. At t=388, the QMS is tuned to mass 33.

The results of Fig. 1(a) show that there is a significant decay of the O_2 signal with time from a freshly-prepared surface. Figure 1(b) shows that the O_2 signal can not be restored by migration of sub-surface oxygen in a few minute period; similar results were obtained if delays of up to 15 hours were made between measurements. The results of Fig. 1(b) also indicate that the O_2 PSD does not result from a thermal process; if it did, there would me a much slower decay of the O_2 signal when the shutter is closed at t = 255 seconds.

The rate of O_2 desorption is a direct monitor of the depletion of oxygen-containing sites. Each chemically distinct site will give rise to an exponentially-decaying signal. The best fit to the data of Fig. 1 was obtained by using a sum of three exponentials, with lifetimes ranging from 3 to 550 sec. The shortest lifetime was found to be inversely proportional to photon flux. From elementary kinetic theory it can be shown that the cross section for PSD should be inversely proportional to the product of the flux density and the lifetime.[9] Using the known flux density of the white-light beamline, we derived a cross section for PSD of weakly-bound O_2 of 3.5 ± 3 Mb. This value seems reasonable in light of previously reported electron and photon stimulated desorption cross sections for neutrals which lie in the range of 0.01 to 100 Mb.[10]

CORE-LEVEL EXCITATION

Results of our experiments using **monochromatized** SR[11] are shown in Fig. 2. This is the **first** data to show desorption of a **neutral** molecule from a surface as the result of core-level excitation. Fig. 2(a) shows the yield of O_2 from a scraped YBCO sample as a function of photon energy in the range 60-140 eV. In Fig. 2(b) is the total-electron yield (TEY, yield of electrons of all kinetic energies) spectrum, which is representative of the bulk photoabsorption,[12] from the same sample. The features in the TEY spectrum are assigned as: Cu(3p) excitation (threshold = 73.8 eV); Ba(4d) to discrete level excitation (peak = 95 eV); Ba(4d) to continuum excitation (threshold = 99 eV, peaks at 107 and 120 eV).[13]

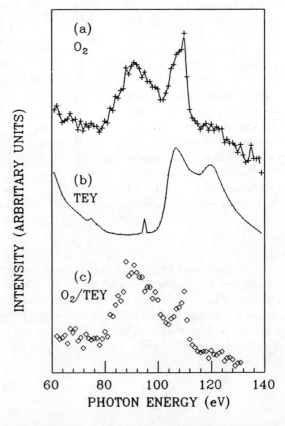

Fig. 2. Comparison of the O_2 PSD spectrum (a) with the TEY (photoabsorption) spectrum (b) of YBCO. In (c) is shown the ratio of the O_2 PSD to the TEY.

Two features dominate the O_2 PSD data: the lower energy feature has a threshold of ~80 eV and a peak at 92 eV, while the higher energy feature has a threshold at 102 eV and a sharp peak at 110 eV. We assign the lower-energy feature to excitation of Cu(3p) electrons and the higher-energy structure is a result of Ba(4d) excitation. In Fig. 2(c) is shown the ratio of the O_2 PSD spectrum to the TEY spectrum. The resulting spectrum is proportional to the number of O_2 molecules produced per photon absorbed. It shows that photoabsorption by Cu atoms is much more effective at producing O_2 than absorption by Ba atoms (Cu peak area/Ba peak area > 3:1). This coincides with the fact that in the ab planes of the YBCO unit cell there are 5 oxygen atoms bound directly to Cu atoms and only 2 oxygen atoms bound directly to Ba atoms.

The nature of the precursor states involved in the core-level O_2 PSD data of Fig. 2 has been discussed previously,[11] so only a brief discussion will be given here. At the Cu(3p) edge, excitation of a Cu^{+2} 3p satellite level results in O_2 PSD. Since this state lies at 7-8 eV higher binding energy than the main Cu^{+2} 3p level,[14] the threshold for O_2 PSD is shifted to higher energies relative to the the Cu(3p) photoabsorption edge (73.8 eV) as given by the TEY data. The satellite level is thought to arise from a bonding to antibonding, σ to σ^*, transition, which in an ionic picture is effectively a metal-to-ligand charge transfer. Auger decay of the satellite state may result in a final state with 2 or 3 holes plus 1 (excited) electron in an antibonding level (2h1e or 3h1e states). Such multi-hole, excited-electron configurations have been found to be very effective pathways for PSD of ions,[15] particularly when the holes are in bonding levels and the excited electron is in an antibonding orbital. In the case of YBCO a 3h1e state localized on an O^{-2} ligand or a 2h1e state localized on an O^- ligand could result in desorption of neutral O, provided the state remained localized for a sufficient time for desorption to occur (~10^{-13} sec). These repulsive states could then lead directly to desorption of O_2 or to production of ground or excited-state oxygen atoms. These oxygen atoms would have a high probability for reactive scattering with surface or near-surface oxygen to form O_2, since the binding energy of oxygen in O_2 (5.1 eV) is much higher than the free energy barrier for oxygen desorption (0.9-1.3 eV).[4]

The structure, with a peak at 110 eV in the O_2 PSD spectrum of Fig. 2 is due to excitation of Ba(4d) electrons. The energy position of the peak in the O_2 PSD spectrum is close to the energy of the first strong peak observed in the photoabsorption (TEY) data; however, there is a noticeable lack of structure in the O_2 PSD spectrum at higher energies (120 eV) corresponding to the second strong peak in the TEY data.

As discussed previously[11] we assign the peak at 110 eV in the O_2 PSD spectrum to be a result of a localized Ba^{+2} (4d→4f) excitation. The energy position of this peak is close to that of the 107 eV peak in the photoabsorption data. There is little or no structure in the O_2 PSD data which corresponds to the 120 eV peak in the photoabsorption data of Fig. 2(b). The reason for this may be that the states reached upon decay of the excitations leading to the 110 eV and 120 eV peaks are different. Some insight into this process may be derived by examining the Ba(5p) CIS data of YBCO.[16] This data gives the relative partial photoionization cross section for production of Ba^+ with a hole in the 5p orbital. The similarity of the O_2 PSD data and the Ba(5p) CIS data suggests that production of a Ba(5p) hole may be an important step in the O_2 desorption process. Data on valence-level O_2 PSD, which will be presented later, in which the Ba(5p) level is **directly** excited gives further credence to this argument.

In Fig. 3 we show the effect of intense irradiation by zero-order light from the 3m TGM beamline on the O_2 PSD spectrum from DBCO. The most dramatic change is the loss of the Ba-related peak at 109 eV. In addition the threshold of the Cu-related feature is shifted to higher energies by 2-3 eV.

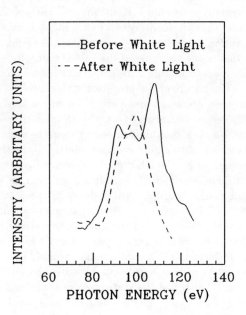

Fig. 3. O_2 PSD spectra from DBCO before(solid line) and after (dashed line) irradiation by intense white light.

We speculate that these changes arise from alterations in the surface or near-surface chemical composition. The loss of the Ba-related peak could be due to conversion of impurity $BaCO_3$ sites to BaO sites, by oxygen loss, as suggested by PES studies.[7] The shift in the Cu edge is also evidence of perturbation of the local electronic structure of the desorbing Cu site. Since the surfaces of these materials are not well characterized, we do not feel inclined to speculate further on the exact nature of these changes.

VALENCE-LEVEL EXCITATION

Figure 4 shows the yield of O_2 as a function of photon energy in the range 10-30 eV. The solid curve is the raw data and the dashed curve shows the data normalized to photon flux. There are two principal features: a weak threshold at ∼15 eV, and a strong structure which peaks at ∼23 eV. The peak at 11.5 eV, which is most evident in the normalized data, is a result of excitation of the 23 eV peak by second-order light.

The first important result obtained from the data of Fig. 4 is that the upper limit for the threshold for oxygen desorption is 15 eV. In order to determine if this is the true threshold, the measurements need to be extended to lower photon energies. This is not easily done due to the presence of large amounts of second-order light at lower energies. If 15 eV is the actual threshold then electronic excitations of energies less than 15 eV will not cause oxygen desorption.

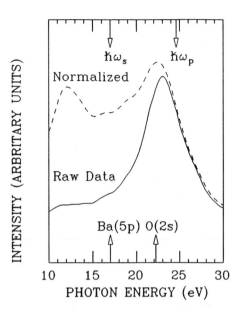

Fig. 4. O_2 PSD spectrum from DBCO in the energy range 10-30 eV presented as raw data (solid line) and normalized to incident photon flux (dashed line). The positions of Ba(5p), O(2s), surface and bulk plasmon excitation energies, as determined by EELS, are marked.[8,15]

In order to explain the structure in Fig. 4, we must try to understand what possible electronic excitations could lead to O_2 desorption. The deep core-level data of Fig. 2 can be succesfully interpreted in terms of excitation of Cu(3p) and Ba(4d) electrons. Similarly it is possible that excitation of shallow core levels may lead to the structure observed in Fig. 4. The most likely candidates in this energy region are the Ba(5p) and O(2s) levels. Arrows on the bottom of Fig. 4 mark the positions of Ba(5p) and O(2s) to conduction band transitions as revealed by EELS.[8,17] Since the energy positions of these transitions coincide with the structure in Fig. 4, it is possible that excitation of Ba(5p) and O(2s) electrons may lead to an antibonding state and desorption of O_2. The argument for participation of a Ba(5p) hole in the desorption process is strengthened by the fact that the Ba(5p) CIS data,[16] is similar to the O_2 PSD yield at the Ba(4d) edge.

There is an alternative explanation for the data of Fig. 4. Recently it has been shown that laser-light excitation of surface plasmons on small metal particles can lead to desorption of atoms.[18] In order to see if such a mechanism could be involved in desorption of O_2 from DBCO, we have marked the energy positions observed for excitation of surface and bulk plasmons in YBCO.[17] The proximity of the surface and bulk plasmon excitation energies to the structure in Fig. 4 gives strength to the argument that plasmon excitation could lead to oxygen desorption. The oxygen observed in our measurements does not necessarily originate from the surface. Atomic or molecular oxygen produced in the near-surface region should have a high probability for escaping the surface. In fact some of the observed O_2 may be formed by reactive scattering of atomic oxygen originating in the bulk. Therefore, excitation of bulk plasmons, as well as surface plasmons, could contribute to O_2 PSD.

CONCLUSIONS

In this paper we have presented results of PSD measurements on high-T_c superconductors. We are able to interpret the O_2 PSD results in terms of core-level excitation, with the possible involvement of bulk and surface plasmons in the lower-energy (10-30 eV) data. At this stage of our understanding it is difficult to gain further insight into the desorption process without theoretical input. A first step in this direction would be theoretical understanding as to why the copper-oxide superconductors have a relatively large propensity to lose oxygen. Obtaining a grasp of this basic concept should allow us to make significant progress in understanding the radiation-induced desorption process.

ACKNOWLEDGMENTS

We would like to thank David Larbalestier and Jeff Seuntjens for supplying the samples used in these studies. The experiments were performed at the Wisconsin Synchrotron Radiation Center, a national facility supported by the National Science Foundation.

REFERENCES

1. *"Chemistry of High Temperature Superconductors"*, D. L. Nelson, M. S. Whittingham, and T. F. George Eds., ACS Symposium Series **351**, Washington, DC (1988).
2. *"Thin Film Processing and Characterization of High-Temperature Superconductors"*, J. A. Harper, R. J. Colton and L. C. Feldman Eds., AIP Conference Proc. **165** (1988).
3. J. D. Jorgensen, M.A. Beno, D.G. Hinks, L. Soderholm, K.J. Volin, R.L. Hitterman, J.D. Grace, I.K. Schuller, C.U. Segre, K. Zhang, and M.S. Kleefisch, Phys. Rev. B **36**, 3608 (1987).
4. H. Strauven, J.P. Locquet, O.B. Verbeke, and Y. Bruynseraede, Solid State Comm. **65**, 293 (1988).
5. N. G. Stoffel, J.M. Tarascon, Y. Chang, M. Onellion, D. W. Niles, and G. Margaritondo, Phys. Rev. B **36**, 3986 (1987).
6. T. J. Wagener, Y. Gao, H.M. Meyer III, I.M. Vitomirov, C.M. Aldao, D.M. Hill, J.H. Weaver, B. Flandermeyer, and D.W. Capone II, in Ref. 2, p. 368.
7. Y. Gao, T.J. Wagener, J.H. Weaver, A.J. Arko, B. Flandermeyer, and D.W. Capone II, Phys. Rev. B **36**, 3971 (1987).
8. J. Yuan, L. M. Brown, and W. Y. Liang, J. Phys. C **21**, 517 (1988).
9. R. A. Rosenberg and C.-R. Wen, Phys. Rev. B **37**, 5841 (1988).
10. M.J. Drinkwine and D. Lichtman, Progress in Surf. Sci. **8**, 123 (1977). A.R. Burns, Phys. Rev. Lett. **55**, 525 (1985). F.L. Tabares, E.P. Marsh, G.A. Bach, and J.P. Cowin, J. Chem. Phys. **86**, 738 (1987).
11. R. A. Rosenberg and C.-R. Wen, Phys. Rev. B **37**, 9852 (1988).
12. A. Bianconi, Appl. Surf. Sci. **6**, 392 (1980).

13. M. M. Hecht and I. Lindau, Phys. Rev. Lett. **47**, 821 (1981). M. M. Hecht, Ph.D. Thesis, Stanford Synchrotron Radiation Laboratory Report No. 82/07, 1982 (unpublished), and references therein.
14. K.S. Kim, J. Elec. Spec. Rel. Phen. **3**, 217 (1974).
15. *"Desorption Induced by Electronic Transitions, DIET I"*, N. H. Tolk, M. M. Traum, J. C. Tully, and T. E. Madey, Eds. (Springer, New York, 1983).
16. M. Onellion, Y. Chang, D.W. Niles, R. Joynt, G. Margaritondo, N.G. Stoffel, and J.M. Tarascon, Phys. Rev. B **36**, 819 (1987).
17. Y. Chang, M. Onellion, D.W. Niles, R. Joynt, G. Margaritondo, N.G. Stoffel, and J.M. Tarascon, Solid State Comm. **63**,717 1987).
18. W. Hoheisel, K. Jungmann, M. Vollmer, R. Weidenauer, F. Träger, Phys. Rev. Lett. **60**, 1649 (1988).

SURFACE AND ELECTRONIC STRUCTURE OF Bi-Ca-Sr-Cu-O SUPERCONDUCTORS STUDIED BY LEED, UPS AND XPS

Z.-X. Shen, P.A.P. Lindberg, B.O. Wells, I. Lindau and W. E. Spicer
Stanford Electronics Laboratories, Stanford University, Stanford, CA 94305

D.B. Mitzi, C.B. Eom, A. Kapitulnik and T.H. Geballe
Department of Applied Physics, Stanford University, Stanford, CA 94305

P. Soukiassian
Department of Physics, Northern Illinois University, Dekalb, Illinois 60115

ABSTRACT

Single crystal and polycrystalline samples of $Bi_2CaSr_2Cu_2O_8$ have been studied by various surface sensitive techniques, including low energy electron diffraction (LEED), ultraviolet photoemission spectroscopy (UPS) and x-ray photoemission spectroscopy (XPS). The surface structure of the single crystals was characterized by LEED to be consistent with that of the bulk structure. Our data suggest that $Bi_2CaSr_2Cu_2O_8$ single crystals are very stable in the ultrahigh vacuum. No change of XPS spectra with temperature was observed. We have also studied the electronic structure of $Bi_2Sr_2CuO_6$, which has a lower superconducting transition temperature T_c. Comparing the electronic structure of the two Bi-Ca-Sr-Cu-O superconductors, an important difference in the density of states near E_F was observed which seems to be related to the difference in T_c.

INTRODUCTION

After the discovery of $YBa_2Cu_3O_{7-\delta}$[1], new superconductors of Bi-Ca-Sr-Cu-O were found recently [2,3]. There are several interesting properties of the Bi-Ca-Sr-Cu-O superconductors as compared to the $YBa_2Cu_3O_{7-\delta}$ compound. Firstly, there are no one dimensional Cu-O chains in Bi-Ca-Sr-Cu-O compounds, in contrast to the $YBa_2Cu_3O_{7-\delta}$ compound where one dimensional Cu-O chains exist as a result of ordered oxygen vacancies. Therefore, there are no intrinsic oxygen vacancies in the Bi-Ca-Sr-Cu-O compounds, and one expects that Bi-Ca-Sr-Cu-O compounds might be more stable than the $YBa_2Cu_3O_{7-\delta}$ compound. Secondly, according to one electron band calculations, the states near E_F for the $Bi_2CaSr_2Cu_2O_8$ compound have contributions from both Cu 3d - O 2p states of $Cu-O_2$ planes and Bi 6p - O 2p states of Bi-O planes.[4, 5] This is different from the $YBa_2Cu_3O_{7-\delta}$ compound where a band calculation predicts that the states near E_F are only Cu 3d - O 2p hybrids from both the $Cu-O_2$ planes and Cu-O chains [6]. Even though there is evidence that band theories can not account for the correlation effects of Cu 3d electrons very well [7, 8, 9], they are expected to give a good description of the Bi 6p - O 2p states. Therefore, it is plausible that the origin of the states near E_F in the Bi-Ca-Sr-Cu-O compounds is somewhat different than that of the $YBa_2Cu_3O_{7-\delta}$ compound. Thirdly, The superconducting transition temperature for Bi-Ca-Sr-Cu-O compounds seems to be related to the number of the $Cu-O_2$ layers in a unit cell. For example, the superconducting transition temperatures for the compounds with one, two and three Cu-

O_2 layers (Corresponding to the nominal compositions of $Bi_2Sr_2CuO_6$, $Bi_2CaSr_2Cu_2O_8$ and $Bi_2CaSr_2Cu_3O_{10}$) are 20, 85 and 105 K, respectively [10]. Given these interesting properties of Bi-Ca-Sr-Cu-O compounds, it is well motivated to study their electronic structures and compare them to $YBa_2Cu_3O_{7-\delta}$ as well as to each other. In this paper, we present our study of single-crystalline and polycrystalline Bi-Ca-Sr-Cu-O samples. We find that Bi-Ca-Sr-Cu-O compounds are very stable in the ultrahigh vacuum (UHV) and their XPS spectra do not change with temperature. There is a clear difference in the density of states near E_F between Bi-Ca-Sr-Cu-O compounds with one and two $Cu-O_2$ layers, which seems to be related to the difference in the superconducting transition temperatures.

II. EXPERIMENT

Single crystals of $Bi_2CaSr_2Cu_2O_8$ as well as the polycrystalline samples of $Bi_2CaSr_2Cu_2O_8$ and $Bi_2Sr_2CuO_6$ were prepared by mixing powders of Bi_2O_3, $Sr(CO_3)_2$, $CaCO_3$ and CuO. All the samples were determined by X-ray diffraction measurements to be almost pure phase. The superconducting transition temperature T_c was determined by magnetic measurements to be 85 K and 10 K for the $Bi_2CaSr_2Cu_2O_8$ and $Bi_2Sr_2CuO_6$ samples, respectively. The single crystalline samples were cleaved in situ, while the polycrystalline samples were scraped in situ by a diamond file. The XPS and LEED data presented in this paper were taken in a Varian photoemission chamber with base pressure better than 2×10^{-10} torr. The UPS experiments were performed using the 3M Toroidal Grating Monochromator (TGM) beam line at the Synchrotron Radiation Center of the University of Wisconsin at Madison, with a base pressure of 4×10^{-11} torr. The photoelectrons were detected by a double pass cylindrical mirror analyzer, with over all resolutions for UPS and XPS of 0.3 and 1.2 eV, respectively.

III. RESULTS AND DISCUSSIONS

Fig. 1 presents a LEED pattern taken from a UHV cleaved single-crystalline surface of $Bi_2CaSr_2Cu_2O_8$ compound. The cleavage plane of the single crystal has been determined by an x-ray diffraction experiment to be a-b planes.[11] This LEED pattern shows that the unit cell vectors along the two vertical axis have a relative ratio of 1 to 5, which is consistent with the fact that there is a superstructure along the b axis so that the b axis is 5 times longer than the a axis[11]. The same LEED pattern can still be obtained after the single crystalline surface has first been exposed to 20 L of pure oxygen and then kept in UHV for

Fig.1 Low energy electron diffraction pattern from a $Bi_2CaSr_2Cu_2O_8$ single crystal.

about 30 hrs, except that the diffraction spots are somewhat fuzzier, demonstrating that the single crystalline surface of the $Bi_2CaSr_2Cu_2O_8$ compound is very stable in the UHV. As we will discuss later, the XPS spectra from single crystalline $Bi_2CaSr_2Cu_2O_8$ samples are also very stable in UHV.

Fig. 2 presents XPS spectra of the O 1s (up panel) and Bi 4f (lower panel) core level from a $Bi_2CaSr_2Cu_2O_8$ single crystal at different temperatures. The O 1s core level is a good criterion to check the quality of the surface. For the $YBa_2Cu_3O_{7-\delta}$ compound, the O 1s core level usually shows two components located at binding energies of -528 and -531 eV. In addition, the relative intensity ratio of the two components changes with time [7]. Now most people believe that the higher binding energy component is due to the surface contaminations [7]. Therefore, the fact we see a singlet O 1s peak (upper panel of Fig. 2) gives us confidence of the quality of the single crystalline surface. In contrast to the $YBa_2Cu_3O_{7-\delta}$ compound, we found that the O 1s core level spectrum is very stable in the $Bi_2CaSr_2Cu_2O_8$ compound. Very similar spectrum can be obtained 2 days after cleaving in UHV. The structure at the lower binding energy side of the O 1s peak is due to a ghost line of the x-ray source, while the weak broad features at higher binding energy side of the O 1s core level are the energy loss features of the O 1s photoelectrons[12]. The Bi 4f core level shown in the lower panel is a doublet without any trace of a second component, consistent with the crystal structure of the $Bi_2CaSr_2Cu_2O_8$ compound. It is clear that the O 1s and Bi 4f core level spectra from the single crystal do not change with the temperature. This again suggests that the samples are stable in the UHV. Comparing with our earlier results from the $Bi_2CaSr_2Cu_2O_8$ polycrystalline samples, we find the XPS spectra taken from the polycrystalline and single crystalline samples are the same, proving that the results from the polycrystalline $Bi_2CaSr_2Cu_2O_8$ samples are meaningful since they agree with those obtained from high quality single crystals[13].

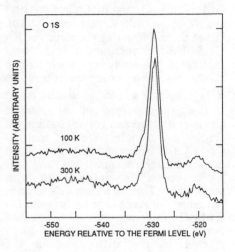

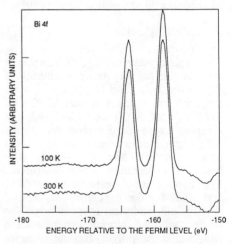

Fig.2 XPS spectra of O 1s (upper panel) and Bi 4f core levels (lower panel) from a $Bi_2CaSr_2Cu_2O_8$ single crystal at two different temperatures.

Taking PES spectra at different photon energies will help us to identify the origin of the states in the valence band. Fig. 3 shows the EDC's of both $Bi_2CaSr_2Cu_2O_8$ (2 Cu L) and $Bi_2Sr_2CuO_6$ (1 Cu L) polycrystalline samples at various photon energies. All the EDC spectra are normalized to have the same maximum intensity. The structure at binding energy about -12.3 eV, which is usually called a d^8 satellite, exhibits a resonant behavior as the photon energy is tuned through the Cu 3p-3d absorption edge near 74 eV. The strength of the resonance, however, appears to be weaker than what is observed in the $YBa_2Cu_3O_{7-\delta}$ and $(LaSr)_2CuO_4$ compounds[8]. The enhancement of the satellite at 74 eV is due to the contributions of Cu states. As previously found in the $YBa_2Cu_3O_{7-\delta}$ compound [7, 8], the existence of the d^8 satellite reveals the strong correlations among the 3d electrons. Fig.3 also displays a spectrum obtained from another, but identical $Bi_2CaSr_2Cu_2O_8$ polycrystalline sample at a photon energy of 1253.6 eV, where the photoionization cross section of Cu 3d states will be at least an order of magnitude larger than that of the states from Bi, O, Ca, or Sr in the valence band. Therefore, the PES spectrum at this higher photon energy essentially gives the Cu partial density of states. We notice that the satellite position, obtained from this spectrum, is different from what one finds from the resonance data (see marks in the figure) and this difference can not be totally explained by the differences in the resolution of the XPS (1.2 eV) and UPS (0.3 eV) measurements. The origin of this discrepancy is not fully understood. Nevertheless, this might due to the fact that the photoemission and resonance photoemission involve different photoionization channels[14]. The d^8 final states can have different multiple components S, P, D, F and G, with the F and G components the strongest[15]. The resonance photoemission involves an Auger process which has very different transition matrix element for the different components. This is probably the reason why the higher binding energy component shows much stronger resonance behavior. Similar effects has been observed by Ghijsen et al. in the CuO system[16]. It is worthwhile to note that the Bi 6s states have binding energy near 11 eV[4], even though the photoionization cross section of these states is small at photon energy of 1253.6 eV.

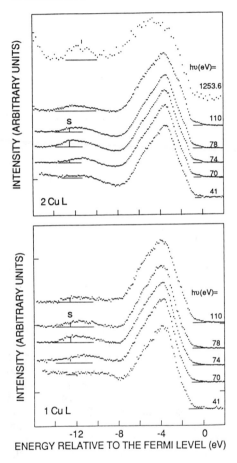

Fig.3 EDC's recorded at different photon energies for polycrystalline $Bi_2CaSr_2Cu_2O_8$ and $Bi_2Sr_2CuO_6$ compounds. The feature marked S is the d^8 satellite which is enhanced slightly at 74 eV.

It is also clear in this figure that the $Bi_2CaSr_2Cu_2O_8$ compound has higher density of states near E_F than the $Bi_2Sr_2CuO_6$ compound. Because the states near the Fermi level are important to the superconductivity, we suggest that the observed difference in the density of states near E_F in the two compounds is related to the difference in the superconducting transition temperature. We have also performed constant-initial-state (CIS) measurements to check the origin of the states near E_F for the Bi-Ca-Sr-Cu-O compounds[17]. We find that the states near E_F in the Bi-Ca-Sr-Cu-O compounds have less Cu character than those of the $YBa_2Cu_3O_{7-\delta}$ compound, which we attribute to the Bi 6p - O 2p states in the Bi-Ca-Sr-Cu-O compounds.

IV. CONCLUSION:

W have performed XPS, UPS and LEED experiments on $Bi_2CaSr_2Cu_2O_8$ and $Bi_2Sr_2CuO_6$ compounds. The $Bi_2CaSr_2Cu_2O_8$ compound appears to be more stable in UHV than the $YBa_2Cu_3O_{7-\delta}$ compound. For a single crystal of the $Bi_2CaSr_2Cu_2O_8$ compound, XPS spectra can reproducibly be obtained 2 days after the cleaving in the UHV. No change of the XPS spectra was observed when the material is cooled down to 100 K. The d^8 like satellite, which undergoes a resonance process at 74 eV, is observed in both the $Bi_2CaSr_2Cu_2O_8$ and $Bi_2Sr_2CuO_6$ compounds, signaling the importance of the correlation effects. A clear difference is observed between the density of states near E_F in $Bi_2CaSr_2Cu_2O_8$ and $Bi_2Sr_2CuO_6$ compounds, which might be related to the difference in the superconducting transition temperature.

ACKNOWLEDGEMENTS

We would like to thank D.S. Dessau for his help in the experiments. The UPS experiments were performed at the Synchrotron Radiation Center of the University of Wisconsin. We acknowledge support from the National Science Foundation - Materials Research Laboratory Program at the Center for Materials Research at Stanford University, the Air Force Contract AFOSR-87-0389 and the JESP contract DAAG 29-85-K-0048.

REFERENCES:

1. M.K. Wu, J.R. Ashburn, C.J. Torng, P.H. Hor, R.L. Meng, L.Gao, Z.J. Huang, Y.Q. Wang and C.W. Chu, Phys. Rev. Lett. **58**, 908 (1987)
2. H. Maeda, Y.Tanaka, M. Fukutami and T. Asano, Jpn. J. Appl. Phys. Lett. **27**, L209 (1988)
3. R.M. Hazen, C.T. Prewitt, R.J. Angel, N.L. Ross, L.W. Finger, C.G. Hadidiacos, D.R. Veblen, P.J. Heaney, P.H. Hor, R.L. Meng, Y.Y. Sun, Y.Q. Wang, Y.Y. Xue, Z.J. Huang, L. Gao, J. Bechtold and C.W. Chu, Phys. Rev. Lett. **60**, 1174 (1988)
4. M.S. Hybertsen and L.F. Mattheiss, Phys. Rev. Lett. **60**, 1661 (1988)
5. H. Krakauer and W.E. Pickett, Phys. Rev. Lett. **60**, 1655 (1988)
6. L.F. Mattheiss and D.R. Hamann, Solid State Communication **63**, 395 (1987)
7. J.C. Fuggle, J. Fink and N. Nücker, to be published in Int. J. Mod. Phys. B (1988); and the references therein
8. Z.-X. Shen, J.W. Allen, J.J. Yeh, J.-S. Kang, W. Ellis, W.E. Spicer, I. Lindau, M.B. Maple, Y.D. dalichaouch, M.S. Torikachvili, J.Z. Sun and T.H. Geballe, Phys. Rev. B **36**, 8414 (1987)
9. Z.-X. Shen, P.A.P. Lindberg, I. Lindau, W.E. Spicer, C.B. Eom and T.H. Geballe, Phys. Rev. B **38**, 7152 (1988)
10. J.M. Tarascon, Y. Lepage, P. Barboux, B.G. Bagley, L.H. Greene, W.R. Mckinnon, G.W. Hull, M. Giroud and D.M. Hwang, preprint, submitted to Phys. Rev. B

11. P.A.P. Lindberg, Z.-X. Shen, B.O. Wells, D.B. Mitzi, I. Lindau, W.E. Spicer and A. Kapitulnik, to be published in Appl. Phys. Lett.
12. Z.-X. Shen, P.A.P. Lindberg, D.S. Dessau, D.B. Mitzi, I. Lindau, W.E. Spicer and A. Kapitulnik, preprint, submitted to Phys. Rev. B
13. Z.-X. Shen, P.A.P. Lindberg, B.O. Wells, D.B. Mitzi, I. Lindau, W.E. Spicer and A. Kapitulnik, Phys. Rev. B, Dec. 1, 1988, in press.
14. L.C. Davis, Phys. Rev. B 25, 2912 (1982)
15. M.R. Thuler, R.L. Benbow and Z. Hurych, Phys. Rev. B 26, 669 (1982)
16. J. Ghijsen, L.H. Tjeng, J. van Elp, H. Eskes, J. Westerink, G.A. Sawatzky and M.T. Czyzyk, preprint.
17. Z.-X. Shen, P.A.P. Lindberg, P. Soukiassian, C.B. Eom, I. Lindau, W.E. Spicer and T.H. Geballe, to be published in Phys. Rev. B

OPTICAL PROPERTIES OF HIGH-T_c SUPERCONDUCTORS: WHO NEEDED THIS ANYWAY

D. E. Aspnes and M. K. Kelly

Bellcore, Red Bank, N.J. 07701-7020 USA

ABSTRACT

We summarize the current status of optical spectroscopy of high-T_c superconductors. Experimental consistency is being approached for single-crystal samples in the infrared, although the interpretation of the data remain controversial. The absence of strong anisotropy in the visible-near UV allows the use of polycrystalline samples to study compositional systematics. Some optical features can be associated with specific structures in the large unit cell, permitting comparisons among materials and with band structure calculations. A strong absorption line at 4.1 eV in oxygen-deficient $YBa_2Cu_3O_{7-\delta}$ is due to a $O\text{-}Cu^{1+}\text{-}O$ local mode and can be used to indicate oxygen composition at surfaces and interfaces.

INTRODUCTION

The discovery of high-T_c superconducting compounds has created new challenges in all areas of solid-state physics.[1] Aside from their spectacularly unusual (and not yet understood) transport properties, these materials exhibit unconventional behavior in a number of areas including optical response. Controversy has arisen not only over the nature of the excitations giving rise to features in optical spectra, particularly in the infrared, but even over which features are actually intrinsic. A major difficulty, characteristic of any new materials field, is sample preparation. In contrast to popular images of high-school students preparing high-T_c compounds in kitchen ovens, these are actually very sophisticated materials, more like semiconductors in their sensitivities to stoichiometry, impurity concentrations, etc. In fact, compared to the early days of semiconductor physics, progress has been nothing less than phenomenal.

Additional complications arise in optical spectroscopy, particularly in the IR, owing to their biaxial, anisotropic nature. Initial data were obtained on sintered samples with randomly oriented grains of characteristic dimensions of the order of the wavelength of light, and mixed contributions from the relatively featureless, highly reflecting metallic basal plane and the highly structured, poorly reflecting c-axis orientation. Consequently, analysis required a combination of crystal optics, effective-medium theory, and scattering theory.[2] Weak structure that could have arisen from the superconducting gap was obscured by much stronger phonon features. However, progress has been rapid, and single crystals of reasonable size and homogeneity are now available in at least some phases. Convergence of experimental results from different samples and laboratories has been satisfactory, even if interpretations remain controversial.

Here, we summarize some aspects of the present situation and indicate potentially fruitful avenues of future development. Our purpose is to summarize understanding of the electronic structure gained from optical investigations. The IR has received the most attention because the IR response should give direct information on superconducting gaps and on the free carriers and interactions that are involved in superconductivity. As surfaces are more of a problem in tunneling experiments, the hope is that IR measurements will more accurately represent intrinsic properties and, especially, unequivocally determine these gaps. The visible-near UV optical properties have similarities with other transition-metal oxides, including relatively small dielectric functions with structures that can be assigned to interband or d-d transitions, or to charge-transfer excitations. The lack of strong anisotropy and the availability of ellipsometric techniques that provide direct information on the dielectric response and are less sensitive to scatter have made possible systematic compositional studies on polycrystalline samples. Some optical features can be associated with specific structures in the large unit cell, permitting comparisons of materials and with band structure calculations.

II. Far- and mid-IR.

In the BCS weak-coupling, T = 0 limit, optical structure arising from the superconducting gap is expected to be found at $2\Delta = 3.53kT_c$ (the factor of 2 arises because carriers can only be created in pairs),[3] although strong-coupling theories predict values of 2Δ as high as $11.5kT_c$.[4] These limits indicate that 2Δ should occur between 100 and 320 cm^{-1} for $La_{1.85}Sr_{0.15}CuO_4$ with T_c = 40 K, 220 and 735 cm^{-1} for $YBa_2Cu_3O_7$ with T_c = 92 K, and 300 and 1000 cm^{-1} for $Tl_2Sr_2Ca_2Cu_3O_{10}$ with T_c = 125 K. However, other theories suggest that the high-T_c superconducting mechanism may be gapless. Consequently, a fairly wide spectral range is involved.

Essentially all presently available IR data are in the form of the normal-incidence R, owing to the absorptive nature of these materials and the difficulty of preparing homogeneous films. As R does not provide direct information on the excitations that give rise to the optical response, these data are analyzed by calculating the optical conductivity, σ, by means of a Kramers-Kronig transform after suitably extrapolating R into experimentally inaccessible spectral regions. While the dielectric function, $\epsilon(\omega)$, is the more fundamental quantity, $\sigma(\omega)$ is preferred because $\sigma(0)$ is finite. The two are related according to $\sigma = -i\omega(\epsilon - 1)/4\pi$.

Surface roughness is the major experimental difficulty in both polycrystalline and single-crystal samples, because it reduces R below intrinsic values. While this is ordinarily not a problem for experiments designed to identify structure in optical spectra, it is significant here because the reflectance in the superconducting state is unity for $\omega < 2\Delta$, so small errors in R lead to large errors in σ. In fact, R measurements would not even be attempted for metals; calorimetric absorptance measurements on thin films are preferred. However, the occurrence of large extrinsic loss mechanisms due to random grain orientations and light scattering among wavelength-sized grains so far has precluded this approach, although it should be very useful as better samples become available. Methods to compensate reflectances for roughness artifacts include the measurement of reference R spectra after the sample has been overcoated with a highly reflecting material such as Pb or Au, and the comparison of data obtained at temperatures above and below T_c.

Examples of IR spectra representative of the reflectance of polycrystalline[5] and single-crystal[6] $YBa_2Cu_3O_{7-x}$ samples are shown in Figs. 1 and 2, respectively. Phonon structure occurs throughout the expected 2Δ range in the polycrystalline sample but is suppressed in the single-crystal data, where the incident electric field is oriented parallel to the metallic basal plane. Crystal-optics calculations show that additional structure can arise from singularities in the dielectric response of uniaxial or biaxial materials when at least one of the principal components of the dielectric tensor is metallic.[7] While considerable effort has been expended in analyzing polycrystalline data for information pertinent to superconductivity,[2] including various cationic substitutions intended to move phonon structure away from the expected gap structure,[8] the single-crystal data provide a better opportunity to identify the gap.

IR data of metallic form are typically analyzed by comparing them to a Drude or a combination of Drude and damped harmonic oscillator responses. The appropriate equation is:

$$\sigma = \frac{\omega_p^2 \omega}{4\pi(\omega\Gamma - i(\omega^2 - \omega_g^2))}, \qquad (1)$$

where ω is the frequency, $\omega_p^2 = 4\pi ne^2/m$ is the square of the plasma frequency, where n and m are the volume density and mass of the contributing carriers, Γ is the broadening parameter, and ω_g is the transition energy. For free carriers, $\omega_g = 0$ and Eq. (1) reduces to the Drude form. At zero frequency $\sigma(\omega)$ reduces to the dc conductance $\sigma_{dc} = \sigma(0) = ne^2/(m\Gamma) = ne^2\tau/m$, where τ is the excitation lifetime. The R response shown in the upper part of Fig. 2 is essentially classic Drude, and was fit to Eq. (1) with $\omega_g = 0$, $\omega_p = 25\,000$ cm^{-1}, and $\Gamma = 7500$ cm^{-1}.[6] The Hagen-Rubens form $R(\omega \to 0) \simeq 1 - 2(2\omega\Gamma/\omega_p^2)^{1/2}$ as $\omega \to 0$ and the characteristic flattening near $\omega = \Gamma$ that results from the associated change in functional dependence are clearly apparent. The fitted expression yielded the σ spectrum shown in the lower part, whose zero-frequency extrapolation was in very good agreement with the measured dc resistivity of 700 $\mu\Omega$-cm, a useful check on both optical results and sample quality. Assuming that

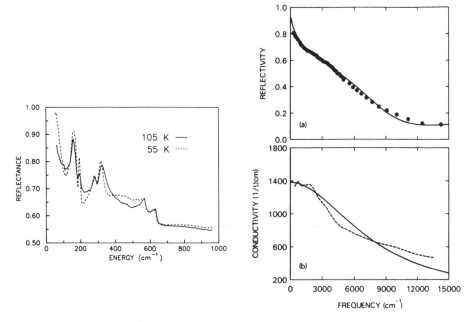

Fig. 1. Reflectance of polycrystalline $YBa_2Cu_3O_{6.85}$ with $T_c = 89$ K in the normal (solid line) and superconducting (dashed line) states (after ref. 5.)

Fig. 2. (top) Room-temperature basal-plane reflectance of single-crystal $YBa_2Cu_3O_{7-\delta}$ with $T_c = 92$ K (points), along with a Drude fit (line). (bottom) Corresponding calculated optical (dashed line) and Drude (solid line) conductivities (after ref. 6.)

$m = m_e$, where m_e is the free electron mass, and a formal valency of 1 free carrier per unit cell ($n = 7 \times 10^{21}$ cm^{-3}) leads to a mean free path $l = 30$ Å and a plasma frequency $\omega_p = 23\,200$ cm^{-1}. The plasma frequency is reasonable, but the mean free path is unusually short.

A more detailed examination of the region below 700 cm^{-1}, with the objective of determining 2Δ, is shown in Fig. 3.[6] To compensate for scattering artifacts, ratios of below-T_c spectra to a characteristic above-T_c spectrum were calculated. The dashed curve indicates the ratio expected from the Mattis-Bardeen theory.[9] The overall lineshape (if not amplitude) agreement suggests that $2\Delta \simeq 500$ cm$^{-1} \simeq 5kT_c$, which is well into the strong-coupling regime. The temperature dependence shown in the inset is consistent with the expected variation of 2Δ.

Virtually the same data were recently reported by a different laboratory, but the conclusions were totally different.[10] These data are shown in Fig. 4. For

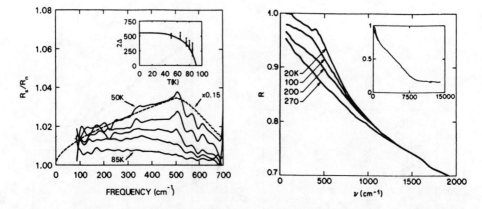

Fig. 3. Ratio of superconducting-to-normal R data for the sample of Fig. 2 for various temperatures (solid lines). The dashed curve shows the spectral dependence predicted by the Mattis-Bardeen theory for $2\Delta = 500$ cm^{-1} (note reduction in amplitude.) The inset shows the temperature dependence of 2Δ obtained in this way vs. the expected dependence (after ref. 6.)

Fig. 4. Reflectance of a YBa$_2$Cu$_3$O$_{7-\delta}$ sample with $T_c = 50$ K at several temperatures (after ref. 10.)

this YBa$_2$Cu$_3$O$_{7-\delta}$ sample, $T_c = 50$ K. The near similarity between the data sets of Figs. 3 and 4 at long wavelengths can be appreciated by mentally taking the difference between the 20 and 270 K spectra of Fig. 4, which shows a maximum near 500 cm^{-1} as in the corresponding data of Fig. 3. The similarity continues into the mid-IR, where a fit to the data of Fig. 4 yielded $\omega_p \simeq 24\,000$ cm^{-1}, $\Gamma \simeq 7500$ cm^{-1}, and $\omega_g \simeq 1700$ cm^{-1} (the finite threshold energy has only an incidental effect on ω_p and Γ_p, and was used for reasons discussed below.) The essential difference occurs at a level that would ordinarily be called detail. The data of Fig. 4 attain a value of unity to within experimental uncertainty at and below 120 cm^{-1}. Consequently, the feature at 120 cm^{-1} was assigned to 2Δ, which leads to $2\Delta \simeq 3.5kT_c$ and suggests a weak-coupling BCS limit. In addition, this seemingly minor distinction has a profound influence on σ for $\omega < 300$ cm^{-1}, as shown in Fig. 5. While σ in Fig. 2 extrapolates smoothly from 2000 cm^{-1} to zero, σ in Fig. 5 contains a peak below 1000 cm^{-1} whose width decreases essentially linearly in temperature. If this structure is separately fit to a Drude expression assuming a free electron mass, the value of ω_p so determined indicates that the carrier density is only 10% of that calculated from the formal valence.

The difficulty here is the appearance of two different types of free carriers, one of which has a very low density. Drawing a parallel with heavy-

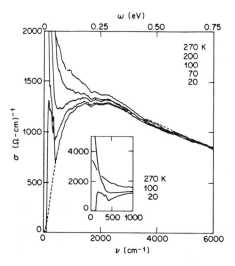 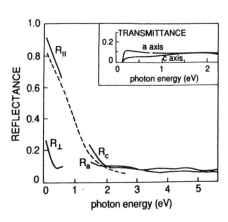

Fig. 5. Conductivity calculated from the data of Fig. 4 (after ref. 10.)

Fig. 6. Reflectances parallel and perpendicular to the c-axis of a $YBa_2Cu_3O_{7-\delta}$ single crystal (after ref. 12.)

Fermion systems and strong-coupling superconductors such as Pb, Thomas et al.[10] interpret these data as different manifestations of the *same* carriers, whose nature is changed as a function of energy due to a strong interaction with a set of (unspecified) optically inactive excitations. The interaction effectively renormalizes the carrier mass at low energies, allowing the anomalously low plasma frequency to be interpreted not as a low carrier density but as an enhanced mass. Consequently, a coherent description of all IR data is obtained. This Holstein mechanism, which is well known in metals physics, is analogous to an indirect transition in a semiconductor, where the absorption of a photon simultaneously creates an excitation along with the electron-hole pair. Higher-order corrections are usually included self-consistently as discussed by Allen.[11] In order to perform calculations to assess this interpretation, Thomas et al. assumed a uniform spectrum of excitations ranging from 250 to 550 cm^{-1}, which indeed gave a good account of the results and which were also supported by an estimate of the coupling constant derived from the jump in specific heat.

These interpretations are mutually inconsistent and need to be tested independently, but to do so will require data that are even better than the standards set by Figs. 2 and 4. Allen[11] has pointed out that the Holstein mechanism should give rise to wavelength-modulation-type difference spectra between superconducting and normal states, since the energy required to create an electron-hole pair increases by 2Δ as the material becomes superconducting. Data sufficiently accurate to see this effect are not yet available.

Recently, single crystals have become large enough to perform R measurements for light polarized along the c axis as well as in the basal plane. Results shown in Fig. 6 indicate that the c-axis reflectance may also be metallic, but with a greatly reduced carrier density.[12] However, the critical test of metallicity is the zero-frequency limit, which has not yet been reached. Values of R above 1.5 eV in Fig. 6 were calculated from ellipsometric data on substantially oriented thin films that were deposited on $SrTiO_3$ substrates, and show that anisotropy is significantly less in the visible-near UV. Representative transmittance spectra for these samples are shown in the inset.

III. Near-IR, visible, near-UV.

While the mid-IR reflectance data of Figs. 2 and 4 appear to be fairly accurately described by the Drude model, this representation fails badly in the near-IR as an expected minimum in R does not occur. Consequently, additional absorption is also present. Because R data are not able to provide definitive information about absorptance, this point has been controversial. However, this spectral range, as well as the visible and near-UV, is accessible to alternative methods of study, ellipsometry and transmittance. Ellipsometry is less sensitive to rough surfaces and scattering from macroscopic inhomogeneities than reflectance and is able to determine both real and imaginary parts ϵ_1, ϵ_2 of the dielectric function in a single non-normal-incidence measurement. Transmittance yields absorptance essentially directly. These approaches facilitate measurements on sintered pellets as well as on crystals and thin films. The relatively weak anisotropy in the visible-near UV also allows easier interpretation of results on unoriented samples. These capabilities are significant because most research on the chemistry of these materials is done using sintered samples, which have made it possible to study a wide range of phases and chemical substitutions. Generally speaking, the dielectric properties of these high-T_c materials are similar to those of other metal oxides, with relatively featureless dielectric functions of approximate value $\epsilon \simeq 4 + i1.5$. This is the origin of their dark grey color.

We continue the discussion with emphasis on the mid-IR. Figure 7 shows spectroellipsometric ϵ_2 data taken from 0.5 to 4.0 eV on a room-temperature polycrystalline $YBa_2Cu_3O_7$ sample; data at a temperature of 80 K were nearly the same.[13] Structure is seen on a relatively flat absorptive background above 1.8 eV, but it is obscured by free-carrier effects at lower energies. To enhance these spectral features, a free-carrier Drude background of strength defined by the dc conductance was subtracted. The result showed a peak near 2.6 eV and evidence of a peak near 0.5 eV, but because 0.5 eV is the limit of the accessible range its identification is uncertain. The origin of the mid-IR peak in sintered samples has been particularly controversial, being attributed to various mechanisms including excitonic absorption and an effective-medium blend of metallic and insulating constituents.[14] Recent theoretical calculations have also attributed this structure to interband transitions from the bridging O that connects the CuO_2 planes to the CuO chains.[15] The apparent absence of this structure in single-crystal samples[6,10] suggests that any interband transition may be weak or strongly polarization dependent.

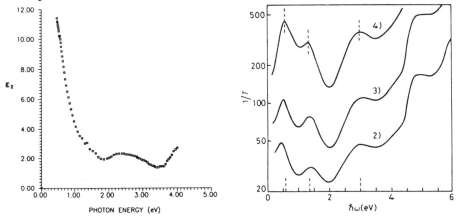

Fig. 7. Imaginary part of the dielectric function of sintered $YBa_2Cu_3O_{7-\delta}$ at room temperature. The data were obtained by ellipsometry (after ref. 13.)

Fig. 8. Reciprocal transmittance of $YBa_2Cu_3O_{7-\delta}$ films. The vertical lines indicate the peaks that are assigned to d-d transitions (after ref. 16.)

Transmittance experiments on thin films provide another perspective of this spectral region. Figure 8 shows inverse transmittance data of a $YBa_2Cu_3O_{7-\delta}$, $\delta \simeq 0$, film deposited on Al_2O_3.[16] X-ray analysis indicated a high degree of orientation with the c axis perpendicular to the surface. These data again reveal structure at 0.6 eV and show additional features at 1.4 and 3.0 eV (in contrast to the essentially featureless data shown in the inset of Fig. 6.) By analogy to similar spectra for NiO, these features were interpreted as d-d transitions within the partially filled Cu3d shell even though the analogous transitions in NiO are about 2 orders of magnitude weaker. The d-d interpretation of the 1.4 eV peak (and perhaps the 0.6 eV peak) is consistent with their apparent absence in reflectance and ellipsometric measurements, because d-d transitions are dipole-forbidden although the selection rules in these materials could be partially relaxed depending on composition. In the presence of a strong Drude response, weak structure is easier to see in transmittance than in reflectance or ellipsometry.

We next consider compositional systematics as studied by optical spectroscopy in the visible-near UV. Having a wide range of specimens with which to work, it becomes possible to draw conclusions based on similarities and differences over a substantial database. A feature common to many of these materials is the broad absorption structure near 2.7 eV, as seen in the ϵ_2 data for $La_{2-x}Sr_xCuO_{4-\delta}$ shown in Fig. 9.[17] It is also present in the 90 K material[16,18,19] and in several nonsuperconducting phases. Changes in composition that strongly affect conductivity have little effect on this feature, and only minor changes are noticed when various cations are substituted for Y and Ba in $YBa_2Cu_3O_{7-\delta}$. It is thus clear that this is attributable to transitions on the CuO_2 plane, which is common to these samples. In fact, most of the optical behavior in the visible-near UV originates with Cu and O. Substitutions for the other cations cause only minor changes due to changes in electronegativity and in lattice parameters from different ionic sizes. An exception is the La ion, which in $La_{2-x}Sr_xCuO_{4-\delta}$ has an unoccupied 4f level that contributes to absorption above 5 eV. In the Bi- and Tl-containing materials, the Bi and Tl planes also appear to contribute to visible-near UV optical spectra.

The nature of the 2.7 eV transition is not completely resolved, but its oscillator strength is consistent with a charge-transfer or interband transition. Calculated optical spectra show some structure in this energy range,[20] but no specific assignments were made.

Interesting results have also been obtained by studying variations that are correlated with changes in doping level and conductivity. Figures 9 and 10 illustrate representative results for Sr in $La_{2-x}Sr_xCuO_{4-\delta}$[17] and O in $YBa_2Cu_3O_{7-\delta}$.[18,21] The latter case is particularly interesting because of the strong, sharp features at 1.7 and 4.1 eV in the low-oxygen, semiconducting phase of $YBa_2Cu_3O_{7-\delta}$. Through chemical variants of this phase and related materials, it was established that the 4.1 eV transition is due to an excitation involving a $O-Cu^{1+}-O$ complex that is spatially localized in the unit cell when oxygen is removed from the chain structure of the material. This feature is thus very specific to this structure and can be used to indicate the presence of the low-oxygen phase, particularly at surfaces and interfaces where it can adversely affect the properties of electrical contacts. It is thus valuable as a processing diagnostic.

The 1.7 eV feature has also received extensive study. Its dependence on pressure[22] and temperature[23] as well as on chemical changes has been investigated. The temperature dependence of both 1.7 and 4.1 eV features is shown in Fig. 11. The lack of pressure dependence, large strength, and strong temperature dependence that involves broadening, weakening, and a shift of energy, have been used to support a charge-transfer interpretation involving the CuO_2 planes. It exists not only for low oxygen content but also in Co-doped material, in which the metallic character is also destroyed. Thus, its presence seems to be related to charge localization in the CuO_2 planes that is eliminated in the metallic state. The 2.1 eV feature of $La_{2-x}Sr_xCuO_{4-\delta}$ may have a similar interpretation as it exists only in the semiconducting limit of low Sr content.[17]

The a-b structural anisotropy in the basal plane is another effect worth studying, although no data have yet been reported. It should be observable since

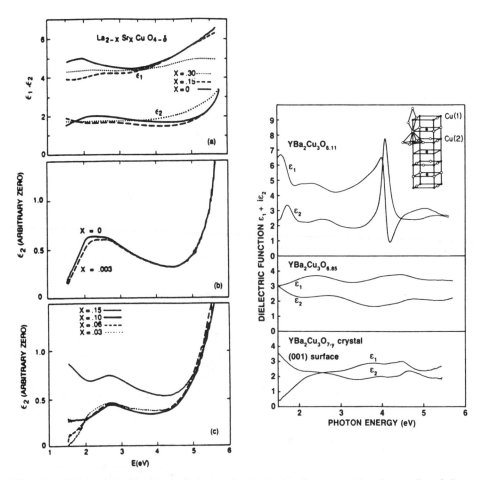

Fig. 9. Dielectric function of $La_{2-x}Sr_xCuO_{4-\delta}$ for several values of x (after ref. 17.)

Fig. 10. Dielectric function of $YBa_2Cu_3O_{7-\delta}$ for various oxygen compositions x (after ref. 18.)

twinning is visible in a polarizing optical microscope.

REFERENCES

1. Recent progress has been summarized in the Proceedings of the International Conference on High Temperature Superconductors and Materials and Mechanisms of Superconductivity, Interlaken, eds. J. Müller and J. L. Olsen, Physica **C153-155** (1988).

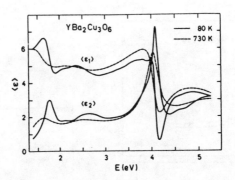

Fig. 11. Comparison of dielectric function spectra of $YBa_2Cu_3O_6$ measured at 80 and 730 K (after ref. 23.)

2. P. E. Sulewski, T. W. Noh, J. T. McWhirter, and A. J. Sievers, Phys. Rev. **B36**, 5735 (1987).

3. See, e.g., J. Callaway, *Quantum Theory of the Solid State* (Academic, New York, 1976), Ch. 7.

4. F. Marsiglio, R. Akis, and J. P. Carbotte, Phys. Rev. **B36**, 5245 (1987).

5. D. A. Bonn, J. E. Greedan, C. V. Stager, T. Timusk, M. G. Doss, S. L. Herr, K. Kamaras, and D. B. Tanner, Phys. Rev. Lett. **58**, 2249 (1987).

6. Z. Schlesinger, R. T. Collins, D. L. Kaiser, and F. Holtzberg, Phys. Rev. Lett. **59**, 1958 (1987).

7. J. Orenstein and D. H. Rapkine, Phys. Rev. Lett. **60**, 968 (1988).

8. A. Wittlin, L. Genzel, M. Cardona, M. Bauer, W. König, E. Garcia, M. Barahona, and M. V. Cabanas, Phys. Rev. **B37**, 652 (1988).

9. D. C. Mattis and J. Bardeen, Phys. Rev. **111**, 412 (1958).

10. G. A. Thomas, J. Orenstein, D. H. Rapkine, M. Capizzi, A. J. Mills, R. N. Bhatt, L. F. Schneemeyer, and J. V. Waszczak, Phys. Rev. Lett. **61**, 1313 (1988).

11. P. B. Allen, Phys. Rev. **B3**, 305 (1971).

12. I. Bozovic, K. Char, S. J. B. Yoo, A. Kapitulnik, M. R. Beasley, T. H. Geballe, Z. Z. Wang, S. Hagen, N. P. Ong, D. E. Aspnes, and M. K. Kelly, Phys. Rev. **B38**, 5077 (1988).

13. A. Bjorneklett, A. Borg, O. Hunderi, and S. Julsrud, Solid State Commun. **67**, 525 (1988).

14. See ref. 7 and K. Kamaras, C. D. Porter, M. G. Doss, S. L. Herr, D. B. Tanner, D. A. Bonn, J. E. Greedan, A. H. O'Reilly, C. V. Stager, and T. Timusk, Phys. Rev. Lett. **60**, 969 (1988) and references therein.

15. S. T. Chui, R. V. Kasowski, and W. Y. Hsu, Phys. Rev. Lett. **61**, 885 (1988).

16. H. P. Geserich, G. Scheiber, J. Geerk, H. C. Li, G. Linker, W. Assmus, and W. Weber, Europhys. Lett. **6**, 277 (1988).

17. S. Etemad, M. K. Kelly, D. E. Aspnes, R. Thompson, J. -M. Tarascon, and G. W. Hull, Phys. Rev. **B37**, 3396 (1988).

18. M. K. Kelly, P. Barboux, J. -M. Tarascon, D. E. Aspnes, W. A. Bonner, and P. A. Morris, Phys. Rev. **38**, 870 (1988).

19. J. Humlicek, M. Garriga, M. Cardona, B. Gegenheimer, and E. Schönherr, Solid State Commun. **66**, 1071 (1988).

20. G. -L. Zhao, Y. Xu, W. Y. Ching, and K. W. Wong, Phys. Rev. **B36**, 7203 (1987).

21. M. Garriga, J. Humlicek, M. Cardona, and E. Schönherr, Solid State Commun. **66**, 1231 (1988).

22. U. Venkateswaran, K. Syassen, H. -J. Mattausch, and E. Schönherr, Phys. Rev. **B38**, 7105 (1988).

23. J. Humlicek, M. Garriga, and M. Cardona, Solid State Commun. **67**, 589 (1988).

PHOTOEMISSION STUDIES OF HIGH TEMPERATURE SUPERCONDUCTORS

Z.-X. Shen, P.A.P. Lindberg, W.E. Spicer and I. Lindau
Stanford Electronics Laboratories, Stanford University, Stanford, CA 94305

J.W. Allen
Department of Physics, University of Michigan, Ann Arbor, Michigan 48109-1120

ABSTRACT

Photoemission studies have been performed on all classes of high temperature superconductors except the Tl-related compounds. Particular attention was paid to the surface cleanliness. Comparison with band calculation shows that the one-electron picture cannot adequately explain the electronic structure of this type of materials. Most important, Cu satellites were observed both in the valence band and the Cu 2p core level for all the samples studied, signaling the importance of the d-d correlation effects. The Cu 3d character of these satellites in the valence band was verified using resonance photoemission. The results have been interpreted in terms of a cluster model derived from the two band Anderson Hamiltonian, which in the past has been used successfully to describe the electronic structure of highly correlated systems. No clear satellite structure was observed in the O 1s core spectrum, which is consistent with the band-like nature of the oxygen states. Examples of changes in the electronic structure, which could be related to T_c, (such as substituting Y by Pr in the Y-Ba-Cu-O system and altering the number of Cu-O layers in the Bi-Ca-Sr-Cu-O system), are also discussed

I. INTRODUCTION

Since the discovery of high-temperature superconductors [1,2] photoelectron spectroscopy has been used extensively to investigate their electronic structures. In the literature there are several review papers that summarize the studies performed in the field [3,4,5,6]. In this paper, we review the work done at Stanford University in collaboration with groups at the University of Michigan, University of California at San Diego, and University of Northern Illinois [7-19]. For more complete reviews of the work by different groups, the reader is referred to the above mentioned review articles[3-6].

One of the great triumphs of solid state physics is the development of the one-electron band theory, which has been widely used to describe the electronic structures of solid materials. For a number of materials, band theory has provided an adequate explanation of the electronic structure. However, it has been found in the past that one-electron band theory cannot describe very well the electronic structure of many transition-metal compounds. For example, one-electron band theory predicts that NiO and CoO should be metals, whereas they are found experimentally to be insulators. Facing this dilemma, Mott and Hubbard proposed that the correlation effects are very important for these elements [20, 21]. In other words, the Coulomb interactions among the localized d-electrons are very strong and many-body effects must be taken into account. Hubbard showed that a band gap forms when the Coulomb interaction U is larger than the band width in the half filled case [20, 21]. Using this approach one could qualitatively explain why NiO and CuO are insulators. This model, named the Hubbard model (or the one-band Hubbard model) after its inventor, takes into account the d-d polar charge transfer but completely ignores the oxygen bands. In the last decade this picture has been modified based on spectroscopic data in which the authors showed the importance of including charge transfer between oxygen and transition metal ions as well as the d-d coulomb interaction in describing the electronic structure of these compounds [22-32]. The approach of these authors is based upon the Anderson Hamiltonian (or two band Hubbard model):

$$H = \varepsilon_d^0 \sum_{i\sigma} d_{i\sigma}^+ d_{i\sigma} + \varepsilon_p^0 \sum_{j\sigma} p_{j\sigma}^+ p_{j\sigma} + \sum_{<ij>\sigma} t_{ij}(d_{i\sigma}^+ p_{j\sigma} + h.c.)$$

$$+ U_d \sum_i n_{i\sigma} n_{i\sigma'} + U_p \sum_j n_{j\sigma} n_{j\sigma'} + U_{pd} \sum_{<ij>} n_i n_j$$

(1)

The first and second terms are for the oxygen and copper bands, respectively. The third term deals with the hybridization of the two bands while the last three terms describe the Coulomb repulsions among the electrons, where U_{dd}, U_{pp}, and U_{pd} are the Coulomb repulsion among the d electrons, p electrons and between p-d electrons respectively. This Hamiltonian is reduced to the one-band Hubbard model when one only considers the Cu bands. The most important parameters in the cluster model are U_{dd} and the charge transfer energy Δ, which is defined as the energy that is needed to transfer one electron from Cu to O. In other words, it is the energy difference of the centroid of the renormalized Cu and O bands. Based on the extensive studies of the transition-metal compounds, Zannan, Sawatzky and Allen (ZSA) proposed a phase diagram for the transition-metal compounds [28]. In the case of $U_{dd} << \Delta$, the d-d polar charge transfer dominates the low energy excitations and the compound can be regarded as a traditional Mott insulator for which the Hubbard model is sufficient. On the other hand, the Anderson Hamiltonian is necessary if $U_{dd} > \Delta$ and the insulators are actually charge transfer insulators. For low energy excitations of the charge transfer insulators, it may be possible to construct an effective one band model from the two band model [26, 27, 45].

One of the important common elements in all classes of high T_c compounds is copper, which is a typical transition metal. Almost immediately after the discovery of the high T_c compounds, possible mechanisms of superconductivity based on the One-band and Two-band Hubbard models were proposed by a number of authors who emphasized the d-d coulomb interactions among the Cu d electrons [33-42]. Anderson predicted that the undoped compound La_2CuO_4 is an antiferromagnetic Mott insulator [33]. This prediction has been confirmed by later experiments which reinforce the importance of the correlations effect [43, 44]. At the same time, extensive band calculations were also performed, and more or less conventional pairing mechanisms were proposed. Given the uncertainty and controversy on the theoretical side, it is important to perform photoemission studies to explore the electronic structure of the high-T_c superconductors.

For the transition metal compounds a typical and important characteristic is the existence of satellite structures in the valence band and core levels. The d-like valence band satellites can be identified by the resonance photoemission (RESPES) technique. We have performed RESPES and X-ray photoemission (XPS) studies of all classes of superconductors except the Tl-related compounds. Clear satellite structures were observed which signal the importance of the correlation effects in the superconductors. We have also compared the photoemission results with the predictions of one-electron band calculations which serve as a good starting point for the understanding of the electronic structure. We systematically find that the experimentally observed states in the valence band are shifted to higher binding energy as compared to the results from one-electron band theory. This result, which is typical of many transition-metal compounds, has been attributed to the renormalization of the correlated bands [14]. Based on these results, we conclude that the one-electron band theory does not describe very well the electronic structures of the high T_c materials. We instead interpret our data in terms of a configuration interaction cluster model derived from the Anderson Hamiltonian. We find that the d-d coulomb interaction U_{dd} is much larger than the charge transfer energy Δ. This validatess the use of a two-band Hamiltonian. Finally, we provide a set of the empirical values for the cluster model Hamiltonian.

This paper is organized in the following way: In section II we discuss the surface preparation and the surface cleanliness, which is a major concern in photoemission spectroscopy. In section III we compare our experimental results with the predictions of one-electron band calculation. In section IV we show the Cu and O core level data, and we use our resonance photoemission results to confirm the Cu 3d nature of the satellite structures in the valence band. Based on these data, in section V we provide an analysis using the configuration interaction cluster (CI) model. Finally, in section VI we discuss the effects on the electronic structure by changing x in the $Y_{1-x}Pr_xBa_2Cu_3O_{7-d}$ system and by altering the number of CuO_2 layers in the Bi-Sr-Ca-Cu-O related High-Tc superconductors.

II. SURFACE PREPARATIONS

Because of the inherent surface sensitivity of photoelectrons, the ability to prepare and maintain a clean surface in ultrahigh vacuum (UHV) is of crucial importance in all photoemission experiments. In the field of High-T_c superconductors, questions have been raised regarding the validity of photoemission results for the elucidation of the mechanism behind the superconductivity. This is mainly because of the ease with which many ceramic materials, including the perovskite-related structures, lose oxygen especially in UHV. Moreover, the inappropriateness of many surface cleaning procedures, such as ion-sputtering and in-situ annealing, because of disrupted valence state bands and loss of oxygen respectively, leaves only "mechanical" treatments as fracturing, in situ scraping (polycrystalline bulk samples) and in situ brushing (thin films) as possible means of producing clean surfaces in UHV. Until recently it has been difficult to know if these treatments produce surfaces characteristic of the bulk High-T_c superconductors. However, with the progressive improvement in the fabrication of High-T_c superconductors the situation has changed. Single crystals of a size large enough for photoemission studies are now available. Furthermore, probably because of the layered character of the high-T_c superconductors, single crystals can readily be cleaved, thus allowing an evaluation of earlier studies performed on polycrystalline materials. Stoffel et al. was the first to present LEED pictures and photoemission spectra from *in situ* cleaved single crystals of $YBa_2Cu_3O_7$ [46].

Recently we were able to cleave single crystals of the $Bi_2Sr_2CaCu_2O_8$ High-T_c superconductors in UHV. The cleaved surfaces were examined using Low-Energy Electron Diffraction (LEED). In Fig. 1 we present one of our LEED patterns recorded at a beam energy of 45 eV [18]. This LEED pattern was reproduced for six different crystals and is completely representative for the results we have obtained. As we described in detail elsewhere [18], the LEED pattern conclusively shows that the cleavage plane is c-axis oriented. Moreover, the surface structure, revealing a superstructure along one of the axes, is in agreement with the bulk crystal structure [47]. Thus, we infer that the cleaved single-crystal produces a surface of high quality with a long-range periodicity representative of the layers in the bulk. Furthermore, 24 hours in UHV (at 1×10^{-10} torr), including many hours of ultraviolet photoemission spectroscopy (UPS) and XPS measurements, did not adversely affect the quality of the LEED pattern, although the diffraction spots became somewhat fuzzier. In light of these facts, we are confident that these surfaces obtained in UHV conditions are clean over a time long enough to perform extensive photoemission studies.

With the above discussion in mind, in this paragraph we will compare photoemission spectra, recorded in both the UPS and the XPS regimes, from *in situ* scraped polycrystalline and *in situ* cleaved single crystalline samples of $Bi_2Sr_2CaCu_2O_8$. In Fig. 2 we show the valence band spectra of the poly- and single-crystalline samples using (a) HeII (40.8 eV) and (b) MgKa (1253.6 eV) radiation. The significantly different energy resolutions of the two photon sources accounts for the different widths of the spectral features in the two panels. Apart from a higher background and slightly broader features for the polycrystalline sample in panel (a), there is a similarity between the spectra of the poly- and single crystalline samples in both panels. The HeII spectra (a) reveal a clear Fermi level cutoff for both samples, and all the peak positions are the same. The similarity between the two types of sample surfaces is further illustrated in Fig. 3 where the O1s core levels of the two samples are shown. Fig. 3 shows that both samples exhibit a single, symmetric O1s core level at the same binding energy and without any sign of oxygen in different chemical states (within the experimental resolution). Note that the appearance of single O1s peaks for

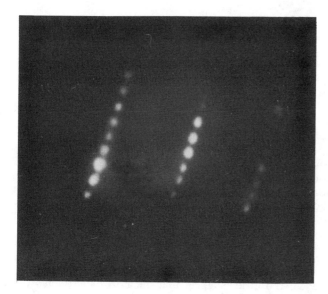

Fig. 1. Low-Energy Electron Diffraction (LEED) pattern from a $Bi_2Sr_2CaCu_2O_8$ single-crystal recorded at 45 eV beam energy.

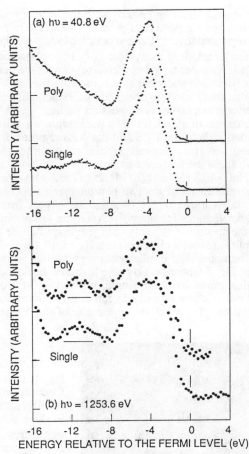

Fig. 2. Comparison of spectra from poly- and single-crystalline samples of $Bi_2Sr_2CaCu_2O_8$ measured at a photon energy of (a) 40.8 eV and (b) 1253.6 eV.

$Bi_2Sr_2CaCu_2O_8$ sharply contrasts with the corresponding core level data for the $YBa_2Cu_3O_7$ system [3-5], with probably only one exception, as far as we are aware[48]. Based on these remarks, we conclude that the valence band states probed by photoemission at room temperature are almost identical for *in situ* scraped polycrystalline and *in-situ* cleaved single crystals of $Bi_2Sr_2CaCu_2O_8$. Consequently, the results previously obtained on *in situ* scraped polycrystalline samples of $Bi_2Sr_2CaCu_2O_8$ appear to reflect the electronic structure of a clean surface, thus validating the use of scraping as a cleaning procedure for this type of ceramic.

It should be pointed out here that the preparation of the $YBa_2Cu_3O_7$ compound surface is more tricky. We found that the valence band and the O 1s core level PES spectra of $YBa_2Cu_3O_7$ samples change with time in UHV at room temperature [7]. Recent works by List and Arko el at. suggest that the PES experiments on the $EuBa_2Cu_3O_7$ (which is the same as $YBa_2Cu_3O_7$) samples must be performed at low temperature [49]. They found that the valence band PES spectra change dramatically upon warming the single crystals from 20 K to room temperature, which they attribute to the loss of oxygen in the surface region. This result may be related with the fact that the O1s core level spectra from different $YBa_2Cu_3O_7$ samples are so inconsistent [3-5]. This difference of the surface stability in the UHV between the $Bi_2Sr_2CaCu_2O_8$ and

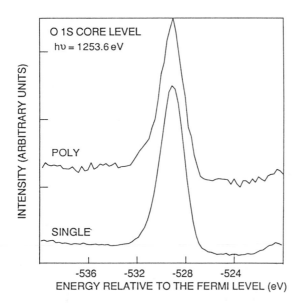

Fig. 3. O1s core level spectra for poly- and single-crystalline samples of $Bi_2Sr_2CaCu_2O_8$ measured at a photon energy of 1253.6 eV.

$YBa_2Cu_3O_7$ is not fully understood at this point, but we think this might be related to the fact that the $YBa_2Cu_3O_7$ compound has an intrisic defect structure (i.e., the oxygen vacancies in the same planes as the one dimensional Cu-O chains), while the $Bi_2Sr_2CaCu_2O_8$ compound does not have such a defect structure.

Because our discussion and modeling are based solely on the two common phenomena observed in all families of superconductors we studied even with the above mentioned uncertainty, we feel confident about our results. As we will discuss later, the two phenomena are: (a) satellite structures in the valence band and Cu core levels, and (b) the shift of the valence band to higher binding energy compared with band theory. These two phenomena were also observed in $EuBa_2Cu_3O_7$ compounds cleaved at 20 K [49].

III. COMPARISON WITH BAND CALCULATIONS

Extensive band calculations have been performed on the high temperature superconductors [6]. These calculations provide much insight into the electronic structures of the high T_c compounds, in particular the energy distribution of different bands and the degree of hybridization in the valence band. However, significant discrepancies were found between the experimental results and the results derived from one-electron band calculations. As an example, we present in Fig. 4 the valence band EDC's of $Bi_2Sr_2CaCu_2O_8$ as a function of the photon energies in comparison with curves obtained from a one electron band calculation. The theoretical curves were obtained by adding the theoretical partial density of states (DOS) from band theory weighted by the photoionization cross section of the various elements at different

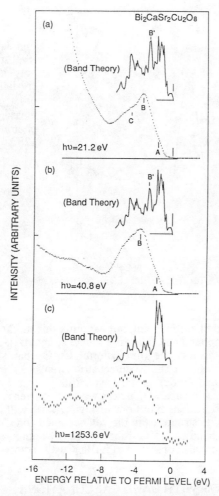

Fig. 4. Comparison between photoemission spectra (dotted curves) from $Bi_2Sr_2CaCu_2O_8$ recorded at three different photon energies and the results of one-electron band calculations weighted by the photoionization cross-sections at the corresponding photon energies.

photon energies [50]. By comparing the experimental and theoretical curves, one finds that the experimental results do not agree with the band theory very well. There are two obvious differences between the experimental and theoretical curves. First of all, the DOS observed at E_F is lower than predicted by the band theory. Second, the centroid of the experimental valence band is shifted to higher binding energy by about 1.5 eV. These two discrepancies are observed in all the superconductors we have studied. As has been pointed out earlier, we will only concentrate on the results which are common to all the superconductors. The shift in the valence band centroid has also been observed in some transition-metal oxides, such as NiO, where Sawatzky and Allen explained this shift as due to the coulomb interaction U_{dd} of the 3d electrons [26]. We believe the shift of the valence band centroid observed in the superconductors is due to a mechanism similar to that in NiO, which is clearly due to correlation effects. Thus, this comparison shows the necessity to take into account many body effects.

IV. RESPES AND XPS EXPERIMENTS

As we have said in the introduction, the photoelectron spectroscopy (PES) features which are important in many transition-metal compounds, are the satellite structures observed in the valence band and the core level data. These satellites clearly demonstrate the existence of correlation effects among the d electrons in these compounds. For all families of the high temperature superconductors we studied, clear satellite structures were observed both in the valence band and the core levels. The assignment of a core level satellite to a particular element, (e.g., Cu in the high T_c material) is fairly straight forward because the core levels of the different elements are usually located at different binding energies. On the other hand, the determination of the valence band features requires more sophisticated experiments. Resonance photoemission (RESPES) has in the past been proven to be an extremely effective tool for the determination of the character of the valence band features in the transition metal compounds[51-59]. For simplicity we will concentrate on the element Cu, even though we performed RESPES experiment for different elements in the compounds. If one tunes the photon energy through the Cu 3p to 3d absorption edge at 74 eV, some electrons will be excited into the empty d states. Because these empty d states are localized, the electrons will decay through a two-electron Koster-Kronig process to the same final states as direct 3d photoemission, causing intensity changes of the Cu features and providing a good way to determine the origin of the valence band features. To our knowledge the first RESPES experiment on the $YBa_2Cu_3O_7$ compound was performed by Kurtz et al [60]. We performed RESPES experiments independently on the $YBa_2Cu_3O_7$ and $(La_{1-x}Sr_x)_2CuO_4$ compounds [7]. In Fig. 5 we present the RESPES data for $(La_{1-x}Sr_x)_2CuO_4$ compound with the photon energy in the vicinity of the Cu 3p ---> Cu 3d absorption edge. Since feature D in Fig. 5 exhibits a resonant behavior at the absorption edge, it is assigned to a Cu 3d satellite. As we pointed out in our previous paper [7], the features A and B show antiresonance behavior, revealing

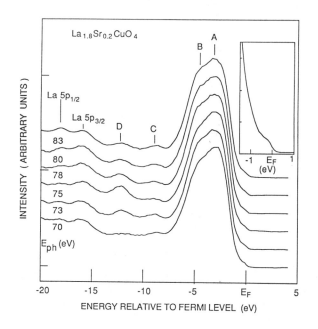

Fig. 5. Photoemission spectra from $La_{1.8}Sr_{0.2}CuO_4$ for various photon energies around the Cu2p absorption threshold around 74 eV. The inset shows the detailed structure of the Fermi edge.

the Cu nature of these states. The origin of feature C, however, is still in question. We will come back to this problem when we discuss the O 1s core level data. Another way to perform RESPES measurement involves the constant-initial-state (CIS) mode, where the photon energy is varied together with the energy of the analyzed electrons to maintain a fixed initial state energy. The intensity modulation in a CIS measurement can yield information about the character of the valence states. In Fig. 6 we show the CIS data for $Bi_2Sr_2CaCu_2O_8$ (2 Cu L) and $Bi_2Sr_2CuO_6$ (1 Cu L) [17], which reveal whether or not Cu states are involved in the features at 0.5, 3.2 and 12.3 eV binding energies in the $Bi_2Sr_2CaCu_2O_8$ compound. This figure clearly demonstrates that the feature at -12.3 eV are due to a Cu 3d satellite. It is interesting to note that the states close to the Fermi level (-0.5 eV) do not show any Cu resonant behavior at all, suggesting that the states near the Fermi level are mainly oxygen related for the $Bi_2Sr_2CaCu_2O_8$ compound, which is consistent with the results from angle-resolved experiments by Takahashi et al [61]. It should be noted that the enhancement of the density of states near E_F as presented in ref. 61 near 18 eV is very strong for oxygen 2s to 2p resonance, stonger than what we observed from $La_{1+x}Ba_{2-x}Cu_3O_7$ compound [15]. Because the Sr 4p edge is also at 18 eV [17], we call for caution to interpret the enhacement of the emission near E_F at 18 eV to be due to the oxygen 2s to 2p resonances. As we will discuss below, the existence of the Cu 3d satellite in the valence band in all the high T_c compounds shows the importance of the correlation effects. The position of the satellite feature in the high T_c compound yields information about the d-d coulomb interaction U_{dd}.

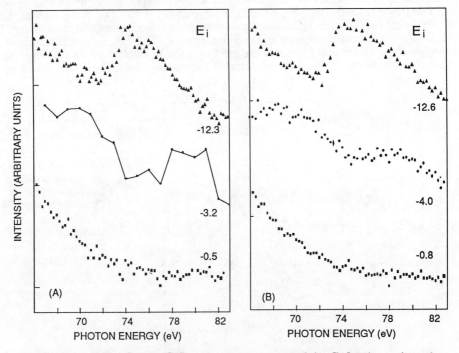

Fig. 6. Constant-Initial-States (CIS) measurements around the Cu2p absorption edge for (a) the 2 Cu L and (b) the 1 Cu L system of the Bi-Sr-Ca-Cu-O superconductors. The initial state energies are given in the right margin of each panel.

As we have mentioned earlier, the Cu satellite structures could exist in the core level as well as in the valence band. Because Cu satellites are observed in the valence band, we expect to see satellites in the core level data. In Fig. 7 we present the Cu 2p core level data from a single-crystalline and a polycrystalline $Bi_2Sr_2CaCu_2O_8$ sample [16]. It is clear from this figure that the main lines of the Cu 2p core level are accompanied by strong satellites. The satellites and the main lines are assigned to d^9 and $d^{10}\underline{L}$ configurations, respectively, using the cluster configuration interaction model [7]. The existence of the d^9 satellite in the Cu 2p core level again suggests the importance of strong correlation. The energy separation between the d^9 and $d^{10}\underline{L}$ configurations, which reflects the coulomb interaction U_{cd} between a Cu 3d hole and a Cu 2p core hole, is about 9 eV.

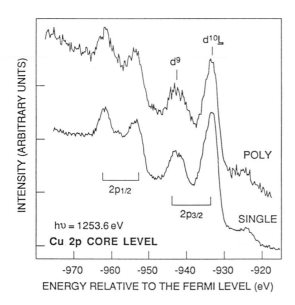

Fig. 7. Cu 2p core level spectra for poly- and single-crystalline samples of $Bi_2Sr_2CaCu_2O_8$ measured at a photon energy of 1253.6 eV.

The O 1s core level data for the high T_c superconductors have been somewhat controversial. The results from most polycrystalline samples of earlier superconductors of $YBa_2Cu_3O_7$ and $(La_{1-x}Sr_x)_2CuO_4$ have two components at different binding energies. [3-5] We have presented in Fig. 3 an O 1s core level spectrum from a well characterized $Bi_2CaSr_2Cu_2O_8$ single crystal surface, together with a spectrum from a polycrystalline sample from our earlier study, in which the O 1s core level is shown to have only one sharp peak near -529 eV. This data conclusively demonstrates that only one component is intrinsic to the $Bi_2CaSr_2Cu_2O_8$ superconductor. Even though the oxygen sites in the $Bi_2CaSr_2Cu_2O_8$ are not equal, the chemical shifts of O 1s core level are smaller than the experimental resolution of ~ 1.2 eV. Because of the similaritity of the different oxygen sites in $Bi_2CaSr_2Cu_2O_8$ and $YBa_2Cu_3O_7$ as revealed by their crystal structure, we suggest that only the lower binding energy component is intrinsic to $YBa_2Cu_3O_7$. The fact that all the published O 1s data for $YBa_2Cu_3O_7$ have two components suggests that the $YBa_2Cu_3O_7$ compound surface is more difficult to prepare, which, as we pointed out earlier, is probably related with the intrinsic defect structure of the $YBa_2Cu_3O_7$ compound. The fact we see a sharp singlet O 1s peak in the $Bi_2CaSr_2Cu_2O_8$ compound gives us confidence about the PES data obtained from $Bi_2CaSr_2Cu_2O_8$ compound. It is clear that the O 1s core level does not have a strong satellite structure as the Cu 2p core level does. On the other hand, there have been suggestions that the coulomb interaction among the oxygen 2p electrons, U_{pp}, is about 5-8 eV[5, 62], not too much smaller than U_{dd}, the coulomb interaction among the Cu 3d electrons. The fact that we observed satellites in the Cu 2p core level spectrum while we did not see satellites in the O 1s core level spectrum cannot be explained within the context of the configuration interaction model, but it may be understood if one takes into the account the difference between the oxygen and copper band widths. Since no strong satellite structure is observed in the O 1s core level, which is consistent with the band-like nature of the oxygen 2p states, it seems unlikely that a clear oxygen two hole satellite will be observed in the valence band spectra. Returning to the issue of feature C in Fig. 4, it has been suggested by Thiry et al.that this is due to the two-oxygen hole satellites in the valence band [63]. In view of the O 1s core level data, we think this is very unlikely to be true.

TABLE I

	$Bi_2CaSr_2Cu_2O_8$	$La_{1.8}Sr_{0.2}CuO_4$	$YBa_2Cu_3O_7$(#1)
$\delta E(2p)$	8.9	8.9	8.8
$I_s/I_m(2p)$	0.33	0.34	0.35
W(VB)	4.6	4.7	4.8
$\delta E(VB)$	8.3	8.5	8.2
$I_s/I_m(VB)$	0.03	0.02	0.03
$E_m(VB)$	-4.0	-3.9	-4.2
$E_s(VB)$	-12.3	-12.4	-12.7

Table I: Experimental quantities for the three materials as defined in the text. All energies are given in eV. Here I_s/I_m (2p) and I_s/I_m (VB) are the satellite and main line intensity ratios for the Cu 2p core level and the valence band, respectively, while $\delta E(2p)$ and $\delta E(VB)$ are the energy separations in the Cu 2p core level and the valence band respectively. W(VB) is the width of the valence band, and the $E_m(VB)$ and $E_s(VB)$ are the binding energies of the main valence band and the valence band satellite respectively.

Finally, we summarize our core level and the valence band experimental data in Table 1. In Table I I_s/I_m (2p) and I_s/I_m (VB) are the satellite and main line intensity ratios for the Cu 2p core level and the valence band respectively, while $\delta E(2p)$ and $\delta E(VB)$ are the energy separations between the main and satellite lines in the Cu 2p core level and the valence band respectively. W(VB) is the width of the valence band and the E_m(VB) and E_s(VB) are the binding energies of the main valence band and the valence band satellite respectively. The detailed analysis of the experimental data with the cluster interaction model (CI) is based solely on these data. We would like to emphasize that, even though some uncertainty exists in the present data, (e.g., feature C in Fig. 4), the data we list in the table or which will be used in our model, are not varying from sample to sample, and similar values have also been reported by other groups [3-6]. The valence band Cu d^8 satellite is also observed in the low temperature cleaved $EuBa_2Cu_3O_7$ single crystal samples [49]. As we pointed out at the beginning of this paper, these are the results we would like to stress.

V. CI MODEL DESCRIPTION OF EXPERIMENTAL DATA

In this section, we analyse our data using a local cluster model which is the simplest version of the Anderson model, that retains the various charge fluctuation processes [7]. In this approach, a $(CuO_6)^{10-}$ cluster is regarded as a separable unit, and its electronic structure is described by configuration interaction. In essence, it is an Anderson impurity Hamiltonian description of the Cu 3d electrons hybridized to the oxygen 2p states but neglecting the 2p bandwidth. The ground state is a linear combination of properly symmetrized d^9 and $d^{10}\underline{L}$ configurations, where $\underline{L}$ denotes a hole of appropriate symmetry relative to filled ligand O 2p states. These two configurations differ by an energy Δ, which is defined as the charge transfer energy needed to move an electron from an oxygen site to a Cu site as shown in the diagram of Fig. 8. The N = 9 configuration is the ground-state, in which the cluster system has one hole at Cu or O sites with different probabilities. After one electron has been removed (N = 8), there are two holes in the cluster. It is clear from this figure that the

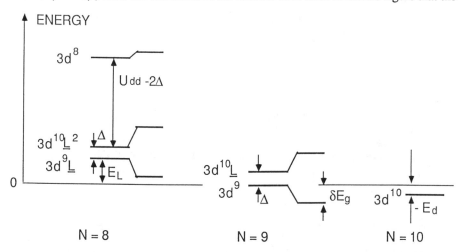

Fig. 8. Energy-level diagram for configurations of the cluster model description of the ground state (N=9), valence band photoemission final states (N=8), and BIS final states (N=10). The changes in splittings are due to hybridization.

TABLE II

	$Bi_2CaSr_2Cu_2O_8$	$La_{1.8}Sr_{0.2}CuO_4$	$YBa_2Cu_3O_7(\#1)$
U_{cd}	7.8	7.8	7.8
T	2.4	2.4	2.5
Δ	0.4	0.3	0.5
U	6.2	6.1	6.5
E_L	2.4	2.3	2.6
E_d	-2.0	-2.0	-2.1
δE_g	2.2	2.25	2.2
E_{m1}(VB)	0.7	0.65	0.9
E_{m2}(VB)	6.8	6.85	7.5
E_s(VB)	12.3	12.25	12.8
I_s/I_m(VB)	0.07	0.07	0.07
E_{BIS}	0.2	0.25	0.2
n_d	9.4	9.5	9.4
E_{gap}	1.0	0.9	1.1

Table II: Model parameters and quantities for the three materials measured. Energies are in eV. The various symbols are defined in reference 7. As can be clearly seen in Fig.8, U_{cd} is the coulomb repulsion between a 2p core hole and a 3d hole, while U is the coulomb repulsion between two 3d holes; T is the charge transfer integral, and Δ is the charge transfer energy by moving one electron from oxygen to copper; E_L is the energy difference between the $3d^9$ and $3d^9\underline{L}$ configurations; E_d is the energy difference between $3d^9$ and $3d^{10}$ configurations; δE_g is the hybridization shift of the ground state; the centroid of E_{m1}(VB) and E_{m2}(VB) can be compared with the E_m (VB) in table I; E_s (VB) is the valence band satellite position obtained by the model parameters T, Δ, and U which can be compared with the E_s (VB) in table I; I_s/I_m (VB) is the valence band satellite and main line intensity ratio from the cluster model which can be compared with the I_s/I_m (VB) in table I; E_{BIS} is the energy of BIS peak expected from the cluster model; n_d is the number of the d electrons; E_{gap} is the theoretical gap energy defined as the summation of E_{BIS} and E_{m1}(VB).

$3d^8$ satellite is pushed to higher energy by the strong coulomb repulsion between the Cu 3d holes. In other words, when an electron is removed from the d^9 configuration, it has to overcome the extra Coulomb interaction between the two holes left in the d^8 configuration, resulting in a satellite at higher binding energy. For the case of $U_{dd} \gg \Delta$, the separation of the satellite and the main band reveals the Coulomb interactions between the d electrons. These satellites, not predicted by one electron band theory but observed in all the high T_c compounds near -12.5 eV, can be explained easily in the context of the cluster model. The 8.5 eV separation between the satellite and the valence band demonstrates that U_{dd} is larger than the valence band width so that the correlated models are necessary. Based on the experimental data listed in Table I, we obtained the values for the model parameters listed in Table II [7]. It should be noted that the parameters given in Table II are at best rough estimates of the physical quantities. Nevertheless, important qualitative information can be extracted from the numbers tabulated. For example, instead of emphasizing the exact value of Δ, we would like to emphasize the fact that Δ is much smaller than U_{dd}.

Therefore, the charge transfer energy dominates the low energy excitations. The main result from this analysis is that both the d-d correlation and the charge transfer are important for the electronic structure of the high T_c superconductors so that the two band Hubbard model is necessary in order to describe the electronic structure of the high T_c materials. It is also worthwhile to point out that it is important to treat the oxygen states as band-like states. As we pointed out earlier in our discussion of the O1s core level data, it is impossible to reconcile the difference in the core level satellite structures between the O 1s and Cu 2p within the scope of the cluster model. Futhermore, a recent angle-resolved study by Shih et al. on NiO showed that the oxygen bands in the NiO system were band like; whereas the d bands are highly correlated [14]. McMahan et al. have sucessfully reproduced the main features in the valence band photoemission spectrum of the superconductors by an Anderson impurity model, where they obtained their model parameters from first principle calculations and treated the oxygen states as extented bands [64]. Using similar approach but with some emperical experimental values of U_{dd}, Δ and T, Eskes et al. were also able to independently simulate the experimental d-electron-removal spectra [65].

VI. CASE STUDIES OF CHANGES IN ELECTRONIC STRUCTURE DUE TO CHANGES WITHIN UNIT CELL

Finally, we will show some examples of how PES experiments can reveal information on changes of the electronic structures of different high T_c compounds, which appear to be related to the superconductivity.

The first example is the recently discovered Bi-Ca-Sr-Cu-O superconductors [66-67], whose superconducting transition temperatures T_c are correlated with the number of Cu-O layers in the system. The T_c for Bi-Sr-Ca-Cu-O superconductors with one, two, and three Cu-O layers are 20K, 85K and 105 respectively [68]. Exploring the differences in the electronic structure of these superconductors is a logical next step, which might give us some clue about the mechanism of high-temperature superconductivity. We have used photoelectron spectroscopy to study and compare the electronic structure of the two CuO_2-layer (2 CuL) sample (with nominal composition of $Bi_2Ca_1Sr_2Cu_2O_8$) and the one CuO_2-layer (1 CuL) sample (with nominal composition of $Bi_2Sr_2CuO_6$) [13]. In Fig. 9, we compare the valence band spectra of the two samples at photon energies of 70 and 74 eV. The enhancement of the emissions from the Cu d^8 satellites, located near -12 eV in the valence band at 74 eV photon energy, can be seen clearly in both compounds. The 2 Cu L sample exhibits a clear Fermi edge, while the density of states near the Fermi level of the 1 Cu L sample is much lower. As presented in Fig. 6, the states near the Fermi level (-0.5 and -0.8 eV) for the Bi-Sr-Ca-Cu-O compounds have almost no Cu character, in strong contrast to that of the $YBa_2Cu_3O_7$ compound [7,13,48]. We suggest that the Bi 6p - O 2p bands are well described by one electron band theories and that the Fermi edge observed in the 2 Cu L compound mainly arises from the Bi-O p bands, as the band theories predicted [12,13]. Therefore, the difference observed in the states near E_F between the 2 Cu L and 1 Cu L compounds is mainly the difference of the Bi-O states, which implies a slightly different occupation of the Bi 6p - O 2p states in the two compounds. This is consistent with the corresponding core level data which show that the Bi ions have slightly higher valency in the 1 Cu L sample than in the 2 Cu L sample. (See the 0.3 eV chemical shift of the Bi 5d core level in the panel 2 of Fig. 9)

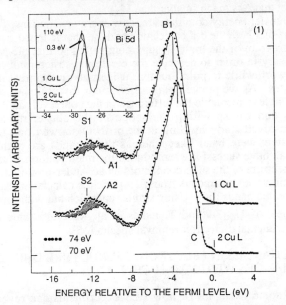

Fig. 9. Comparison of valence band spectra of the 2 Cu L and the 1 Cu L system of the Bi-Sr-Ca-Cu-O superconductors. Note the change of the density of states at the Fermi energy and the intensity changes of the satellite structure at -12 eV. The inset shows the shift of the Bi 5d core levels.

The same difference in the occupation of the Bi-O states in the two compounds is also found from a one-electron band calculation [69], suggesting again that the band theory gives a good description of the Bi-O p bands despite the strong correlation among the Cu 3d electrons. Given the importance of the states near E_F to the superconductivity, we suggest that the observed difference in the density of states near E_F is probably related to the different superconducting transition temperatures of the different compounds [13].

The second example we want to show here is how electronic structure studies can provide some insights into the quenching of superconductivity in the $Y_{1-x}Pr_xBa_2Cu_3O_7$ system [11]. It is found that the superconductivity in the $Y_{1-x}Pr_xBa_2Cu_3O_7$ system is quenched as Y is replaced by Pr.[70-72]. This result is unusual because most of the $RBa_2Cu_3O_7$ compounds, where R is a rare earth element, are superconducting with T_c near 90K except for the cases of Ce, Pr and Tb. (For Ce and Tb, the 1-2-3 compounds do not form.[73]) $Y_{1-x}Pr_xBa_2Cu_3O_7$ compounds form the same orthorhombic crystal structure as $YBa_2Cu_3O_7$, but the degree of orthorhombic distortion relative to the corresponding tetragonal structure is diminished with the increasing substitution of Pr. A model for the Tc - quenching could be that the valence of Pr is 4+ so that extra charge is transferred to the Cu-O planes and fills the holes that are widely believed to be the superconducting carriers. We have performed XPS, Bremsstrahlung isochromat spectroscopy (BIS), and RESPES studies near the Cu (3p-->3d), Pr (4d--->4f) and O (2s--->2p) thresholds. Fig. 10 shows the Pr 4f spectrum of the Pr ions in the $Y_{1-x}Pr_xBa_2Cu_3O_7$ compounds in comparison with that of the Pr metal. It is clear that the Pr 4f states of the Pr ions in $Y_{1-x}Pr_xBa_2Cu_3O_7$ are changed considerably. As has been discussed in detail in the paper by Kang et al. [11], our data imply that Pr 4f / O 2p / Cu 3d hybridization alters the electronic or the magnetic structure of the x = 0 material, which results in the quenching of the superconductivity. This results is consistent with conclusions drawn from the pressure dependence of the transport properties of these alloys [76].

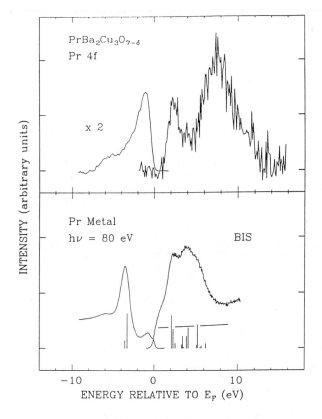

Fig. 10. Top panel : The complete Pr4f spectral weights for $PrBa_2Cu_3O_{7-d}$, obtained by combining RESPES and BIS spectra. The RESPES spectrum is scaled relative to the BIS spectrum so that 1/7 of its total area is below E_F.

Bottom panel: The cooresponding spectral weights for Pr metal, by combining a PES spectrum at hv = 80 eV [74] and a BIS spectrum [75].

These two examples show that PES experiment can provide very direct information about the electronic structure which is very important for the understanding of the possible mechanisms of the high temperature superconductivity.

VII. SUMMARY

We have performed PES studies of the electronic structures for all the classes of the high temperature superconductors except the Tl related compounds. One important common feature for all the high T_c compounds is the Cu 3d-O 2p bands, which are highly correlated we often has been observed in the transition metal compounds. The one-electron band theory is not sufficient to describe these bands. Our results suggest that both the d-d coulomb interaction and the p-d charge transfer are important for the electronic stucture of the superconductors, with the charge transfer process dominating the low energy excitations. On the other hand, there are also differences in the different compounds. The Bi-O bands derived from the Bi-O planes in the $Bi_2Ca_1Sr_2Cu_2O_8$ compound, for example, are very important for the states near E_F. Even though the Cu-O hybrids are best described by the two band Hubbard model, the details of the electronic structure, especially for the states near E_F still remain largely unknown. There is evidence from the $Bi_2Ca_1Sr_2Cu_2O_8$ compound that the states near the Fermi level, which presumably are most important for the superconductivity, are mostly of oxygen character [13, 61]. Our results show that the dispersive nature of the oxygen band is important. However, how to take the translational symmetry and correlation into account at the same time is still unexplored. Therefore, more experimental and theoretical work is needed to elucidate the electronic structure of the high T_c superconductors and many transition-metal oxides.

ACKNOWLEDGEMENTS

We are very grateful to many of our colleagues who participated in our experiments as reflected in the references 7 to 19. We especially thank M.B. Maple for sharing his insight and for providing the original motivation to study the $Y_{1-x}Pr_xBa_2Cu_3O_7$ alloys. We thank R.S. List and A.J. Arko for communicating the results on low temperature-cleaved single crystals before publication. The RESPES experiments were performed at Stanford Synchrotron Radiation Laboratory (SSRL) and Synchrotron Radiation Center of the University of Wisconsin. SSRL is funded by the DOE under contract DE-AC03-82ER-13000, Office of Basic Energy Sciences, Division of Chemical / Material Sciences. We gratefully acknowledge support from the National Science Foundation - Materials Research Laboratory Program at the Center for Materials Research at Stanford University, the Air Force Contract AFSOSR-87-0389, JESP contract DAAG 29-85-K-0048, and at the University of Michigan the U.S. National Science Foundation through Low Temperature Physics Grant No. DMR-87-21654 (J.W.A.).

REFERENCES

1. J. G. Bednorz and K. A. Müller, Zeitschrift für Physik B **64**, 189 (1986).
2. M. K. Wu, J. R. Ashburn, C. J. Torng, P. H. Hor, R. L. Meng, L. Gao, S. J. Huang, Y. Q. Wang, and C. W. Chu, Phys. Rev. Lett. **58**, 908 (1987).
3. Göran Wendin, J. de Physique, Colloque **C9**, and the references therein.
4. R.L. Kurtz, AIP Conf. Proc. No. **165**, p. 222, AIP, New York, 1988, and the references therein.
5. C. Fuggle, J. Fink and N. Nücker, to be published as Int. J. Mod. Phys. B (1988), and the references therein.
6. K.C. Hass, to appear in Solid State Physics. (H. Ehrenreich and D. Turnbull, eds.) Vol. **42**, Academic Press, Orlando, 1989

7. Z.-X. Shen, J.W. Allen, J.-J. Yeh, J.-S. Kang, W. Ellis, W. E. Spicer, I. Lindau, M. B. Maple, Y. D. Dalichaouch, M. S. Torikachvili, J.Z. Sun and T.H. Geballe, Phys. Rev. B, Vol. **36**, 8414, 1987
8. Z.-X. Shen, J.-J. Yeh, I. Lindau, W.E. Spicer, J.Z. Sun, K. Char, N. Missert, A. Kapitulnik, T.H. Geballe, M.R. Beasley, SPIE Symp. Proc. Vol. **948-10**, 41-48 1987
9. Z.-X. Shen, J.W. Allen, J.-J. Yeh, J.-S. Kang, W. Ellis, W. E. Spicer, I. Lindau, M. B. Maple, Y. D. Dalichaouch, M. S. Torikachvili, J.Z. Sun and T.H. Geballe, MRS Symp. Proc. Vol. **99**, P349, 1987
10. P.A.P. Lindberg, Z.-X. Shen, I. Lindau, W. E. Spicer, C.B. Eom and T.H. Geballe, Appl. Phys. Lett. **53**, 529 (1988)
11. J.-S. Kang, J. W. Allen, Z.-X. Shen, W.P. Ellis, J.J. Yeh, B.-W. Lee, M.B. Maple, W. E. Spicer and I. Lindau, Journal of Less Common Metals, 24 & 25, (1988)
12. Z.-X. Shen, P.A.P. Lindberg, I. Lindau, W.E. Spicer, C.B. Eom, T.H. Geballe, Phys. Rev. B, **38**, 7152 (1988).
13. Z.-X. Shen, P.A.P. Lindberg, P. Soukiassian, I. Lindau, W.E. Spicer, C.B. Eom, T.H. Geballe, Phys. Rev. B. to be published.
14. C. K Shih, Z.-X. Shen, P.A.P. Lindberg, I. Lindau, W.E. Spicer, S. Doniach, J.W. Allen, Submitted to Phys. Rev. Lett., and the reference therein.
15. P.A.P. Lindberg, Z.-X. Shen, J. Hwang, C.K. Shih, I. Lindau, W. E. Spicer, D.B. Mitzi and A. Kapitulnik, Solid State Communication, to be published.
16. Z.-X. Shen, P.A.P. Lindberg, B.O. Wells, D.B. Mitzi, I. Lindau, W.E. Spicer and A. Kapitulnik, Phys. Rev. B, Dec. 1 (in press).
17. P.A.P. Lindberg, P. Soukiassian, Z.-X. Shen, C.B. Eom, I. Lindau, W.E. Spicer, T.H. Geballe, to be published in Appl. Phys. Lett., in press.
18. P.A.P. Lindberg, Z.-X. Shen, B.O. Wells, D. Mitzi, I. Lindau, W.E. Spicer, A. Kapitulnik, Appl. Phys. Lett. to be published.
19. P.A.P. Lindberg, Z.-X. Shen, B.O. Wells, D. Dessau, D. Mitzi, I. Lindau, W.E. Spicer, preprint, submitted to Phys. Rev. B.
20. N.F. Mott, Proc. Phys. Soc., Sect A **62**, 416 (1949)
21. J. Hubbard, Proc. Roy. Soc. Ser A, **276**, 238 (1963); **277**, 237 (1964); **281**, 401 (1964).
22. G. van der Laan, C. Westra, C. Haas, and G. A. Sawatzky, Phys. Rev. B **23**, 4369 (1981).
23. G. van der Laan, Solid State Commun. **42**, 165 (1982).
24. G. A. Sawatzky, in *STUDIES IN INORGANIC CHEMISTRY*, Vol. 3 (Elsevier, Amsterdam, 1983) p. 3.
25. A. Fujimori and F. Minami, Phys. Rev. B **30**, 957 (1984).
26. G. A. Sawatzky and J. W. Allen, Phys. Rev. Lett. **53**, 2339 (1985).
27. J. W. Allen, J. Mag. Mag. Mat. **47-48**, 168 (1985).
28. J. Zaanen, G. A. Sawatzky, and J. W. Allen, Phys. Rev. Lett. **55**, 418 (1985).
29. G. van der Laan, J. Zaanen, and G. A. Sawatzky, Phys. Rev. B **33**, 4253 (1986).
30. J. Zaanen, C. Westra, and G. A. Sawatzky, Phys. Rev. B **33**, 8060 (1986).
31. L. Ley, M. Taniguchi, J. Ghijsen, R. L. Johnson, and A. Fujimori, Phys. Rev. B **35**, 2839 (1987).
32. A. Fujimori, M. Saeki, N. Kimizuka, M. Tanigukchi, and S. Suga, Submitted to Phys. Rev. B **35**, 8814 (1987).
33. P. W. Anderson, Science **235**, 1196 (1987).
34. G. Baskaran, Z. Sou, and P. W. Anderson, Solid State Commun., **63**, 973 (1987).

35. P. W. Anderson, G. Baskaran, Z. Zou, and T. Hsu, Phys. Rev. Lett. **58**, 2790 (1987).
36. H. B. Schüttler, M. Jarrell, and D. J. Scalapino, Solid State Commun., to be published
37. C. M. Varma, S. Schmitt-Rink, and E. Abrahams, Solid State Commun., **62**, 681 (1987).
38. S. Robaszkiewicz, R. Micnas, and J. Ranninger, Phys. Rev. B**36**, 180 (1987).
39. H. Aoki and H. Kamimura, Solid State Commun., 8
40. V. J. Emery, Phys. Rev. Lett. **58**, 2794 (1987).
41. J. E. Hirsch, Phys. Rev. B**35**, 8726 (1987).
42. D. H. Lee and J. Ihm, Solid State Commun.**62**, 811 (1987).
43. D. Vaknin et al., Phys. Rev. Lett. **58**, 2802 (1987); Mitsuda et al., Phys. Rev. B**36**, 822 (1987); T. Freltoft et al. , Phys. Rev. B **36**, 826 (1987).
44. M.A. Beno, et al., Appl. Phys. Lett. **51**, 57 (1987); D.C. Johnston et al., in *Chemstry of High-Temperature Superconductors*, edited by D.L. Nelson, M.S. Whittingham, and T.F. George, ACS symposium Series **351**, 1987; J.D. Jorgensen et al., Phys. Rev. B**36**, 3608 (1987)
45. F.C. Zhang and T.M. Rice, Phys. Rev. B **37**, 3759 (1988).
46. N.G. Stoffel, P.A. Morris, W.A. Bonner, D. LaGraffe, Ming Tang, Y. Chang, G. Margaritondo, M. Onellion, Phys. Rev. B **38**
47. H.W. Zandbergen, Y.K. Hwang, M.J.V. Menken, J.N. Li, K. Kadowaki, A.A. Menovsky, G. van Tendeloo and S. Amelinckx, Nature **332**, 620 (1988).
48. J. H. Weaver, H. M. Meyer III, T. J. Wagener, D. M. Hill, Y. Gao, D. Peterson, Z. Fisk and A. J. Arko, Phys. Rev. B **38**, 4558 (1988).
49. R.S. List and A.J. Arko, private communication.
50. M.S. Hybertsen and L.F. Matheiss, Phys. Rev. Lett. **60**, 1661 (1988) (band calculation) , J.-J. Yeh and I. Lindau, At. Data Nucl. Data Tables **32** (1985) (Photoionization cross section).
51. J. W. Allen, S.-J. Oh, O. Gunnarsson, K. Schönhammer, M. B. Maple, M. S. Torikachvili, and I. Lindau, Adv. in Phys. **35**, 275 (1986).
52. L. C. Davis, Phys. Rev. B **25**, 2912 (1982).
53. U. Fano, Phys. Rev. **124**, 1866 (1961).
54. J.-I. Igarashi and T. Nakano, submitted to J. Phys. Soc. Japan.
55. M. Aono, T. C. Chiang, F. J. Himpsel, and D. E. Eastman, Solid State Commun. **37**, 471 (1981).
56. J. Sugar, Phys. Rev. B **5**, 1785 (1972).
57. O. Gunnarsson and K. Schönhammer, in *Giant Resonances in Atoms, Molecules, and Solids,* edited by J. P. Connerade, J.-M. Esteve, and R. C. Karnatak, Proceedings of NATO Summer School, Les Houches, June 1986 (New York, Plenum, 1987), P. 405.
58. O. Gunnarson and T. C. Li, Phys. Rev. B **36**, 9488 (1987).
59. M. H. Hecht and I. Lindau, Phys. Rev. Lett. **47**, 821 (1981).
60. R.L. Kurtz, R.L. Stockbauer, D. Muller, A. Shih, C.E. Toth, M. Osofsky and S.A. Wolff, Phys. Rev. B **35**, 8818 (1987)
61. T. Takahashi, H. Matsuyama, H. Katayama-Yoshida, Y. Okabe, S. Hosoya, K. Seki, H. Fujimoto, M. Sato and H. Inokuchi, Nature Vol. **334**, 691 (1988)
62. D. van der Marel, J. van Elp, G.A. Sawatzky and D. Heitmann, Phys. Rev. B **37**, 5136 (1988)
63. P. Thiry, G. Rossi, Y. Petroff, A. Revcolevschi and J. Jegondez, Europhys. Lett., in press.
64. A.K. McMahan, R.M. Martin and S. Stapathy, Phys. Rev. B **38**, 5560 (1988).
65. H. Eskes and G.A. Sawatzky, Phys. Rev. Lett. **61**, 1415 (1988)

66. H. Maeda, Y. Tanaka, M. Fukutomi, and T. Asano, Jap. Journ. Appl. Phys. Lett. **27**, 1209 (1988)
67. R.M. Hazen, C.T. Prewith, R.J. Angle, N.L. Ross, L.W. Finger, C.G. Hadiaiacos, D.R. Veblen, P.J. Heaney, P.H. Hor, R.L. Meng, Y.Y. Sun, Y.Y. Xue, Z.J. Huang, L. Gao, J. Bechtold and C.W. Chu, Phys. Rev. Lett. **60**, 1174
68. J.M. Tarascon, Y. Le Page, P. Barboux, B.G. Bagley, L.H. Greene, W.R. Mckinnon, G.W. Hull, M. Giroud and D.M. Hwang, preprint, Submitted to Phys. Rev. B
69. R.V. Kasowski, private communication.
70. Y. Dalichaouch, M. S. Torikachvili, E.A. Early, B.W. Lee, C.L. Seaman, K.N. Yang, H. Zou and M.B. Maple, Solid State Commun. **65**, 1001 (1987)
71. L. Soderholm, K. Zhang, D.G. Hinks, M.A. Beno, J.D. Jorgensen, C.U. Segre, and I.K. Schuller, Nature **328**, 604 (1987)
72. J.K. Liang, X.T. Xu, S.S. Xie, G.H. Rao, X.Y. Shao and Z.G. Duan, Z. Phys. B **69**, 137 (1987)
73. K.N. Yang, B.W. Lee, M.B. Maple and S.S. Laderman, Applied Physics (in press)
74. D.M. Wieliczka, C.G. Olson and D.W. Lynch, Phys. Rev. Lett. **52**, 2180 (1984)
75. J.K. Lang, Y. Baer and P.A. Cox, J. Phys. F **11**, 121 (1981)
76. J.J. Neumeier, M.B. Maple and M.S. Torikochvili, Physica C (to be published)

SURFACE AND INTERFACE PROCESSES

Characterization of the Ceramic: Substrate Interface of Plasma Sprayed High-Temperature Superconductors 352

Reaction between $YBa_2Cu_3O_{7-x}$ and Water 360

Interaction of CO, CO_2, and H_2O with Ba and $YBa_2Cu_3O_7$ 368

A Nonaqueous Chemical Etch for $YBa_2Cu_3O_{7-x}$ 376

Effect of Silver on the Water-degradation of YBaCuO Superconducting Films ... 384

Aluminum and Gold Deposition on Cleaved Single Crystals of $Bi_2CaSr_2Cu_2O_8$ Superconductor 391

Surface and Interface Properties of High Temperature Superconductors 399

Characterization of the Ceramic:Substrate Interface of Plasma Sprayed High-Temperature Superconductors

A. S. Byrne, W. F. Stickle
Perkin-Elmer, Physical Electronics Laboratory, Mountain View, CA 94043

C. Y. Yang, T. Asano, M. M. Rahman
Microelectronics Laboratory, Santa Clara University, Santa Clara, CA 95053

ABSTRACT

The adhesion of thin films of high-temperature superconducting ceramics to various substrates is of great importance for the successful application of superconductors in microelectronic devices. As the first step toward constructing semiconductor devices, thin films of the $YBa_2Cu_3O_{7-x}$ perovskite-based superconductor were plasma sprayed onto silicon wafers and aluminum coated silicon wafers. The substrate-thin film interfaces have been characterized using X-ray Photoelectron Spectroscopy, XPS, and Auger Electron Spectroscopy, AES. The interfacial chemistry and composition which exist before and after annealing are compared and contrasted. The XPS and AES data reveal differences in the chemistry and composition between the interface and the bulk material. Interfacial segregation of various components of the plasma sprayed material is observed following subsequent heat treatment. These differences may account for varying degrees of interfacial stability.

INTRODUCTION

The formation of $YBa_2Cu_3O_{7-x}$ thin films on silicon is of great importance for the application of high temperature superconductors in microelectronics devices. This group of high T_c films is well known to be reactive in ambient air.[1] This reactivity will affect bulk properties of the film and the chemistry and mechanical properties of the interface. It has been reported that poor adhesion of these films to silicon substrates may be related to chemical reactions at the film-substrate interface.[2] The use of intermediate layers such as MgO[3], ZrO_2[4], and aluminum[5] have been shown to improve adhesion performance and film stability.

© 1989 American Institute of Physics

The ceramic deposited directly on bare silicon wafers exhibits very poor adhesion. A study of the interface on samples prepared with an aluminum layer shows differences which can be correlated to the improved film adhesion. This investigation is concerned with the chemistry of the ceramic-substrate interface including the role of the aluminum layer.

The role of the annealing cycle and its effect on the chemistry of the interface is also investigated. It is well known that proper annealing is essential for achieving superconducting Y-Ba-Cu-O films. Correct choice of annealing conditions should optimize the film's superconducting properties without adversely affecting the film's mechanical properties.

EXPERIMENTAL

The samples were prepared using a Metco plasma spray system. The spray coating was performed in open air using argon carrier gas with a Metco 7MB spray gun. In order to allow a direct evaluation of the role of aluminum, silicon wafers with native oxide at the surface were half coated with aluminum by vacuum evaporation. A Y-Ba-Cu-O film approximately 30 microns thick was then deposited on the entire surface of the wafer by the plasma spray method. Figure 1 shows the schematic cross section of the test structures. The source materials, Y_2O_3, $BaCO_3$, and CuO were mixed together in a slurry, spray dried and sintered before spraying. The Y-Ba-Cu-O powder was prepared to yield 1-2-3 stoichiometric films after spraying. The wafers were preheated to 500°C for the deposition. The samples were annealed in dry O_2 at 950°C for 5 hours. The cooling rate was 2°C per minute to 700°C followed by 1°C per minute to room temperature.

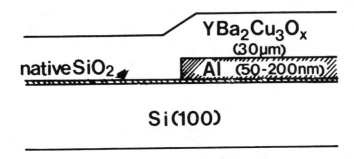

Figure 1. Schematic cross section of test structure.

The sample matrix consisted of aluminum coated (500, 1000, and 2000Å thick layers) and uncoated silicon wafers upon which the films were deposited. The samples were analyzed before and after the annealing cycle. Films which had little or no mechanical stability on the substrate immediately after spraying were not annealed. On samples where delamination occurred after annealing, the resulting surface was analyzed as it was exposed. Adhesive tape was used to remove the film on samples which had a more stable coating.

X-ray Photoelectron Spectroscopy, XPS, also known as Electron Spectroscopy for Chemical Analysis, ESCA, and Auger Electron Spectroscopy, AES, have been used to characterize the sprayed ceramic material and the interface where delamination has occurred. The XPS data were taken with a Perkin-Elmer model 5400 ESCA spectrometer using monochromated Al radiation. Depth profile data were obtained by ion milling using 4kV Ar^+ ions removing material at a rate of 35Å per minute relative to Ta_2O_5. XPS was used to analyze the exposed surface of samples which peeled unassisted and those which were manually peeled using adhesive tape. Atomic concentration data were obtained using peak areas and elemental sensitivity factors. The AES data were taken with a Perkin-Elmer model 660 Scanning Auger Microprobe.

RESULTS and DISCUSSION

The use of an aluminum interlayer has been successful in achieving a more stable film on silicon wafers. Figure 2 shows the effectiveness of an aluminum layer in preventing peeling and cracking of the plasma films. The region of the wafer without the aluminum layer shows cracks and peeling over the entire surface, while the region with aluminum adheres to the substrate.

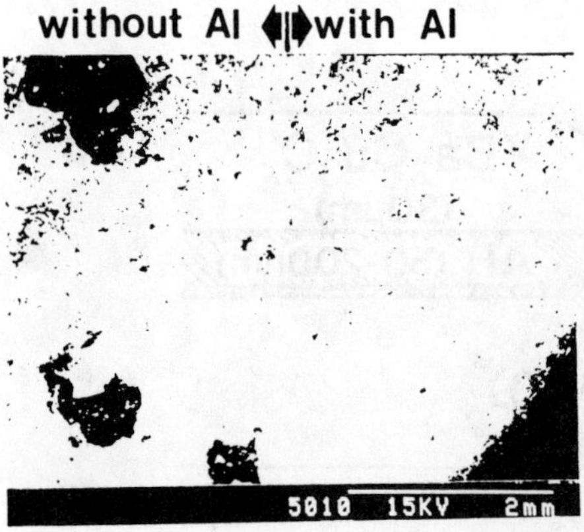

In Figure 3, a typical resistance vs temperature curve is shown for material sprayed on an aluminum coated silicon wafer. A critical temperature of about 87 K is indicated for material prepared using the plasma spray technique.

Both uncoated and aluminum coated samples show very poor adhesion before annealing when ex-

Figure 2. SEM image of plasma sprayed wafer surface.

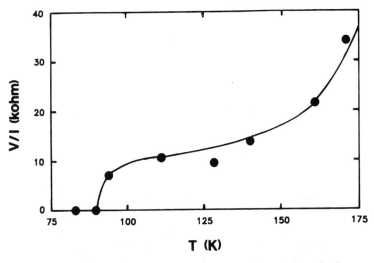

Figure 3. Resistance vs. Temperature curve for plasma sprayed material on aluminum coated silicon wafer.

posed to ambient air. On samples coated with aluminum, the film peels off at the ceramic-aluminum interface after exposure to air for several weeks. The depth profile in Figure 4 indicates the distribution of elements where the film peels. The exposed surface is primarily oxidized aluminum with small amounts of barium, yttrium, copper, and carbon. The aluminum, Figure 5, appears oxidized at the surface (curve a). Sputter etching the layer shows that more of the aluminum in unreacted as indicated by the increased intensity of the low binding energy peak (curves b and c). The aluminum layer is probably oxidized during the deposition process. Barium carbonate and hydroxide form because of reaction of BaO with the air. Adventitious hydrocarbon is also present.

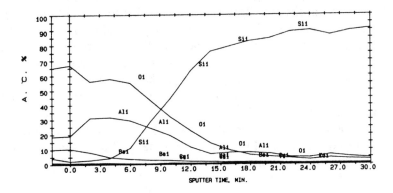

Figure 4. XPS depth profile of delaminated area on unannealed aluminum coated wafer.

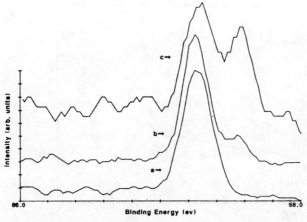

Figure 5. *Aluminum 2p region showing oxide at the surface (curve a) and increasing metal component with depth into the layer (curves b and c).*

The samples made without an aluminum layer will peel within days after spraying. The exposed surface is the silicon substrate with patches of the Y-Ba-Cu-O film remaining. The islands which remain are depleted in yttrium and copper in comparison to their concentrations in the original film[6]. Table I shows the atomic concentrations of yttrium, barium and copper to be markedly different when comparing the delaminated areas to the original film.

The annealing cycle changes the interface to improve the film adhesion on the aluminum coated samples. On these samples, the film material is initially stable to an adhesive tape pull test. After several weeks of exposure to air, the pull test shows separation occurs in the film itself with a relatively thick (>500Å) continuous layer of Y-Ba-Cu-O remaining on the substrate. Atomic concentration data for this layer is shown in Table I. This layer does not have the same stoichiometry as the peeled film, indicating that segregation is occurring.

Secondary electron images of interfaces on samples prepared without and with aluminum are shown in Figures 6a and b respectively. Elemental maps of the micrograph in Figure 7a are shown in Figures 7b, c, and d for Cu, Ba, and Si respectively. On both samples a 5 micron layer has formed between the ceramic and the substrate. Auger analysis indicates this layer is rich in barium and probably is a Ba-Si compound. Samples prepared without an aluminum layer develop pockets of copper between the barium rich layer and the silicon (see arrow B in Figure 6). The formation of the copper precipitates is inhibited by the aluminum layer. No aluminum is detected on the annealed aluminum coated wafers, suggesting that it has percolated into the film and is present in concentrations below the technique's detection limits. The use of Secondary Ion Mass Spectrometry, SIMS, is planned to better understand the distribution of aluminum in this system.

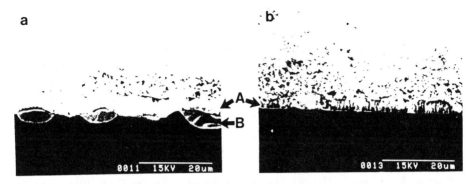

Figure 6. SEM images of samples a) without aluminum layer, b) with aluminum layer. Arrow A indicates the Ba-Si layer. Arrow B shows the copper precipitates.

Table I: Atomic Concentration (%)

	Y	Ba	Cu	O	C	Al	Si
Unannealed Interface (500Å)	1	7	3	51	20	18	-
Unannealed Interface (No Al)	<1	8	2	49	24	-	16
Unannealed Ceramic Bulk[7]	16	17	16	53	-	-	-
Annealed Interface (500Å)	5	11	5	53	25	-	-

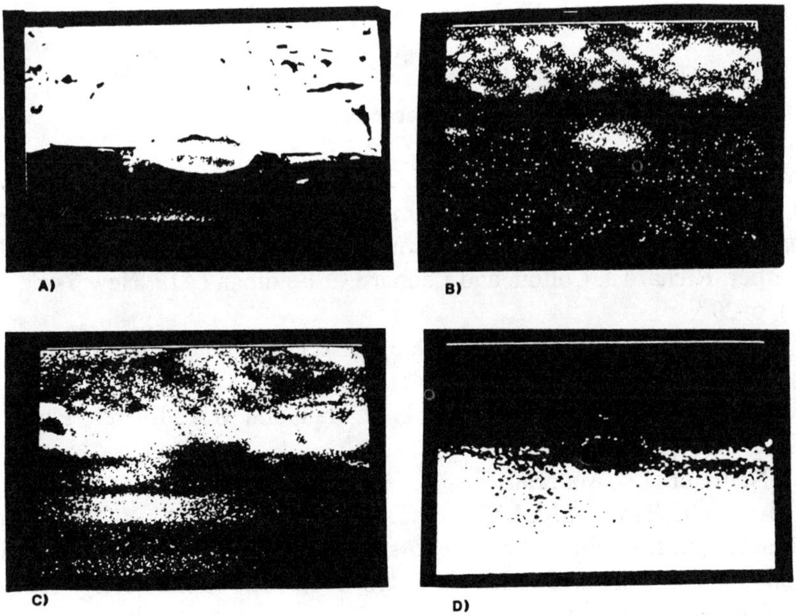

Figure 7. A) SEM image of interface with copper precipitate. B) Cu Auger map. C) Ba Auger map. D) Si Auger map.

CONCLUSION

We have looked at plasma spray deposited Y-Ba-Cu-O films on silicon wafers where delamination occurs quite readily. The use of an aluminum barrier layer improves the adhesion and prevents formation of copper precipitates at the interface. A Ba-Si layer was also observed at the interface independent of the presence of aluminum. Segregation was observed in the film at the interface where the film peeled. The aluminum layer was not detected after annealing and apparently percolates into the ceramic film. Future work using SIMS could yield more information on how the aluminum behaves in this system. The long term stability of the Y-Ba-Cu-O films is still a problem, which the material will continue to have until the parts are encapsulated.

ACKNOWLEDGEMENTS

We wish to thank J. D. Reardon of Perkin-Elmer Metco Division for providing the plasma spray services and coatings expertise in this work. We also wish to thank K. D. Bomben of Perkin-Elmer Physical Electronics for useful discussions. K. Tran and T. Y. Yao of Santa Clara University provided valuable assistance in testing and processing the samples.

References

1. W. F. Stickle, K. D. Bomben, A. M. Turner, R. C. Budhani, and R. F. Bunshah, in *Thin Film Processing and Characterization of High Temperature Superconductors, Anaheim, CA 1987*, AIP Conf. Proc. No. 165, edited by James M. E. Harper, Richard J. Colton, and Leonard C. Feldman (AIP, New York, 1988), p. 297.
2. P. Madakson, J. J. Cuomo, D. S. Yee, R. A. Hoy, and G. Silla, J. Appl. Phys. **63**, 2046 (1988).
3. A. Mogro-Campero, L. G. Turner, E. L. Hall, and M. C. Burrell, Phys. Lett. **52**, 1185 (1988).
4. M. Naito, et al, J. Materials Res. **2**, 713 (1987).
5. T. Asano, K. Tran, M.M. Rahman, T. Y. Yau, A. Byrne, C. Y. Yang, and J. D. Reardon, in *Extended Abstracts of the 20th (1988 International) Conference on Solid State Devices and Materials, Tokyo, 1988* (The Japan Society of Applied Physics, Tokyo), pp. 603-604.

6. A. S. Byrne, C. Y. Yang, M. M. Rahman, M. Gao, W. F. Stickle, D. W. Harris, and S. H. Chiao, *The Second Topical Conference on Quantitative Surface Analysis*, AVS, Monterey, 1987 (unpublished).
7. These data are based on sputtered samples.

REACTION BETWEEN $YBa_2Cu_3O_{7-x}$ AND WATER

M. W. Ruckman, S. M. Heald, D. Di Marzio, H. Chen
and A. R. Moodenbaugh
Brookhaven National Laboratory, Upton, NY 11973

C. Y. Yang
University of Michigan, Ann Arbor, MI 48109

ABSTRACT

Reaction between water at 80°C and $YBa_2Cu_3O_{7-x}$ (0.8<x<0.0) is studied using x-ray absorption fine structure (EXAFS) and x-ray diffraction (XRD). Oxygen deficient $YBa_2Cu_3O_{7-x}$ reacts with water and decomposes into $BaCO_3$, CuO and at least one other Cu containing phase. $YBa_2Cu_3O_7$ reacts slowly with water and the bulk material is modified before it decomposes. The structural and chemical modifications of this material resemble those reported for $YBa_2Cu_3O_7$ reacted with hydrogen gas. We conclude that either hydrogen or water enters the bulk before decomposition.

INTRODUCTION

An important limitation hindering some applications of high T_c copper oxide superconductors is their sensitivity to atmospheric moisture and water.[1] It has been shown that $YBa_2Cu_3O_7$ samples decompose when mixed with water and x-ray diffraction patterns show that CuO, $BaCO_3$, $Y(OH)_3$ and other phases containing Y, Ba, Cu and O are the reaction products.[2,3] Oxygen gas is also liberated and volumetric studies have been reported.[4] The reactivity of the 1-2-3 phase with water is not unexpected because Cu and/or O may be in oxidation states (i.e. Cu^{3+} or O^-) which are susceptible to reduction and both Y and Ba form oxides that are easy to hydrate. The kinetics of $YBa_2Cu_3O_7$ corrosion is likely to depend on the oxygen content. Harris and Nyang[5] report that oxygen deficient material loses its superconductivity faster than $YBa_2Cu_3O_7$ and suggest that the presence of oxygen vacancies accelerates the reaction between 1-2-3 and water by providing channels for easy transport of water into the lattice. They also conclude that water reacts with Ba to create a bulk lattice defect.

In this paper, XRD and EXAFS and x-ray absorption near-edge structure (XANES) measurements are used to study the interaction of $YBa_2Cu_3O_{7-x}$ with water. The data show that oxygen deficient 1-2-3 material reacts completely with water during the time it is present for reaction and that a highly disordered reaction product forms. $YBa_2Cu_3O_7$ retains the perovskite structure when exposed to water for the same duration but some of the Cu ions are reduced. Comparing the data with results reported by Yang et al.[6] for $H_xYBa_2Cu_3O_7$ and Harris and Nyang[5] for the 1-2-3 material exposed to water vapor suggest that hydrogen or water is present in the bulk material after exposure to water.

EXPERIMENTAL

$YBa_2Cu_3O_7$ and oxygen deficient samples were prepared by the well-known technique and the details are reported elsewhere.[7] For orthorhombic $YBa_2Cu_3O_7$, the final oxygen firing was 40 hours at 970°C and 8 hours at 700°C followed by furnace cooling to room temperature. Oxygen deficient samples were prepared by reheating $YBa_2Cu_3O_7$ at 900°C for a predetermined time. This second firing was terminated by quenching the sample in liquid nitrogen. The oxygen content was deduced from sample weight change relative to $YBa_2Cu_3O_7$ and samples produced by these techniques gave the expected single phase tetragonal or orthorhombic XRD patterns.[8]

Water exposure was accomplished by mixing distilled water with powdered samples that had been screened through a 400 mesh sieve. Sufficient water was added to form a liquid resembling black paint and the solution was placed on a glass evaporation dish. To accelerate the rate of reaction,[5] the solutions were heated to 80°C on a hot plate and the reaction ended by the complete evaporation of the water. The typical reaction time was approximately 10 minutes. The experimental methodology was designed, in part, to maximize the volume fraction of bulk material reacted with water and the superconductor water mixture was heated to 80°C to reduce reaction time from hours to minutes. For x-ray absorption studies, the reacted material was reground and spread out on scotch tape. Cu k-edge XANES and EXAFS studies were done on the NSLS X-11A beam line at Brookhaven National Laboratory using Si(111) monochromator crystals. A Cu foil standard was run simultaneously with each edge measurement for use as an energy calibration. All near edge spectra are energy referenced to a small peak halfway up the Cu metal edge. Full details of the background removal and EXAFS data analysis procedures employed are contained in Ref. (9). The EXAFS data have been filtered with a Gaussian window and have been weighted to place emphasis on the low z elements. XRD patterns were recorded using a GE XRD-5 x-ray diffractometer equipped with a Cu K-α x-ray source.

RESULTS AND DISCUSSION

XRD patterns (Fig. 1) recorded over a 2θ range 26-40° show the structure of $YBa_2Cu_3O_7$ and $YBa_2Cu_3O_{6.2}$ before, marked "a", and after, marked "b", reaction with water. During contact with water, gas bubbles which were probably oxygen are seen. Eickenbush et al.[4] conducted volumetric measurements of the gases liberated during the water reaction with $YBa_2Cu_3O_7$ which show that this gas is oxygen and indicate that sufficient oxygen is released to suggest the reduction of one Cu^{3+} or O^- ion per unit cell. Both Cu^{3+} and O^- have electrochemical potentials sufficient to disrupt H_2O and release hydrogen and oxygen. Bansal and Sandkuhl[3] and Barkatt et al.[10] report that the pH of the water superconductor solution increases during reaction indicating that a strong

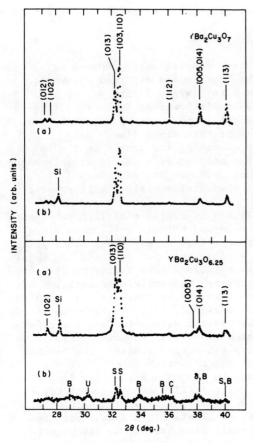

Fig. 1. XRD patterns for $YBa_2Cu_3O_7$ and $YBa_2Cu_3O_{6.25}$ (a) before and (b) after reaction with water.

base, possibly $Ba(OH)_3$, forms and goes into solution. Our XRD data show that $YBa_2Cu_3O_7$ retains the orthorhombic structure after exposure to water while $YBa_2Cu_3O_{6.2}$ loses the tetragonal structure. For $YBa_2Cu_3O_{6.2}$, traces of the 1-2-3 phase, marked "S", remain and weak reflections assigned to $BaCO_3$, marked "B", and an unknown phase, marked "U" are seen. During reaction with water, this material changed color from black to brownish green, whereas, $YBa_2Cu_3O_7$ remained black. Examination of all the other oxygen deficient samples show that the same decomposition took place. The presence of $BaCO_3$ in the samples is due to the reaction of $Ba(OH)_3$ with atmospheric CO_2.

Cu k-edge XANES spectra for all the samples are shown in Fig. 2a and b. Before reaction with water (Fig. 2a), the Cu k-edge data shows a feature which undergoes a systematic variation with oxygen concentration. A peak near the onset of the Cu k-edge, $(E=E_o=0)$, assigned to a 1s-4pπ transition, grows with decreasing oxygen concentration. Several earlier studies[11] indicate that the appearance of this state is related to the formation of Cu^{1+} in the Cu(1), (Chain) site when oxygen is removed from the O(1) site. For water exposed $YBa_2Cu_3O_7$(Fig. 2b), the 1s-4pπ state also appears on the Cu k-edge and we interpret this as indicating that a partial reduction of some Cu^{2+} ions to Cu^{1+} has occurred. The shape of the edge above this feature remains unchanged. For oxygen deficient material (Fig. 2b), the Cu k-edge undergoes a complete change after reaction with water. The 1s-4pπ transition associated with Cu^{1+} in the Cu(1) site, either does not increase or disappears into a new edge feature found at higher photon energy. This new edge feature, marked "1", is near the energy found for the 1s-4p transition in CuO reference compound.[11] This may suggest that a compound similar in bonding to CuO has formed but this conclusion must be qualified by the observation that this system is not homogeneous. Small peaks above the main edge attributable to multiple

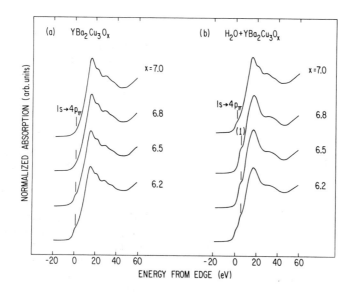

Fig. 2. Cu k-edge XANES for the 1-2-3 material before and after reaction with water.

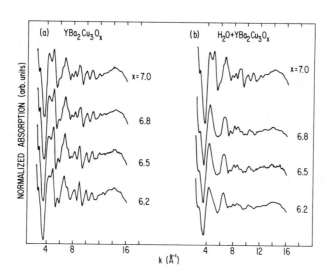

Fig. 3. Cu k-edge EXAFS for the same materials.

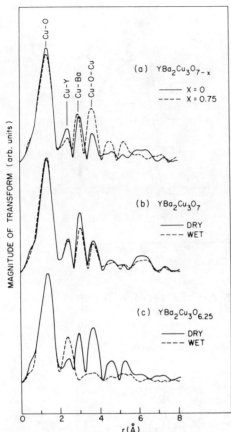

Fig. 4. k-weighted EXAFS for $YBa_2Cu_3O_7$ and $YBa_2Cu_3O_{6.25}$ before and after reaction with water.

scattering events also disappear and this suggests that the compound becomes more disordered.

Cu k-edge EXAFS, shown in Figs. 3 and 4, show that structural changes accompany the modification of the Cu k-edge. For $YBa_2Cu_3O_7$, the EXAFS signals (Fig. 3) appear unchanged before and after water exposure. However, Fourier transforms (Fig. 4) show that changes have occurred in the radial distribution function. The positions of the atomic shells show little or no change, but the weight of the atomic shells containing barium which are closest to the Cu(1) and Cu(2) sites is reduced. The ability of EXAFS to analyze the structure of disorderd materials is of use in studying the oxygen deficient material after water exposure. All of the EXAFS signals for $YBa_2Cu_3O_{7-x}$ (0<x<0.8) show the same basic modification. Fewer EXAFS oscillations are seen and they damp out more quickly in k-space. The Fourier transforms for $YBa_2Cu_3O_{6.2}$ show the removal of R-space structure due to the higher order shells and this suggests that the reactedmaterial is highly disordered. The copper-oxygen bond length remains almost unchanged because the peak in radial distribution function assigned to Cu-O nearest neighbors does not shift.

Examination of the Fourier transforms for $YBa_2Cu_3O_7$ after reaction with water and $YBa_2Cu_3O_{6.25}$shows that structural changes accompanying the partial reduction of Cu^{2+} for both cases are not equivalent. For $YBa_2Cu_3O_7$ (unmodified) and $YBa_2Cu_3O_{6.25}$, going from the orthorhombic to the tetragonal structure causes a shift in the position of higher order shells and changes in the intensity from the shells marked Cu-Y (decreases) and Cu-O-Cu (increases). The shell marked Cu-Ba also undergoes a shift to smaller R number and a slight increase in intensity. However for the water modified $YBa_2Cu_3O_7$, the shells do not move in R-space and the Cu-Ba shell becomes weaker. In conjunction with the XRD result, we conclude that water has not caused the 1-2-3 material to undergo the orthorhombic to tetragonal phase transition upon Cu reduction.

XRD and x-ray absorption spectroscopy show that the structure of the oxygen deficient material is completely disrupted by water exposure and the material incongruently decomposes into $BaCO_3$, CuO, and other disordered Y-Ba-Cu-O phases containing Cu^{2+}. $YBa_2Cu_3O_7$ undergoes a much slower reaction and most of the material retains the orthorhombic structure. Bulk changes involving the copper oxidation state, namely, the reduction of some Cu^{2+} to Cu^{1+} and a change in the structure of the barium site are detected. Our data clearly indicate that the bulk superconductor is being modified during the interfacial reaction with water.

Similarity between the XANES and EXAFS for the water modified 1-2-3 material and 1-2-3 material reacted with hydrogen gas suggests the possible entry of hydrogen into the bulk superconductor. Yang et al.[6] find that hydrogen readily dissolves in the 1-2-3 material up to ~0.2 atoms/unit cell. Beyond that concentration, a nonsuperconducting hydride forms. They also find that hydrogen does not form a water molecule or hydroxyl radical in the bulk because infrared absorption bands in the 3000-4000 cm^{-1} range associated with these components are absent. However, bands at 841 and 1467 cm^{-1} associated with a Cu-H bond were reported. XANES measurements found that hydrogen incorporation causes reduction of some of the Cu^{2+} to Cu^{1+} and EXAFS results showed the same type of structural modification illustrated in Fig. 4 for $YBa_2Cu_3O_7$ after exposure to water. Yang et al. report that the Y K and Ba $L_{2,3}$ edges are unmodified by hydrogen uptake.

Loss of oxygen could also cause a reduction of some of the 2+ copper. However, specific changes in the transformed EXAFS spectra which are related to structure for the 1-2-3 phase exposed to water and the tetragonal oxygen deficient phase when compared to $YBa_2Cu_3O_7$ suggest that this is an unlikely explanation for our experimental results. Ba removal from the unit cell does occur during some stage of the reaction of $YBa_2Cu_3O_7$ with water. Ba leaching leading to vacancies in the atomic shell containing barium could account for the reduction in those shells, but removal of Ba ions should destabilize the whole $YBa_2Cu_3O_7$ unit cell. Since little change is seen in the position or intensity of nearby atomic shells, we believe the removal of Ba from the bulk is unlikely at this early stage of reaction between water and the 1-2-3 material.

Reaction of $YBa_2Cu_3O_7$ with water causes a loss of superconductivity. Harris and Nyang[5] studied the change in the Meissner effect upon exposure to water and found that oxygen deficient material lost the superconducting phase faster than $YBa_2Cu_3O_7$. They concluded that oxygen vacancies provide pathways which allow the easy transport of water molecules into the bulk and report that water goes into vacancies in the Cu-O chains near the Ba ion. Reaction with the Ba ion leads to the formation of a lattice defect which at high concentration causes the disruption of the $YBa_2Cu_3O_7$ lattice. Such lattice defects have been observed in transmission electron micrographs of $YBa_2Cu_3O_7$ samples exposed to water vapor.[12] Our results are in good agreement with these earlier findings. Entry of water and its interaction with barium would change Cu 1s photoelectron scattering from the shells containing barium and the insertion of hydrogen in an O(1) site for O^{2-} should reduce the copper valence. We do not see the orthorhombic to tetragonal phase transition reported by Harris and Nyang upon the loss of superconductivity in the 1-2-3 samples exposed to water. Vacancies play a key role in the diffusion of oxygen[13,14] and apparently water in the high T_c material. Lack of these vacancies in the Cu-O chains reduces the mobility and probably explains the greater stability of $YBa_2Cu_3O_7$ relative to the oxygen deficient material.

CONCLUSIONS

We find that the rate of reaction between water and $YBa_2Cu_3O_7$ at 80°C is strongly dependent on the presence of oxygen vacancies. Oxygen deficient material with as little as 0.2 oxygen vacancies per unit cell is completely decomposed during the ten minute exposure to water and the reaction produces a highly disordered material containing $BaCO_3$, CuO and at least one other phase containing Y, Ba, Cu and/or O. $YBa_2Cu_3O_7$ is more stable and is not decomposed. XRD and x-ray absorption data show that the orthorhombic structure remains but some of the Cu ions are reduced from 2+ to 1+ and Cu 1s photoelectron scattering from the first atomic shell containing Ba is modified. Our data suggest that hydrogen or water is intercalating into $YBa_2Cu_3O_7$ prior to the decomposing reaction. If hydrogen or water is being intercalated, comparison with data for $H_xYBa_2Cu_3O_7$ and 1-2-3 exposed to water indicates that T_c and the Meissner effect would be reduced.

ACKNOWLEDGEMENTS

This work is supported by the U.S. Department of Energy, Division of Materials Sciences under Contract No. DE-AC02-76CH00016. The X-11 beam line and National Synchrotron Light Source operations are supported by the U.S. Department of Energy under Contract Nos. DE-AS05-80-ER10742 and DE-AC02-76CH00016, respectively.

REFERENCES

1. H. S. Horowitz, R. K. Bordia, R. B. Flippen, R. E. Johnson, U. Chowdhry, Mater, Res. Bull. 23, 821 (1988); J. G. Thompson, B. G. Hyde, R. L. Withers, J. S. Anderson, J. D. Fitzgerald, J. Bitmead, and M. S. Paterson, Mater. Res. Bull. 22, 1715 (1987); I. Nakada, S. Sato, Y. Oda, T. Kohara, Jpn. J. Appl. Phys. 26, L697 (1987).
2. M. F. Yan, R. L. Barns, H. M. O'Bryan Jr., P. K. Gallagher, R. C. Sherwood, and S. Jin, Appl. Phys. Lett. 51, 532 (1987).
3. N. P. Bansal and A. L. Sandkuhl, Appl. Phys. Lett. 52, 324 (1988).
4. H. Eickenbush, W. Paulus, R. Schölhorn, and R. Schlögl, Mater. Res. Bull. 22, 1505 (1987).
5. L. B.. Harris and F. K. Nyang, Solid State Commun. 67, 359 (1988).

6. C. Y. Yang, X-Q. Yang, S. M. Heald, J. J. Reilly, T. Skotheim, A. R. Moodenbaugh, and M. Suenaga, Phys. Rev. B $\underline{36}$, 8798 (1987).
7. J. M. Tranquada, S. M. Heald, A. R. Moodenbaugh, Youwen Xu, Phys. Rev. B (to be published).
8. R. J. Cava, B. Batlogg, R. B. van Dover, D. W. Murphy, S. Sunshine, T. Siegrist, J. P. Remeika, E. A. Reitman, S. Zahurak, and G. P. Espinosa, Phys. Rev. Lett. $\underline{58}$, 1676 (1987); J. P. Jorgensen, M. A. Beno, D. G. Hinks, L. Soderholm, K. J. Volin, R. L. Hitterman, J. D. Grace, I. K. Schuller, C. U. Segre, K. Zhang, and M. S. Kleefisch, Phys. Rev. B $\underline{36}$, 3608 (1987).
9. E. A. Stern and S. M. Heald in <u>Handbook of Synchrotron Radiation</u>, E. E. Koch, Editor, Vol. 1b, p. 995, North-Holland, Amsterdam, 1983.
10. A. Barkatt, H. Hojaji, and K. A. Michael, Adv. Ceram. Mater. $\underline{2}$, 701 (1987); K. G. Frase, E. G. Linger, and D. R. Clarke, Adv. Ceram. Mater. $\underline{2}$, 698 (1987).
11. S. M. Heald, J. M. Tranquada, A. R. Moodenbaugh, and Youwen Xu, Phys. Rev. B $\underline{38}$, 761 (1988); K. B. Garg, A. Bianconi, S. Della Longa, A. Clozza, M. de Santis, and A. Marcelli, Phys. Rev. B $\underline{38}$, 244 (1988); F. W. Lytle, R. B. Greegor, and A. J. Panson, Phys. Rev. B $\underline{37}$, 1550 (1988).
12. B. G. Hyde, J. G. Thompson, R. L. Withers, J. G. Fitzgerald, A. M. Stewart, D. J. M. Bevan, J. S. Anderson, J. Bitmead, and M. S. Paterson, Nature $\underline{237}$, 402 (1987).
13. P. R. Fretais and T. S. Plaskett, Phys. Rev. B $\underline{36}$, 5723 (1987).
14. H. Bakker, J. P. A. Westerfeld, D. M. R. LoCascio, and D. O. Welch, Physica C (to be published).

Interaction of CO, CO_2 and H_2O with Ba and $YBa_2Cu_3O_7$

S. L. Qiu, C. L. Lin, M. W. Ruckman, J. Chen, D. H. Chen
Youwen Xu, A. R. Moodenbaugh, and Myron Strongin
Brookhaven National Laboratory
Upton, NY 11973-6000

D. Nichols and J. E. Crow
Temple University
Philadelphia, PA 19122

ABSTRACT

Photoemission from high Tc superconductors crucially depends on surface stoichiometry as well as the interaction of impurities with the reactive surface. In this paper we discuss the photoemission spectra of clean $YBa_2Cu_3O_7$ and $YBa_2Cu_3O_{6.25}$ as well as the spectra after the interaction of CO, CO_2 and H_2O with both the high Tc 1-2-3 compound and Ba metal.

INTRODUCTION

Because of the well-known spontaneous modifications of the high Tc 1-2-3 surface, controversy exists over the interpretation of photoemission results obtained for both the valence band and core level spectra, as well as how these spectra change with temperature.

In the data reported here, we observe no change with temperature if impurities are carefully eliminated, or if the samples are cleaned at low temperature. It is our view that early results[1] reporting such effects are basically due to accidental contamination of the surface during the experiment. In the O 1s core level data for $YBa_2Cu_3O_7$ shown here, and from the work of other groups[2] it is known that the photoemission peak due to the bulk oxide occurs at about 529 eV. Further evidence is presented here which confirms that this is the only feature attributable to the 1-2-3 phase. Other features seen by various authors [1,3,4] are extrinsic and most likely due to impurities. The interaction of CO, H_2O, and CO_2 with Ba and $YBa_2Cu_3O_7$ produces changes in the O 1s spectra as well as a positive Ba 4d core level shift due to Ba carbonate formation which makes the Ba 4d core level spectra appear asymmetric. Another feature found near 9.5 eV in the valence band spectrum has a component which cannot be readily assigned to the 1-2-3 material. The nature of this feature and its relationship to impurities and the formation of Ba surface compounds will be discussed.

The submitted manuscript has been authored under contract DE-AC02-76CH00016 with the Division of Materials Sciences, U. S. Department of Energy. Accordingly, the U. S. Government retains a non-exclusive, royalty-free license to publish or reproduce the published form of this contribution, or allow others to do so, for U. S. Government purposes.

The data were taken at the Brookhaven National Synchrotron Light Source (NSLS) U7 beamline, and also by using a conventional Al K-α x-ray source. The ambient pressure in the experimental chamber was 2×10^{-10} Torr. The surface of the 1-2-3 compounds was cleaned by scraping with a stainless steel file.

RESULTS

A. Clean Samples

In Figure 1, we show O 1s core level spectra for freshly scraped O_7 and $O_{6.25}$ samples. The spectra look similar except for a small shift of the lower binding energy edge towards the Fermi level for the O_7 sample. A difference curve is also plotted in Figure 1. The

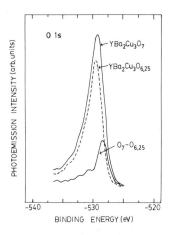 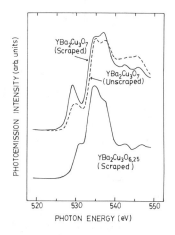

Figure 1 - XPS spectra for O 1s core level of the freshly scraped $YBa_2Cu_3O_7$ and $YBa_2Cu_3O_{6.25}$. The lower solid curve is the difference between these two spectra.

Figure 2 - Oxygen 1s absorption edges of clean $YBa_2Cu_3O_7$, dirty $YBa_2Cu_3O_7$ and clean $YBa_2Cu_3O_{6.25}$. Data beyond 540 eV may be affected by the Au N_3 adsorption edge at about 545 eV from the Au detector and the optics.

peak in the difference curve at low binding energy may be caused by an O 1s core level shift related to better final state screening in the metallic O_7 sample. However, we point out that this peak in the difference curve could also be an unresolved O 1s core level component that is decreased in intensity when oxygen is removed from the 1-2-3 lattice.

In Figure 2, oxygen K-edge data for $YBa_2Cu_3O_7$ (upper solid curve) shows a peak at about 529 eV which is assigned to transitions from O 1s core level to empty 2p states, Nücker et al.[5] attribute these empty O 2p states to oxygen holes in the conduction band. It is interesting to note that in the unscraped sample this peak is small (dashed curve) and there is only a residual feature in the insulating $O_{6.25}$ sample (lower solid curve in Figure 2).

Valence band features, not shown here, for the O_7 and $O_{6.25}$ compounds show more O 2p weight near E_F for the O_7 sample. Cu 2p spectra, also not shown here, show a larger satellite peak for the O_7 sample than for the $O_{6.25}$ sample. This is consistent with the greater Cu 2+ and smaller Cu 1+ concentration in the O_7 sample.

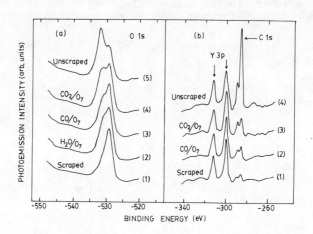

Figure 3 - XPS spectra for (a) O 1s and (b) C 1s core levels of $YBa_2Cu_3O_7$. The exposure of H_2O, CO, and CO_2 is 400L, 200L, and 2000L respectively.

B. Impurities on $YBa_2Cu_3O_7$

Figures 3 (a) and (b) show the effect of selected impurities on the O 1s and carbon 1s core level spectra. A huge carbon 1s peak was observed for the unscraped O_7 sample (spectrum (4) in (b)) and the O 1s core level shows two peaks located at 529 eV and about 532 eV. The 532 eV O 1s core level component along with the carbon 1s peak can be significantly reduced by scraping as shown in spectrum (1) which indicates that the dominant feature at about 532 eV in the O 1s core level region (spectrum (5) in (a)) is mainly due to the carbon related contamination. The interaction of H_2O, CO, and CO_2 with the 1-2-3 compound at room temperature produces species with binding energies ranging from 531 to 533 eV as shown in Figure 3 (a). The intensity of the carbon 1s peak increases as the exposure of CO or CO_2 is increased (Figure 3 (b)).

Figure 4 shows the Ba 3d and 4d core level spectra of $YBa_2Cu_3O_7$. Spectra (2) in both (a) and (b) was taken from the unscraped O_7 sample while the spectrum from a freshly scraped sample is shown as spectrum (1). Similar to the case of the O 1s spectra shown in Figure 3, the dominant feature seen in Ba 3d and 4d spectra of the dirty O_7 sample is directly related to the carbon contaminated surface.

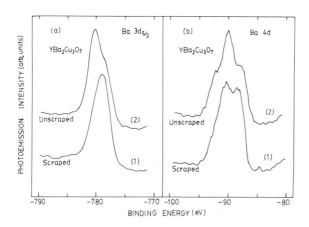

Figure 4 - XPS spectra of (a) Ba 3d and (b) Ba 4d core levels for clean and dirty $YBa_2Cu_3O_7$.

Figure 5 shows the emergence of various valence band features as different impurities are put on the O_7 surface. Figure 5 (a) shows the features taken at 76 eV photon energy and (b) at 105 eV. Although a weak feature at 9.5 eV is observed on the freshly scraped sample at the copper resonance energy of 76 eV, this feature is not observed at 105 eV. In the data shown at both 105 eV and 76 eV the 9.5 eV peak is always increased by the absorption of these common contaminating gases.

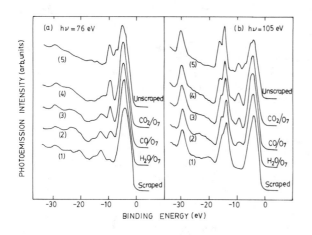

Figure 5 - Valence band spectra taken at a photon energy of (a) 76 eV and (b) 105 eV. The exposure of H_2O, CO, and CO_2 is 400L, 200L, and 2000L respectively.

C. Impurities on Ba

To better understand photoemission from the species that could form on the surface of the 1-2-3 compounds during the interaction with various gases we have studied the O 1s, valence band and Ba 3d and 4d spectra for O_2, CO, and H_2O on Ba metal. The O 1s spectrum for O_2 on Ba at 20K is shown in Figure 6 (a) (1). The feature near 529 eV is thought to be Ba oxide (BaO) with a formal valence of -2. At low coverage of O_2 this BaO feature is at 530 eV and shifts to lower binding energy with increasing the oxygen exposure. This shift is probably due to the incorporation of oxygen below the metal surface. From other work we have done[6], the peak at 532 eV is thought to be the peroxide, BaO_2 and the peak at 541 eV is solid O_2. Figure 6 (a) (2) shows H_2O on Ba at room temperature where the broad feature with binding energy ranging from 531 to 533 eV is due to hydroxide. Spectrum (3) in Figure 6 (a) for CO on Ba shows evidence for the formation of Ba carbonate. The shoulder at 530 eV is the BaO feature for low coverage of oxygen as mentioned above. The dominant peak at about 532.5 eV is accompanied by a huge carbon 1s feature (data are not shown) which resembles the case of the dirty O_7 sample. This indicates that the carbon related contamination on the surface of the 1-2-3 compounds is a Ba carbonate (possibly $BaCO_3$).

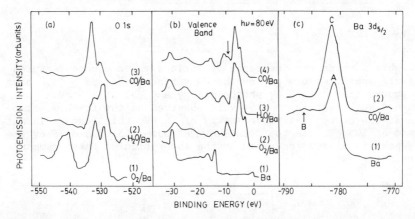

Figure 6 - (a) O 1s XPS spectra for (1) O_2 on Ba at 20K, (2) H_2O on Ba at 300K, (3) CO on Ba at 300K.
       (b) Valence band spectra for (1) Ba metal, (2) O_2 on Ba at 20K, (3) H_2O on Ba at 300K, (4) CO on Ba at 300K.
       (c) Ba 3d XPS spectra for (1) Ba metal, (2) CO on Ba.

The spectra in Figure 6 (b) show the dramatic affects of impurities on the Ba metal valence band. For a high O_2 coverage, spectrum (2) shows again a systematic shift of the Ba 5s and 5p core level to lower binding energy relative to that of Ba metal (spectrum (1)) due to the incorporation of oxygen below the metal surface. In contrast, the Ba 5s and 5p states shift to higher binding energy after the interaction of H_2O and CO with Ba metal (spectra (3) and (4)). Only the peak at 9.5 eV in spectrum (4), denoted by an arrow, is accompanied by a huge carbon 1s peak and therefore it can be assigned to a feature of Ba carbonate. The features at 4.5, 6.5, and 11 eV in spectrum (4) as well as the features at 3, 5, and 10 eV in spectrum (2) of Figure 6 (b) will be discussed in detail elsewhere[6].

Peak A in Figure 6 (c) represents the metallic Ba $3d_{5/2}$ core level and the broad shoulder denoted by B is a plasmon. After the interaction of CO with Ba metal, the plasmon feature is gone and a shift of 0.8 eV between peaks C and A is observed. Figure 6 (c) suggests that the higher binding energy components of Ba 3d and 4d core levels observed on the 1-2-3 compound are due to Ba carbonate contamination. Both Figure 6 (b) (4) and (c) show that there is a positive Ba core electron chemical shift due to Ba carbonate which makes the Ba core levels asymmetric.

DISCUSSION

A. O 1s Core Level Spectra

There have been reports that an intrinsic feature exists near 532 eV[3,4] and also at 533 eV[1] at low temperature. We find no evidence for intrinsic features at these energies except for a possible weak shoulder at 531 eV. Furthermore, upon scraping, the O K-edge feature at 529 eV shown in Figure 2 increases in intensity upon removal of these higher binding energy O 1s core level features. It has been claimed by Sarma et al.[7] that the scraping process removes the species of O⁻ on the surface and this explains the lack of change with temperature. The data in Figure 2 argues against this conclusion since removing oxygen would be expected to reduce the 1s-2p near-edge feature and make the spectrum look like the $O_{6.25}$ sample shown in Figure 2, which does not show this feature. Growth of the 1s-2p peak with scraping, in fact, indicates that the surface band structure originally has less oxygen 2p holes, and scraping produces a more stoichiometric O_7 surface where there are more oxygen holes.

B. Valence Band Spectra

Various investigators[8,9] claim the 9.5 eV peak is intrinsic and caused by a double hole final state on oxygen similar to that for Cu. Our data is not conclusive on this point, but it is clear that impurities can produce a feature in this regime. CIS spectra for the 9.5 eV feature (not shown here) show that there is a small resonance at the Cu 3p core level threshold, similar to that found for the well known Cu shake-up feature at 12.5 eV. The 9.5 eV peak is absent at 105 eV photon energy, and it is quite small at 76 eV as shown in Figure 5. We emphasize "small," since in most work,[8,10] the 9.5 eV feature is larger than or equal to the adjacent Cu shake-up peak. In these cases the 9.5 eV peak is probably caused by impurities. Tang et al.[8] have observed this feature at 100 eV photon energy for a single crystal sample and have argued that this feature is intrinsic. Our data leads us to conclude that the 9.5 eV feature taken at 100 eV photon energy is not an intrinsic feature and is probably due to residual impurities. Our examination of the interaction of H_2O and

CO with Ba illustrates that some of the likely Ba related surface compounds have the necessary valence band and core level features.

C. Ba 3d and 4d Core Level Spectra

The Ba 3d and 4d core level spectra shown in Figure 4 (2) for the dirty O_7 sample resemble those reported by Werfel et al.,[11] and Ford et al.[12] From such asymmetric spectra they concluded that there are two differently charged Ba ions, one is in an almost perfect oxygen coordination, the other with a distorted oxygen neighborhood. Our results show clearly that with the decreasing intensity of carbon 1s peak due to scraping the surface, the spectra of the Ba 3d and 4d core levels become less asymmetric as shown in Figure 4 (1). Furthermore, the interaction of CO with Ba metal produces positive chemical shift of Ba core levels which makes the Ba 3d and 4d spectra asymmetric. Therefore, the asymmetry of the Ba 3d and 4d spectra is due to Ba carbonate contamination rather than two different Ba sites.

In conclusion, this work and other recently published results[2,13] show that the intrinsic photoemission features of the high Tc superconductors are distinguishable from those caused by impurities. This implies that surface spectroscopies like photoemission and SEXAFS can provide important information about the electronic structure of these materials.

ACKNOWLEDGMENTS

We are grateful to J. Davenport, P. D. Johnson, V. J. Emery, G. Wendin, and J. Tranquada for their helpful comments. We thank the staff of the National Synchrotron Light Source at Brookhaven National Laboratory, where the photoemission measurements were performed. We also thank F. Loeb, J. Raynis, and R. Raynis for their excellent technical support. Special thanks are also due N. Michelsen who suffered greatly during the preparation of this manuscript. This work was supported by the U. S. Department of Energy under Contract No. DE-AC02-76CH00016.

REFERENCES

1. D. D. Sarma, K. Sreedhar, P. Ganguly, and C. N. R. Rao, Phys. Rev. B36, 2371, 1987; C. N. R. Rao, P. Ganguly, M. S. Hedge, and D. D. Sarma, J. Am. Chem. Soc. 109, 6893, 1987.
2. See for example, J. H. Weaver, H. M. Meyer III, P. J. Wagner, D. M. Hill, Y. Gao, D. Peterson, Z. Fisk, and A. J. Arko, Phys. Rev. B 38, 4668, 1988; D. C. Miller, D. E. Fowler, C. R. Brundle, and W. Y. Lee in: American Institute of Physics Conference Proceedings, No. 165, New York, pg. 336, 1988.
3. Y. Dai, A. Manthiram, A. Campion, J. B. Goodenough, Phys. Rev. B 38, 5091, 1988.
4. S. Horn, J. Cai, S. A. Shaheen, Y. Jeon, M. Croft, C. L. Chang, and M. L. denBoer, Phys. Rev. B36, 3895, 1987.
5. N. Nücker, J. Fink, J. C. Fuggle, P. J. Durham, and W. M. Temmerman, Phys. Rev. B37, 5158, 1988.
6. S. L. Qiu, C. L. Lin, J. Chen, and Myron Strongin, to be published.

REFERENCES (cont.)

7. D. D. Sarma, K. Prabhakaran and C. N. R. Rao, Solid State Commun., 67, 263, 1988.
8. M. Tang, N. G. Stoffel, O. B. Chen, D. LaGraffe, P. A. Morris, W. A. Bonner, G. Mararitondo, and M. Onellion, Phys. Rev. B 38, 897, 1988.
9. G. Wendin, in proceedings of the 14th International Conference on X-ray and Inner-shell Processes (J. Phys. (Paris) to be published.
10. J. A. Yarmoff, D. R. Clarke, W. Drube, U. O. Karlsson, A. Taleb-Ibrahimi, and F. J. Himpsel, in: American Institute of Physics Conference Proceedings, No. 165, New York, pg. 264, 1988.
11. F. Werfel, M. Heinonen, and E. Suoninen, Z. Phys. B Condensed Matter 70, 317, 1988.
12. W. K. Ford, C. T. Chen, J. Anderson, J. Kwo, S. H. Liou, M. Hong, G. V. Rubernacker, and J. E. Drumheller, Phys. Rev. B 37, 7924, 1988.
13. R. L. Kurtz, R. Stockbauer, T. E. Madey, D. Mueller, A. Shih, and L. E. Toth, Phys. Rev. B. 37, 7936, 1988.

A NONAQUEOUS CHEMICAL ETCH FOR $YBa_2Cu_3O_{7-x}$

R. P. Vasquez, M. C. Foote, and B. D. Hunt
Jet Propulsion Laboratory, California Institute of Technology
Pasadena, California 91109

ABSTRACT

A nonaqueous chemical etch, with Br as the active ingredient, is described which removes the insulating hydroxides and carbonates that form on high temperature superconductor surfaces as a result of atmospheric exposure. X-ray photoemission spectra have been recorded before and after etching $YBa_2Cu_3O_{7-x}$ films. It is found that, after the etch, the photoemission peaks associated with surface contaminants are greatly reduced, the Y:Ba:Cu ratio is close to the expected 1:2:3, and the oxidation state of the Cu (2+) is not affected. The temperature at which the film reaches zero resistance (~80 K) is not affected by the etch.

INTRODUCTION

The discovery of superconducting cuprate compounds[1,2] has stimulated a great deal of work in this area.[3,4] The new high temperature superconductors include compounds such as (i) $RBa_2Cu_3O_{7-x}$ where R is a rare earth (most commonly yttrium) and x < ~0.45, (ii) $(La_{1-x}M_x)_2CuO_{4-y}$ where M = Ba, Sr, or Ca and x < ~0.2, y < ~0.1, (iii) $BiCaSrCuO_{7-x}$ (and similar compounds containing the same elements plus aluminum), and (iv) Tl compounds such as $TlBaCu_3O_{5.5+x}$ and $Tl_2Ca_2Ba_2Cu_3O_{10+x}$.

Superconducting surfaces are required for thin film tunneling devices. Such surfaces would also be required for scientific studies involving surface-sensitive spectroscopies such as x-ray photoelectron spectroscopy (XPS) and scanning tunneling microscopy (STM), and may be important in obtaining experimental information on such scientifically important problems as determining the mechanism of high temperature superconductivity. However, the high temperature superconductors react with atmospheric H_2O and CO_2, forming insulating hydroxides and carbonates on the surface and at grain boundaries.[5-12] The surfaces of these materials are thus typically not superconducting, and the surface studies performed to date are generally affected by, and sometimes dominated by, the nonsuperconducting surface species. The removal of this nonsuperconducting surface layer is a necessary starting point for thin film studies and device processing.

The most common physical surface cleaning method, ion milling, is known[13] to cause reduction of the Cu and hence is not suitable for surface studies of these materials. The best surfaces for such applications have been obtained by scraping in vacuum.[9,10,14] However, this is not a practical device processing technique, and extreme care must be taken in thin film studies. A chemical cleaning technique for these applications would be highly desirable, and has recently been described.[15] A chemical etchant for the bulk superconductor would also be desirable for applications such as depth profiling and device patterning.

Due to the reactivity of the high temperature superconductors to water,

© 1989 American Institute of Physics

a chemical etch which minimizes surface damage should include nonaqueous solvents or gas phase reactants, and the reaction products should be gaseous or soluble in nonaqueous solvents. It has been proposed[15] that halogenation of the surface meets these requirements. The same treatment is also likely to be useful as an etchant of the bulk superconductor. As previously discussed,[15] the oxides, hydroxides, and carbonates of Ba, Cu, Y and La (as well as the other rare earths), Sr, Bi, Tl, and Ca are all likely to be etched by halides. The bromides and iodides of Cu, Ba, La and Y (as well as the other rare earths), Al, Bi, Sr, and Ca, and the bromides of Tl are all soluble in polar nonaqueous solvents such as alcohols, acetone, or ether. The formation of fluorides and chlorides is a less promising approach to achieving a clean surface due to their limited solubility in suitable solvents. Passivation of the surface to avoid degradation due to atmospheric exposure is also technologically important. Surface halides may also make an effective passivating layer. In this case, the solubility of the halide byproducts in a suitable solvent is not necessary and surface chlorides or fluorides may prove useful.

In order to test these ideas, a continuing series of experiments was initiated in this lab. The feasibility of using halide etchants on high temperature superconductors has been demonstrated[15] by using a nonaqueous Br solution to etch $YBa_2Cu_3O_{7-x}$. Experiments which verify and extend the earlier results are briefly described below. The Br etchant is shown to greatly reduce the amount of surface carbonates, resulting in surfaces which are close to the ideal stoichiometry, without affecting the oxidation state of the Cu (2+).

EXPERIMENTAL

$YBa_2Cu_3O_{7-x}$ films were prepared by multilayer deposition[16,17] on yttria- stabilized cubic zirconia substrates. Typically, a total of 21 to 27 layers of Y, BaF_2, and Cu were deposited using resistively heated sources of BaF_2 and Cu, and an electron beam-heated Y source. Individual layer thicknesses were 200, 800, and 225 Å for the Y, BaF_2, and Cu layers, respectively. Samples were annealed at ~865° C for 3 hours in 1 atmosphere of oxygen bubbled through deionized water, followed by slow-cooling in dry oxygen at a rate of 3° C/min.

The XPS measurements were done on a Surface Science Instruments SSX-501 small spot XPS spectrometer with monochromatized Al K_α x-rays (1486.6 eV). A dry box flushed with ultrahigh-purity nitrogen is attached to the sample load lock. The etching solution consisted of 1% by volume Br in absolute ethanol. Samples were etched in a nitrogen-flushed glove bag in a fume hood by dipping in the etchant solution, rinsed by dipping in absolute ethanol, and blown dry with nitrogen. Ten minutes etching time was found sufficient to completely remove a 1 μm-thick film. The samples were transferred from the glove bag to the dry box in a sealed dessicator; the samples were thus never exposed to atmospheric water vapor or carbon dioxide after the etch. XPS spectra were recorded with a resolution of 1.0 eV. In addition to the superconducting films, XPS spectra were also obtained from sputter-cleaned Cu, and from CuO and Cu_2O powders pressed into In. It was not found necessary to correct the binding energies for charging for the powders mounted in this manner.

RESULTS AND DISCUSSION

The formation of Y, Ba, and Cu bromides by the etching solution was verified with XPS measurements on unrinsed samples. Rinsing in absolute ethanol removed most of the bromides, as expected, although Br is still detectable on rinsed surfaces at a level of ~2 atomic percent. An oxygen peak in the O 1s spectrum (~531.1 eV), which is characteristic of surface contaminants, was found to be greatly reduced in intensity relative to the oxygen peak attributed to the bulk superconductor (~528.5 eV).[9,10,14,18] This effect is shown in Fig. 1. Fifteen to thirty seconds of etching was found to be sufficient to reduce the high binding energy O 1s peak to near its minimum intensity, with relatively little additional reduction in intensity observed for longer etch times. The spectrum shown in Fig. 1(c) compares favorably with published spectra of samples scraped in vacuum.[9,10,14] The reduced intensity of the high binding energy O 1s peak correlates with the reduced intensity of a Ba 3d doublet which is shifted 2.2 eV to higher binding energy relative to the Ba 3d doublet attributed to the bulk superconductor.[15] The removal of a surface Ba compound is also evident in the valence band spectra in Fig. 2. Before etching, the Ba 5p doublet near 12 eV is not clearly resolved, as shown in Fig. 2(a), indicating the presence of more than one chemical species. After etching, the spin-orbit split Ba 5p peaks are clearly evident (see Fig. 2(b)) with an intensity ratio which is close to the 2:1 which is expected for a single chemical species. A high binding energy C 1s peak characteristic of carbonates is also reduced in intensity after the etch, as shown in Fig. 3. These results are consistent with findings[7-10,12] that the surface and grain boundary material is primarily $BaCO_3$. The residual high binding energy O 1s and C 1s peaks apparent in Figs. 1(c) and 3(b), as well as a residual high binding energy Ba 3d doublet,[15] are presumably due to grain boundary material, an interpretation that is consistent with the fact that

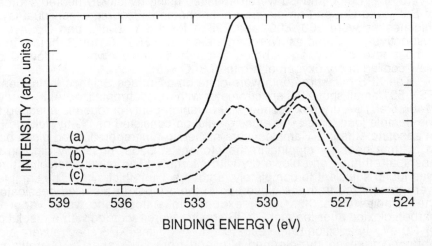

Fig. 1. O 1s XPS spectra from $YBa_2Cu_3O_{7-x}$ (a) before etching, (b) etched in 1% Br in absolute ethanol for 15 seconds, and (c) etched in 1% Br in absolute ethanol for 60 seconds.

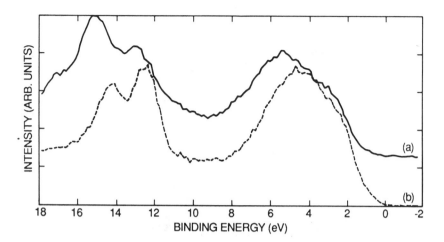

Fig. 2. XPS valence band spectra of $YBa_2Cu_3O_{7-x}$ (a) before etching, and (b) etched in 1% Br in absolute ethanol for 15 seconds.

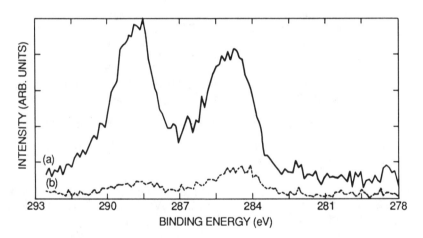

Fig. 3. C 1s XPS spectra from $YBa_2Cu_3O_{7-x}$ (a) before etching, (b) etched in 1% Br in absolute ethanol for 15 seconds.

these peaks are still present even after ion etching through 75% of the film.

The Cu 2p core level peaks of the superconductor occur at nearly the same binding energy but are broader compared to spectra from CuO, and no gross changes are observed as a result of the Br etch, with the Cu clearly remaining in the 2+ oxidation state.[15] The Cu L_3VV Auger peaks shown in Fig. 4, however, do show a clear difference between the superconductor and CuO. The superconductor L_3VV Auger peak occurs at ~1 eV higher kinetic energy than does that of CuO, and also appears to be broader as a result of increased satellite intensity. This has previously

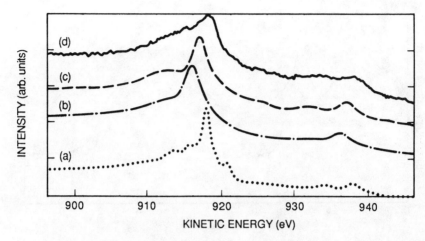

Fig. 4. XPS Cu L_3VV spectra from (a) sputter-cleaned Cu, (b) Cu_2O, (c) CuO, and (d) $YBa_2Cu_3O_{7-x}$ etched for 15 seconds in 1% Br in absolute ethanol.

been interpreted[19] as evidence of increased covalency of the Cu-O bonding in the superconductor. The XPS Cu 2p core level and L_3VV Auger data are summarized in Table I. Before etching, the surface composition (normalized to Ba), as determined from the XPS spectra of a typical $YBa_2Cu_3O_{7-x}$ thin film sample, is found to be $Y_{0.37}Ba_2Cu_{1.5}O_{5.2}$. After etching for 15 seconds, the average surface composition measured on four samples is found to be $Y_{0.99}Ba_2Cu_{2.62}O_{6.87}$, close to that which is ideally expected.

Measurement of resistance vs. T shows that a 1 μm-thick film etched for 30 seconds (removing ~500 Å) has zero resistance at 78.5 K, while a similar unetched film, grown and annealed at the same time, has zero resistance at 81.9 K. This difference is within the sample-to-sample

Table I. Summary of the measured Cu 2p binding energies and peak full-widths at half-maximum (ΔE_{FW}), Cu L_3VV kinetic energies, and the Auger parameters, all in eV.

	Cu 2p	ΔE_{FW}	Cu L_3VV	Auger parameter
Cu	932.6	1.26	918.7	364.7
Cu_2O	932.5	1.57	916.8	362.7
CuO	933.7	2.78	917.8	364.9
$YBa_2Cu_3O_{7-x}$ before etching	933.8	3.38	918.9	366.1
$YBa_2Cu_3O_{7-x}$ after etching	933.6	3.42	919.0	366.0

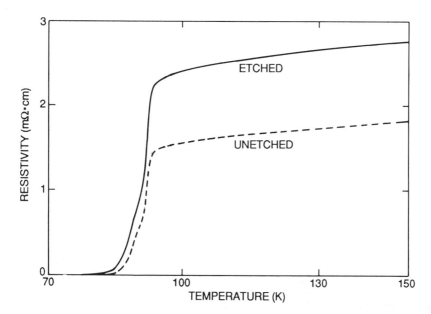

Fig. 5. Comparison of the resistivity vs. T of an etched and unetched film which were grown and annealed at the same time.

variability. Thus, the etchant appears to act only on the surface without adversely affecting the zero resistance temperature of the bulk of the film. Fig. 5 shows the resistivity vs. temperature for these samples, where the resistivity has been calculated from the measured resistance and from the estimated film thickness (±10%) and the length and width of the area of the film between the silver paint electrodes. The apparent increased resistivity of the etched film compared to the unetched film (Fig. 5) may simply indicate that the superconducting region of the film is thinner than the total film thickness. Such an effect might be due to incomplete oxidation of the film near the substrate, as has been reported by other researchers,[9] or to interaction of the film with the substrate. Either of these effects would result in a non- superconducting region in the interior of the film and therefore an increase in apparent resistivity if the total film thickness is used in the calculation. The possibility that the increased resistivity may be due to etch-induced damage, however, cannot be ruled out at this time.

Preliminary results show that the surface resistance of Pb contacts evaporated onto an etched film is <600 $\mu\Omega$-cm^2 at room temperature and <150 $\mu\Omega$-cm^2 at liquid He temperature, two orders of magnitude less than the values measured on an unetched film. While techniques for obtaining lower resistance contacts have been previously reported,[20-23] the present results have been obtained without resorting to mechanical abrasion or sputter etching and without using a noble metal contact. These results have been obtained in spite of the fact that XPS spectra of Pb deposited on a superconducting film show a partial reduction of Cu^{2+} to Cu^{1+}. The electrical measurements, combined with the XPS results, indicate that the nonaqueous Br solution is a promising approach to etching and cleaning high temperature superconductor surfaces. Qualitatively similar results would be expected if the bromine were substituted with iodine, or if the absolute ethanol were substituted with other polar nonaqueous solvents.

As previously mentioned, the limited solubility of the fluorides and chlorides in nonaqueous solvents makes them unsuitable as intermediates in etching and cleaning processes. It should be noted, however, that the fluorides and chlorides (as well as the bromides and iodides) may be useful as passivating layers. The fluorides could be formed on the surface using wet chemical techniques, such as with HF in ethanol, or using gas phase reactants, such as F_2 or XeF_2. Similar techniques could be used to form the chlorides, bromides, and iodides. Since the chlorides, bromides, and iodides are easily hydrolized, the fluorides appear more promising for this purpose.

CONCLUSIONS

A nonaqueous chemical etch for high temperature superconductors has been described and demonstrated on $YBa_2Cu_3O_{7-x}$ films. The etch was found to greatly reduce the XPS signals attributed to surface and grain boundary material, leaving the stoichiometry of the surface close to that which is ideally expected. The etch does not affect the oxidation state of the Cu (2+). Electrical measurements show that the temperature at which the film reaches zero resistance (~80 K) is not affected by the etch, though the resistivity at higher temperatures is increased. Deposited Pb contacts on etched films were found to have a resistance two orders of magnitude lower than that measured on similar unetched films. The same etchant should also be useful for etching bulk superconductors and for other high temperature superconductor compounds. Similar treatments may also be promising for passivation.

ACKNOWLEDGEMENTS

The authors gratefully acknowledge useful discussions with Dr. W. J. Kaiser and Dr. H. G. LeDuc, and the technical assistance of S. Cypher. The research described in this paper was carried out by the Jet Propulsion Laboratory (JPL), California Institute of Technology, and was supported by the Strategic Defense Initiative Organization, Innovative Science and Technology Office, and by the Defense Advanced Research Projects Agency, through an agreement with the National Aeronautics and Space Administration. The work was performed as part of JPL's Center for Space Microelectronics Technology.

REFERENCES

1. J. G. Bednorz and K. A. Muller, Z. Phys. B $\underline{64}$, 189 (1986).
2. M. K. Wu, J. R. Ashburn, C. T. Torng, P. H. Hor, R. L. Meng, L. Gao, Z. J. Huang, Y. Q. Wang, and C. W. Chu, Phys. Rev. Lett. $\underline{58}$, 908 (1987).
3. *Thin Film Processing and Characterization of High-Temperature Superconductors*, Eds. J. M. E. Harper, R. J. Colton, and L. C. Feldman (American Institute of Physics, New York, 1988) Conference Proceedings Vol. 165, Proceedings of the American Vacuum Society Topical Conference, November 6, 1987, Anaheim, CA.
4. *High Temperature Superconductors*, Eds. M. D. Brodsky, R. C. Dynes, K. Kitazawa, and H. L. Tuller (Materials Research Society, Pittsburgh, PA, 1988), Symposium Proceedings Vol. 99, Proceedings of the Materials Research Society meeting November 30 - December 4, 1987, Boston, MA.
5. M. F. Yan, R. L. Barns, H. M. O'Bryan, Jr., P. K. Gallagher, R. C. Sherwood, and S. Jin, Appl. Phys. Lett. $\underline{51}$, 532 (1987).
6. R. L. Barns and R. A. Landise, Appl. Phys. Lett. $\underline{51}$, 1373 (1987).
7. Z. Iqbal, E. Leone, R. Chin, A. J. Signorelli, A. Bose, and H. Eckhardt, J. Materials Res. $\underline{2}$, 768 (1987).
8. N. P. Bansal and A. L. Sandkuhl, Appl. Phys. Lett. $\underline{52}$, 323 (1988).
9. D. C. Miller, D. E. Fowler, C. R. Brundle, and W. Y. Lee, in Ref. 3, p. 336.
10. A. G. Schrott, S. L. Cohen, T. R. Dinger, F. J. Himpsel, J. A. Yarmoff, K. G. Frase, S. I. Park, and R. Purtell, in Ref. 3, p. 349.
11. K. Kitazawa, K. Kishio, T. Hasegawa, A. Ohtomo, S. Yaegashi, S. Kanbe, K. Park, K. Kuwahara, and K. Fueki, in Ref. 4, p. 33.
12. H. S. Horowitz, R. K. Bordia, R. B. Flippen, and R. E. Johnson, in Ref. 4, p. 903.
13. G. Panzner, B. Egert, and H. P. Schmidt, Surf. Sci. $\underline{151}$, 400 (1985).
14. P. Steiner, V. Kinsinger, I. Sander, B. Siegwart, S. Hufner, and C. Politis, Z. Phys. B $\underline{67}$, 19 (1987).
15. R. P. Vasquez, B. D. Hunt, and M. C. Foote, submitted to Appl. Phys. Lett.
16. B.-Y. Tsaur, M. S. Dilorio, and A. J. Strauss, Appl. Phys. Lett. **51**, 858 (1987).
17. A. Mogro-Campero, B. D. Hunt, L. G. Turner, M. C. Burrell, and W. E. Balz, Appl. Phys. Lett. $\underline{52}$, 584 (1988).
18. P. Steiner, R. Courths, V. Kinsinger, I. Sander, B. Siegwart, S. Hufner, and C. Politis, Appl. Phys. A $\underline{44}$, 75 (1987).
19. D. E. Ramaker, N. H. Turner, J. S. Murdy, L. E. Toth, M. Osofsky, and F. L. Hutson, Phys. Rev. B $\underline{36}$, 5672 (1987).
20. Y. Tzeng, A. Holt, and R. Ely, Appl. Phys. Lett. $\underline{52}$, 155 (1988).
21. J. W. Ekin, A. J. Panson, and B. A. Blankenship, Appl. Phys. Lett. $\underline{52}$, 331 (1988).
22. R. Caton, R. Selim, A. M. Buoncristiani, and C. E. Byvik, Appl. Phys. Lett. $\underline{52}$, 1014 (1988).
23. A. D. Wieck, Appl. Phys. Lett. $\underline{52}$, 1017 (1988).

Effect of Silver on the Water-Degradation of YBaCuO Superconducting Films

Chin-An Chang, J. A. Tsai, and C. E. Farrell

IBM T. J. Watson Research Center, Yorktown Heights, N. Y. 10598

Abstract

Degradation of YBaCuO superconducting films by water is studied using different starting structures with or without a silver layer. Several layer structures are compared, including Cu/BaO/Y_2O_3/substrate, Ag/Cu/BaO/Y_2O_3/substrate, Cu/BaO/Y_2O_3/Ag/substrate, and Ag/Cu/BaO/Y_2O_3/Ag/substrate, deposited either at room temperature or at high temperatures. Superconducting films were formed from these structures, with a zero resistance between 81 and 88 K. Upon exposing to moisture or immersing in water, a large difference in film degradation is observed, being much less for the structures containing a silver layer. Changes in film properties are analyzed, including decreasing critical current densities at 77 K, decreasing temperatures for zero resistance, increasing contact resistance at low temperatures, increasing film resistance at room temperatures, and changes in film structures. The possible mechanisms involved are also discussed.

INTRODUCTION

The sensitivity of YBaCuO superconductors to moisture has been a general concern for the application to various uses. The superconductors are reported to react readily with water, forming semiconducting and nonconducting products, responsible for the degradation and instability observed (1-4). Efforts in understanding and improving film degradation by water have been reported, with encouraging results in some cases (5). Recently, we have studied the degradation by both moisture in air and by water of the superconducting films. A large difference in degradation is observed between structures containing a silver layer and those without a silver layer (6, 7). The degraded properties include decreasing critical current densities at 77 K, decreased temperatures for zero resistance, increased contact resistance at low temperatures, and increased film resistance at room temperatures. In this paper, we report the degradation studies on structures containing a silver layer. The superconducting films were immersed in water for different lengths of time until the disappearance of superconducting characteristics. Both electrical and structural aspects are explored to understand the changes in the superconducting films.

EXPERIMENTAL AND RESULTS

Several structures with and without silver have been studied. Results on the $Cu/BaO/Y_2O_3/YSZ$, $Ag/Cu/BaO/Y_2O_3/YSZ$, and $Ag/(Cu/BaO/Y_2O_3)_3/Ag/YSZ$ structures have been reported elsewhere (6, 7), YSZ standing for yttria-stabilized zirconia substrate. The results on the structures without a silver layer are briefly reviewed here to contrast those using structures containing silver as reported here. For the $Cu/BaO/Y_2O_3/YSZ$ structure, immersion in water for 5 min led to the loss of zero resistance at 77 K. This was accompanied by a large increase in contact resistance by several orders of magnitude at 77 K, and a large rise in film resistance at room temperature. Such a degradation by water for the superconducting YBaCuO films is shown to be greatly reduced using silver-containing structures. In this paper we present results on the $Ag/Cu/BaO/Y_2O_3/BaO/Y_2O_3/Cu/Ag/YSZ$ structure. By comparing this and other structures, the effect of silver on reducing the degradation of YBaCuO superconducting films by water becomes a clear trend, and can be of technological importance to such superconducting films. The layers were deposited in sequence, without breaking the vacuum, using single electron-beam evaporation technique (8, 9). Both the BaO and Y_2O_3 layers were deposited by evaporating Ba and Y metals in the presence of a partial pressure of oxygen of 1×10^{-4} torr. Cu and Ag layers were deposited in the absence of such an oxygen partial pressure. Layer thickness used was 800, 450, 500 and 1200Å for each of the Ag, Cu, BaO and Y_2O_3 layers, respectively. The structures were first annealed at 900°C in helium for 5 min, followed by a slow cooling in oxygen. Electrical measurement was made using four-point probe technique by lowering the sample holder into a liquid nitrogen dewar, ending at 77 K. Critical current densities were measured using samples of narrow stripe geometry. The samples were immersed in water at room temperature for various lengths of time up to 60 hrs, and their electrical and structural characteristics were studied.

All the freshly annealed samples show zero resistance around 87 K and higher, with a critical current density around several hundred A/cm^2 at 77 K. Upon immersion in water, several changes were observed, including increasing film resistances at room temperature, reducing temperatures for zero resistance, reducing critical current densities at 77 K, and increasing contact resistance at low temperatures, and broadening in transition into the superconducting state. Fig. 1 shows the electrical measurement for the samples with immersion times up to 40 hrs. Zero resistance at 77 K is observed up to an immersion time of 5 hrs. The resistance-temperature relation changes from metallic to semiconductive at long immersion times, starting to be noticeable at an immersion time of 40 hrs. The relation becomes clearly semiconductive at still longer immersion time, as shown in Fig. 2 for an immersion of 60 hrs. Afterward, the same sample was annealed a second time by residing at 900°C in helium for 5 min, followed by a slow cooling in oxygen. It is interesting to notice that this second thermal treatment changed the electrical behavior back to a superconductive one, also shown in Fig. 2. The film did not reach zero resistance at 77 K using 1 μamp, yet it shows a clear transition into the superconductive state. The room temperature resistance also decreases after the second heat treatment from that after the 60 hrs immersion treatment.

We have analyzed the structural changes of the films upon different water and thermal treatments. X-ray diffraction study shows the formation of the 123 superconducting phase for the freshly annealed film, with strong intensities for the 020/006, 001 and 010/003 peaks. The peak intensities decrease with increasing immersion times in water. Fig. 3 shows the diffraction pattern after 60 hrs of immersion. Only a medium intensity is observed for the 020/006 peak. The silver peak remains very strong throughout the immersion treatment. Upon a second anneal at 900°C, x-ray diffraction, shown in Fig. 4, shows the formation of increasing amount of the 123 phase, as evidenced by the increased intensities for all the major 001 peaks. This is consistent with the recovering of the superconductive transition shown in Fig. 2.

The result presented above shows a clear reduction in the water-induced degradation of the YBaCuO superconducting film by the use of a silver layer in the starting structures. It agrees with those using different silver-containing structures. Silver also improves the contact resistance to the film, even with prolonged immersion in water. We believe that silver distributes itself both on the surface and along the grain boundaries. The latter is especially beneficial to the superconducting films by improving the weak links among the superconducting grains. We are currently investigating the detailed distribution of silver and its changes upon water immersion. The fact that a much reduced degradation is observed for several structures containing silver is a very encouraging result to the application of YBaCuO superconducting films to various areas. Further work along this line should provide more understanding and improvement of the superconducting films needed for technologies.

SUMMARY

Degradation of YBaCuO superconducting films by water is shown to be greatly reduced using starting structures containing silver. Both the film resistance and contact resistance remain low after several hours of immersion in water, while retaining the superconductivity above 77 K. Structural analysis provides further understanding of the reactions taking place due to the water immersion and additional annealing.

REFERENCES

1. M. F. Yan, R. L. Barns, H. M. O'Bryan, P. K. Gallagher, R. C. Sherwood, and S. Jin, Appl. Phys. Lett. 51, 532 (1987).
2. R. L. Barns and R. A. Laudise, Appl. Phys. Lett. 51, 1373 (1987).
3. K. Kitazawa, K. Kishio, T. Hasegawa, A. Ohtomo, S. Yaegashi, S. Kanabe, K. Park, K. Kuwahara, and K. Fueki, mat. Res. Soc. Symp/ Proc. Vol. 99, (1988), p. 33.
4. P. M. Mankiewich, J. G. Scofield, W. J. Skocpol, R. E. Howard, A. H. Dayem, and E. Good, Appl. Phys. Lett. 51, 1753 (1987).
5. D. E. Farrell, M. R. DeGuier, B. S. Chandrasekhar, S. A. Alterovitz, P. R. Aron, and R. L. Gagaly, Phys. Rev. B 35, 8797 (1987).
6. Chin-An Chang, Appl. Phys. Lett. 53, 1113 (1988).
7. Chin-An Chang and J. A. Tsai, to be published.
8. Chin-An Chang, Appl. Phys. Lett. 52, 924, (1988).
9. Chin-An Chang, C. C. Tsuei, C. C. Chi, and T. R. McGuire, Appl. Phys. Lett. 52, 72 (1988).

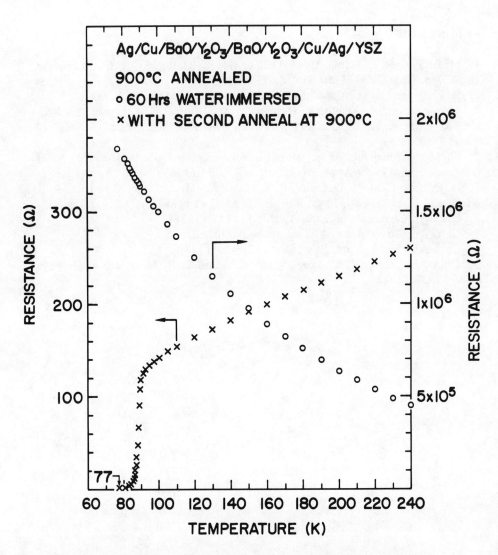

Figure 2. Resistance measurements of the same structure shown in Fig. 1 with 60 hrs of immersion in water, and with an additional annealing at 900°C. The transition into a superconductive state is redeveloped after the second heat treatment.

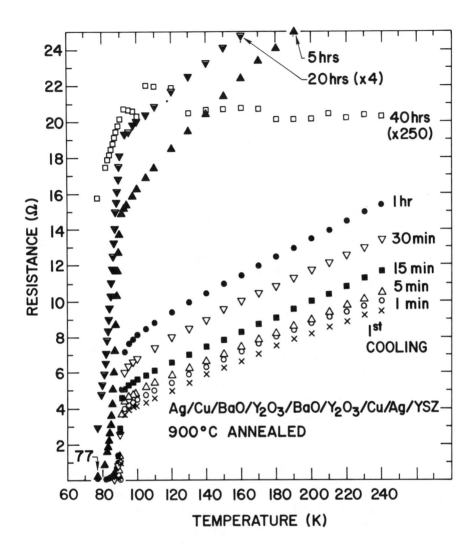

Figure 1. Electrical resistance measurements of an Ag/Cu/BaO/Y$_2$O$_3$/BaO/Y$_2$O$_3$/Cu/Ag/YSZ structure annealed at 900°C, and immersed in water for different times. The current used for the resistances shown is 10μamp. Using a current of 1 μamp, zero resistance is observed after a total immersion time of 5 hrs.

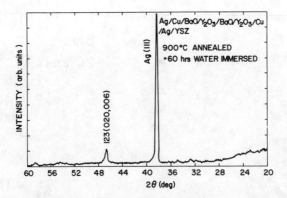

Figure 3. X-ray diffraction of the same structure shown in Fig. 3 after 60 hrs of immersion in water, showing largely diminished signals of the 123 superdonducting phase.

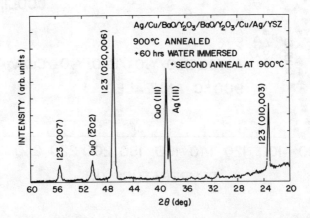

Figure 4. X-ray diffraction pattern of the structure shown in Fig. 2 after the 60 hrs water immersion and a second anneal at 900°C, showing strong peaks of the 123 phase.

ALUMINUM AND GOLD DEPOSITION ON CLEAVED SINGLE CRYSTALS OF $Bi_2CaSr_2Cu_2O_8$ SUPERCONDUCTOR

B. O. Wells, P. A. P. Lindberg, Z-X. Shen, D. S. Dessau, I. Lindau,
W. E. Spicer
Stanford Electronics Laboratories, Stanford, CA 94305
D. B. Mitzi, A. Kapitulnik
Department of Applied Physics, Stanford University,
Stanford, CA 94305

ABSTRACT

We have used photoelectron spectroscopy to study the changes in the electronic structure of cleaved, single crystal $Bi_2CaSr_2Cu_2O_8$ caused by deposition of aluminum and gold. Al reacts strongly with the superconductor surface. Even the lowest coverages of Al reduces the valency of Cu in the superconductor, draws oxygen out of the bulk, and strongly modifies the electronic states in the valence band. The Au shows little reaction with the superconductor surface. Underneath Au, the Cu valency is unchanged and the core peaks show no chemically shifted components. Au appears to passivate the surface of the superconductor and thus may aid in the processing of the Bi-Ca-Sr-Cu-O material. These results are consistent with earlier studies of Al and Au interfaces with other, polycrystalline oxide superconductors. Comparing with our own previous results, we conclude that Au is superior to Ag in passivating the Bi-Ca-Sr-Cu-O surface.

INTRODUCTION

The Bi-Ca-Sr-Cu-O materials have shown promising superconducting properties and generated a large amount of research interest.[1,2] Some practical uses for these materials seems probable in the future. One obstacle to the use of these materials is that it is difficult to prepare a superconducting surface region and maintain the good surface upon deposition of metallic contacts or the construction of any surface structure. Some previous studies have been performed on metal overlayers on polycrystalline samples of $(La,Sr)_2CuO_4$ [3-5], $YBa_2Cu_3O_7$ [6-9], and Bi-Ca-Sr-Cu-O[10,11] materials. Since none of these studies have been performed on single crystal samples, there has been some question as to the quality of the surfaces and therefore the validity of the results. Our earlier study showed that single crystal and polycrystalline samples had the same electronic structure.[12] Similarly, in this study we find that our results are consistent with earlier reports, thus validating the use of polycrystalline samples. Finally, we are able to show that Au is superior to other noble metals such as Ag in passivating th e Bi-Ca-Sr-Cu-O surface.

MATERIALS PREPARATION AND EXPERIMENTAL PROCEDURE

Photoemission experiments were performed in a Varian photoemission chamber. He II (40.8 eV) and MgKa (1253.6 eV) radiation were used to excite the samples. The photoelectrons were analyzed by a Cylindrical Mirror Analyzer (CMA). The energy resolution was 0.3 and 1.2 eV for the He II and MgKa radiation, respectively. Details of the preparation and characterization of the $Bi_2CaSr_2Cu_2O_8$ single crystals is given elsewhere.[13] They showed zero resistance at 85 K by magnetic susceptiblity measurements. The samples were transferred into the chamber on a transfer arm and cleaved in vacuum. Low Energy Electron Diffraction (LEED) showed sharp diffraction spots characteristic of single crystalline surfaces and indicated a cleave along the a-b plane.[13] After careful outgassing, Al was evaporated using Al deposited on a tungsten coil, and Au was evaporated from a bead. Evaporation rates were measured with a crystal monitor.

Two experiments were performed. In both we started with a single crystal of $Bi_2CaSr_2Cu_2O_8$ cleaved in vacuum. We then deposited a metal overlayer a few angstroms at a time, Al for the first sample and Au for the second. At each stage both core level and valence band spectra were recorded.

RESULTS AND DISCUSSION

Aluminum Overlayers

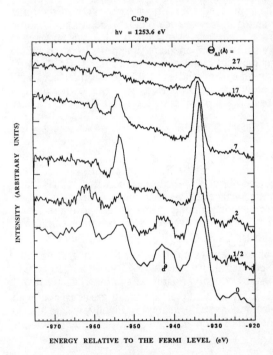

Figure 1
The Cu2p core level for the evolving Al/Bi-Ca-Sr-Cu-O interface showing the disappearance of the Cu satellite

Fig. 1 shows the Cu 2p core levels of the $Bi_2CaSr_2Cu_2O_8$ superconductor for different aluminum coverages (Θ). The d^9 satellite is reduced at even 1/2 Å coverage and completely gone by 2 Å coverage of Al. At this point the main $Cu2p_{3/2}$ line gets sharper. Since the ratio of the d^9 satellite to the d^{10} main line is a measure of the valency of Cu,[14] we infer that the copper valency is immediately reduced from "Cu2+". The Bi-Ca-Sr-Cu-O crystal is believed to cleave between the Bi layers.[13] The top two copper layers are approximately 5 and 15 Å below the surface.[15] The fact that the Cu reacts so strongly with the Al and that this happens at such low coverages suggests that Al is diffusing into the superconductor, breaking Cu-O bonds. This contrasts with our work on rubidium overlayers.[11,16] The large electropositive Rb ions presumably cannot penetrate into the superconductor, and indeed we see no disruption of the Cu satellite structure by the rubidium overlayer.

The O 1s and Bi 4f peaks show complimentary behaviour as shown in Fig. 2. The clean surface shows a single O1s peak and a single Bi4f spin-orbit split doublet. As Al is deposited a shoulder builds on both the O1s and Bi4f peaks and eventually dominates the spectra at higher coverages. The shoulder appears at higher

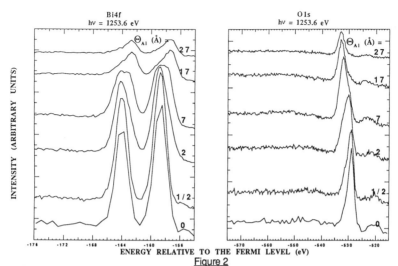

Figure 2
The O1s and Bi4f core levels for the Al/Bi-Ca-Sr-Cu-O interface. Note both peaks show chemically reacted components.

binding energy for O1s and lower binding energy for the Bi4f. The clean O1s peak is at -529.0 eV and the new shoulder is 3.9 eV higher binding energy at -532.9 eV. The clean Bi4f$_{7/2}$ peak is at -158.9 eV, the same as found in Bi$_2$O$_3$, and the shoulder appears 1.4 eV lower at -157.3 eV. We compare the relative intensities of the O and Bi peaks in each spectrum and note that while the intensity of the Bi peak drops monotonically with increasing Al, the intensity of the O1s peak remains essentially the same until $\Theta \geq 17$Å Al and then decreases with increasing Al coverage. We suggest that Al breaks Bi-O bonds. The oxygen then diffuses upwards and bonds with the surface Al forming a layer of Al$_2$O$_3$. As a result, Bi is left in a reduced state, leaving a more negative environment near the Bi ions, and thus shifting the core levels to lower binding energy. The two peak positions correspond to the two states of Bi: the original Bi-O state and the final, post reaction reduced state. Almost all of the surface Bi is eventually reduced. The oxygen, which tends to bond more strongly with Al than it does with the superconductor, is apparently left in a more stable electronic configuration as signalled by the O1s state at a much higher binding energy.

The corresponding valence band spectra are shown in Fig. 3. The clean spectrum has two main features. The main feature which extends from -1 eV to about -7 eV consists of Cu-O bonds mixed with some Bi-O states. The broad structure from -9 to -13 eV may have more than one component but includes the Cu d^8 satellite.[17] Upon deposition of Al, the number of states near the Fermi level decreases and the main feature moves toward higher binding energy. The d^8 satellite disappears with Al coverage. This is further evidence that the Cu valency is reduced and that Bi-O bonds are broken by the Al.

Fig. 4 shows the Al2p peak at different Al coverages. The small Cu 3p peak is at a similar energy. The Cu peak obscures the Al2p peak at low Θ, leading to a complex structure in the spectrum. However, at higher coverages the Cu peak has a negligible contribution to the spectrum. For $\Theta \geq 7$ Å, the Al peak narrows and centers

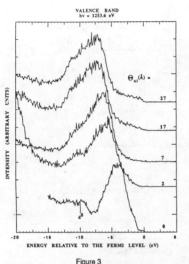
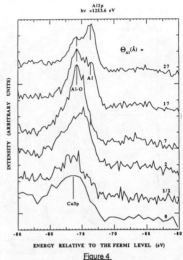

Figure 3
The valence band for the evolving Al/Bi-Ca-Sr-Cu-O interface

Figure 4
The Al2p core level for the Al/Bi-Ca-Sr-Cu-O interface showing the original, lower binding energy Al-oxide peak and the appearance of the metallic Al peak.

at about 75.8 eV. Also at $\Theta = 7$ Å, we first see a lower binding energy shoulder at 73.5 eV. This shoulder is more prominent at $\Theta = 17$ Å and is the largest feature in the spectrum at our maximum coverage of 27Å. As the Al is deposited, it reacts with oxygen in the superconductor to form a layer of Al_2O_3. Once we have deposited 7 Å of Al, this layer of Al_2O_3, thickness about twice that of the deposited Al, acts as a barrier to any further diffusion of oxygen out of the substrate. Thus we see the formation of the metallic Al peak at 73.5eV. This also explains why the O1s core level remains approximately constant in intensity up to $\Theta = 7$Å, but then starts to decrease monotonically with increasing Al coverage. The slow growth of this Al peak and the fact that it does not dominate the spectrum at our largest coverage, indicate that the metallic Al forms islands or at least does not grow smoothly layer by layer. This is consistent with other studies of Al on polycrystalline $YBa_2Cu_3O_7$.[7]

We see that the Al/Bi-Ca-Sr-Cu-O interface is complex and not clean and abrupt. The outermost surface consists of islands of metallic Al or regions of Al mixed with Al_2O_3. This outer-most Al is on top of a layer of Al_2O_3, which in turn is on top of a region of disrupted superconductor with some diffused Al. Presumably, there is still some good superconductor at some distance beneath the surface. Clearly it is impossible to make a well defined tunnel junction with Al oxide or Al directly on Bi-Ca-Sr-Cu-O. This may also indicate a problem in making Al contacts to the high temperature superconductor, even on a clean, superconducting surface.

Gold Overlayers

The copper 2p core levels of the superconductor for different gold coverages are shown in Fig. 5. The d^9 satellite structure is clearly visible at all gold coverages. In fact the ratio of the intensity of the satellite to the main peak is essentially constant. This indicates that the gold does not affect the copper valency as most metal

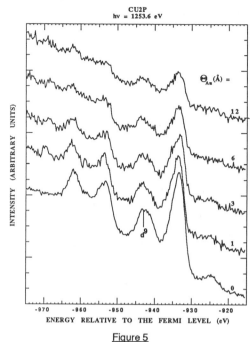

Figure 5
The Cu2p core level for the Au/Bi-Ca-Sr-Cu-O interface showing that the satellites do not disappear.

overlays do. The d^9 satellites are destroyed or degraded by reactive metals such as Al, Ti[9], or Fe[3], and also by more inert metals such as Cu[4] and even Ag[10]. Since superconductivity is believed to take place in the Cu-O planes, we assume that maintaining the Cu valency is essential for superconductivity. Other groups found that Au overlayers preserved the Cu satellites for polycrystalline $(LaSr)_2CuO_4$[5] and $YBa_2Cu_3O_7$.[8]

The O1s peak tends to be sensitive to any chemical activity in the substrate. In Fig. 6a we see that for both the clean surface and at each gold coverage, the shape and position of the O1s peak is unaltered. Fig. 6b shows the Bi4f peak. Again, no change in peak shape with Au coverage is observed. There is a small shift to lower binding energy in the peak position, no more than 0.2 eV. Since this shift is small and does not start as a shoulder and build with Au coverage, we expect that it represents a change in the electronic environment of the Bi rather than a chemical reaction involving the Bi and the Au. The Bi-O plane most likely forms the surface layer of the cleaved

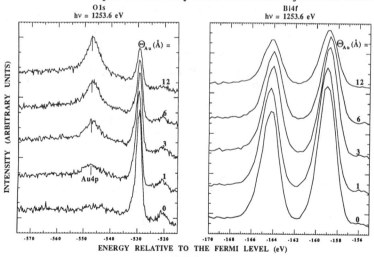

Figure 6
The O1s and Bi4f core levels for the Au/Bi-Ca-Sr-Cu-O interface showing no chemical reaction for either species.

crystal,[13] so it is reasonable to expect that there would be some change in the electronic environment of the Bi by placing it in contact with Au instead of vacuum.

The corresponding valence band spectra are shown in Fig. 7. In contrast to the case with the Al overlayers, we do not see the main feature shift toward higher binding energy. This is further evidence that Au does not cause any breaking of the Cu-O or Bi-O bonds. All we see is a Au valence band spectrum superimposed on a clean $Bi_2CaSr_2Cu_2O_8$ valence band spectrum, with Au features dominating to a greater extent with greater Au coverage.

All of the superconductor substrate peaks fall off rapidly with Au coverage. Fig. 8 shows a graph of the logarithm of the ratio of $Bi4f_{7/2}$ (and O1s) substrate peak intensity at a coverage Θ to the same peak for a clean surface, for overlayers of Au and Ag. The Ag data is from our earlier paper, and was taken on polycrystalline $Bi_2CaSr_2Cu_2O_8$.[10] Clearly the Bi4f and O1s core level intensities fall off fastest for the Au overlayer. The escape depth of electrons are similar in Au and Ag. Since the substrate peak falls off the fastest under the Au overlayer, this is evidence that the Au covers the surface more uniformly than Ag. The Ag may be islanding more or may be diffusing into the superconductor. Others reports state that for $YBa_2Cu_3O_7$, Ag will diffuse into the substrate.[18]

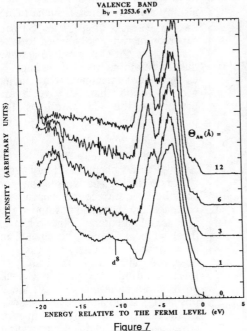

Figure 7
The valence band for the Au/Bi-Ca-Sr-Cu-O interface

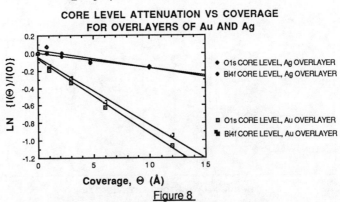

Figure 8
Attenuation curves for both the $Bi4f_{7/2}$ and O1s spectral features under overlayers of Ag and Au. The curves represent the logarithm of the normalized intensity versus coverage. The signals are clearly attenuated faster under the Au overlayer.

CONCLUSIONS

We have found that Al forms a very complex interface with the Bi-Ca-Sr-Cu-O superconductor. The Al penetrates into the superconductor, lowers the valency of the Cu, and draws oxygen out of the bulk to form Al_2O_3. Eventually, regions of metallic Al form on top of Al_2O_3. In contrast, Au will form an overlayer without disrupting the superconductor. The Au/Bi-Ca-Sr-Cu-O interface is much simpler. A protective Au layer may be essential to form well characterized structures on top of the oxide superconductors. Further we found that our data on single crystals is consistent with the results from earlier work on polycrystalline superconductor samples. Also we find that Au is superior to Ag for passivating the $Bi_2CaSr_2Cu_2O_8$ surface because Au does a better job of preserving the Cu valency and seems to cover the surface better.

ACKNOWLEDGEMENTS

Three of the authors would like to acknowledge support received through an NSF Ph.D. Fellowship (D.S.D.), an AT&T Ph.D. Fellowship (D.B.M.), and Alfred P. Sloan and P.Y.I. Fellowships (A.K.). The financial support from the U. S. National Science Foundation through the NSF-MRL program at the Center for Materials Research at Stanford University, and Air Force Contract AFSOR-87-0389 is gratefully acknowledged.

REFERENCES

1. H. Maeda, Y. Tanaka, M. Fukutomi, and T. Asano, Jap. Journ. Appl. Phys. Lett. (to be published).

2. R. M. Hazen, C. T. Prewitt, R. J. Angel, N. L. Ross, L. W. Finger, C. G. Hadidiacos, D. R. Veblen, P. J. Heaney, P. H. Hor, R. L. Meng, Y. Y. Sun, Y. Q. Wang, Y. Y. Xue, Z. J. Huang, L. Gao, J. Bechtold, and C. W. Chu, Phys. Rev. Lett. **60**, 1174 (1988).

3. D. M. Hill, H. M. Meyer III, J. H. Weaver, B. Flandermeyer, and D. W. Capone II, Phys. Rev. B **36**, 3979 (1987).

4. D. M. Hill, Y. Gao, H. M. Meyer III, T. J. Wagener, J. H. Weaver, and D. W. Capone II, Phys. Rev. B **37**, 511 (1988).

5. H. M. Meyer III, T. J. Wagener, D. M. Hill, Y. Gao, S. G. Anderson, S. D. Krahn, J. H. Weaver, B. Flandermeyer, and D. W. Capone II, Appl. Phys. Lett. **51**, 1118, (1987).

6. Y. Gao, T. J. Wagener, J. H. Weaver, B. Flandermeyer, and D. W. Capone II, Appl. Phys. Lett. **51**, 1032 (1987).

7. Y. Gao, I. M. Vitomirov, C. M. Aldao, T. J. Wagener, J. J. Joyce, C. Capasso, J. H. Weaver, and D. W. Capone II, Phys. Rev. B **37**, 3741 (1988).

8. T. J. Wagener, Y. Gao, I. M. Vitomirov, C. M. Aldao, J. J. Joyce, C. Capasso, J. H. Weaver, and D. W. Capone II, Phys. Rev. B **38**, 232 (1988).

9. H. M. Meyer III, D. M. Hill, S. G. Anderson, J. H. Weaver, and D. W. Capone II, Appl. Phys. Lett. **51**, 1750 (1987).

10. P. A. P. Lindberg, Z-X. Shen, I. Lindau, W. E. Spicer, C. B. Eom, and T. H. Geballe, Appl. Phys. Lett. **53**, 529 (1988).

11. P. A. P. Lindberg, P. Soukiassian, Z-X. Shen, S. I. Shah, C. B. Eom, I. Lindau, W. E. Spicer, and T. H. Geballe, Appl. Phys. Lett. (in press).

12. Z-X. Shen, P. A. P. Lindberg, B. O. Wells, D. B. Mitzi, I. Lindau, W. E. Spicer, and A. Kapitulnik, Phys. Rev. B, December 1, 1988 (in press).

13. P. A. P. Lindberg, Z-X. Shen, B. O. Wells, D. B. Mitzi, I. Lindau, W. E. Spicer, and A. Kapitulnik, Appl. Phys. Lett. (to be published).

14. Z-X. Shen, J. J. Yeh, I. Lindau, W. E. Spicer, J. Z. Sun, K. Char, N. Missert, A. Kapitulnik, T. H. Geballe, and M. R. Beasley, SPIE Symposium Proceedings **948-10**, 41 (1988).

15. J. M. Tarascon, W. R. McKinnon, P. Barboux, D. M. Hwang, B. G. Bagley, L. H. Greene, G. Hull, Y. LePage, N. Stoffel, and M. Giroud, Preprint.

16. P. A. P. Lindberg, Z-X. Shen, B. O. Wells, D. S. Dessau, D. B. Mitzi, I. Lindau, W. E. Spicer, and A. Kapitulnik, Phys. Rev. B (to be published).

17. Z-X. Shen, P. A. P. Lindberg, I. Lindau, W. E. Spicer, C. B. Eom, and T. H. Geballe, Phys. Rev. B **38**, 7152, 1988.

18. J. W. Ekin, T. M. Larson, N. F. Bergen, A. J. Nelson A. B. Swartzlander, L. L. Kazmerski, A. J. Panson, and B. A. Blankenship, App. Phys. Lett. **52**, 1819 (1988).

SURFACE AND INTERFACE PROPERTIES OF HIGH TEMPERATURE SUPERCONDUCTORS

J.H. Weaver, H.M. Meyer III, T.J. Wagener, D.M. Hill, and Y. Hu

Materials Science, University of Minnesota, Minneapolis, MN 55455

ABSTRACT

The energy distribution of the occupied and unoccupied electronic states of the high temperature superconductors $La_{1.85}Sr_{0.15}CuO_4$, $YBa_2Cu_3O_{7-x}$, $Bi_2Ca_{1+x}Sr_{2-x}Cu_2O_{8+y}$ and related compounds have been investigated using x-ray and inverse photoemission spectroscopy. Results from polycrystalline and single crystal samples show very similar valence band characteristics, and comparison with theory shows the importance of correlation effects. Core level results reveal a nominally 2+ valence state for copper and inequivalent chemical environments for oxygen. The empty state spectra show structure that can be identified with La 5d and 4f levels for $La_{1.85}Sr_{0.15}CuO_4$, Ba 5d and 4f and Y 4d levels for $YBa_2Cu_3O_{7-x}$, and Bi 6p, Sr 4d and Ca 3d levels for $Bi_2Ca_{1+x}Sr_{2-x}Cu_2O_{8+y}$. Inverse photoemission studies also show that the excitation of O 2s core levels near threshold results in resonant behavior of the unoccupied O levels just above E_F. Interface studies which include overlayers of metals (Au, Ag, Ti, Fe, Cu, Pd, La, Al, In, and Bi), semiconductors (Si and Ge) and dielectric materials (CaF_2, Al_2O_3, SiO_2, Bi_2O_3) are discussed in terms of adatom reactivity with, and modification of, the superconductor surface region.

INTRODUCTION

In the year since the first AVS symposium on High Temperature Superconductivity,[1] the community has witnessed the discovery of another class of superconducting copper-oxides and a 30% increase in the superconducting transition temperature, T_c. The superconducting copper oxides are now of the form La_2CuO_4, $YBa_2Cu_3O_{7-x}$, and Bi-Ca-Sr-Cu-O and Tl-Ba-Ca-Cu-O, where the latter exist in more than one superconducting phases, including the 2-1-2-2 and 2-2-2-3 phases, and there are variable numbers of Cu-O planes. (We will refer to the materials by their metal stoichiometries, 2-1-4, 1-2-3, 2-1-2-2, and 2-2-2-3.) The members of this extended family show superconducting transition temperatures ranging from ~35 K for the 2-1-4's to ~120 K for the Tl-based 2-2-2-3's. Common challenges continue to involve underlying mechanisms for superconductivity, the clear delineation of their various physical and chemical properties, and their integration with new and existing technologies.

The results of photoemission and inverse photoemission studies offer important insight into the fundamental properties of the copper oxide-based superconductors.[2-17] We have used these techniques to reveal the occupied and empty electron states for theory and to examine chemical changes during interface formation, surface degradation, and surface passivation.[16-20] In this paper we first review our photoemission and inverse photoemission results for a number of copper oxide-based polycrystalline and single crystal 2-1-4, 1-2-3, 2-1-2-2, and 2-2-2-3 compounds. These samples are almost free of second phases, and the results allow us to identify features intrinsic to the superconductor. We will discuss the variation of the superconductor features as interfaces are formed with representative overlayers for metallization, encapsulation, or passivation.

© 1989 American Institute of Physics

EXPERIMENTAL

The polycrystalline $La_{1.85}Sr_{0.15}CuO_4$ samples were precipitated from a solution of La-, Sr-, and Cu- oxalate salts that was pressed into a pellet and sintered at 1100 °C for ~12 hours.[21] The resulting samples were ~95% dense, single phase, and exhibited a superconducting transition temperature of 35 K. Polycrystalline $YBa_2Cu_3O_{7-x}$ was formed from a fine stoichiometric powder (average particle size of ~1.4 µm) obtained from Rhone-Poulenc. The powders were three-times ground and sintered at 900°C, pressed into bars, subjected to a step-wise heating schedule in O_2 (maximum temperature of 950 °C), and then cooled slowly to room temperature (measured density ~93%).[22] Electron microscopy showed that fractured surfaces of these polycrystalline samples were characterized by transgranular fractures with minimal grain boundary exposures. Single crystals of $YBa_2Cu_3O_{7-x}$ were grown from a stoichiometric melt of the parent oxides and then cooled in air. Crystals extracted from the melt were ~2mm x ~2mm x ~0.5mm and exhibited a superconducting transition temperature of 85 K. The slightly depressed transition temperature was caused by trace amounts of Al incorporated from the crucible during firing.

X-ray photoemission, XPS, studies were performed in an ultrahigh vacuum spectrometer described in detail elsewhere.[23] A monochromatized photon beam (Al K_α, hv=1486.6 eV) was focussed onto the sample surface and the emitted electrons were energy dispersed with a Surface Science Instruments hemispherical analyzer onto a position-sensitive detector.[24] The size of the x-ray beam was 300 µm, the pass energy of the analyzer was 50 eV, and instrumental resolution was 0.6 eV. For the inverse photoemission, IPES, measurements, a monochromatic low energy electron beam was directed onto the sample at normal incidence and the emitted photons were monochromatized by a grating and then analyzed by a position sensitive detector. The incident electron beam was ~1 x 5 mm. The overall resolution was 0.3-0.6 eV for photon energies 12-44 eV. Sample surfaces were prepared by fracturing or cleaving *in situ* at pressures of 8 x 10^{-11} Torr.

For interface studies, the overlayers were deposited from well degassed thermal evaporation sources at pressures below 5 x 10^{-10} Torr and evaporation rates of ~1 Å/min. In those experiments that involved the formation and deposition of oxide materials, the evaporation of metal or semiconductor materials was performed in an ambient of activated oxygen. The activation was performed by placing a hot, biased thoriated Ir filament between the substrate and the O_2 gas source (bias -250 eV relative to the chamber, emission current of 50 mA, O_2 pressure 1 x 10^{-5} Torr). These conditions provided a source of neutral oxygen atoms (~5 x 10^{-6} Torr), doubly ionized O_2 (~1 x 10^{-6} Torr), and singly ionized O_2 (~5 x 10^{-6} Torr).

RESULTS AND DISCUSSION

In Fig. 1 we show experimental and calculated electronic states for $La_{1.85}Sr_{0.15}CuO_4$, $YBa_2Cu_3O_{7-x}$, and $Bi_2Ca_{1+x}Sr_{2-x}Cu_2O_{8+y}$. The valence bands were obtained with hv = 1486.6 eV and the empty state spectra were obtained with 30.25, 36.25 and 38 eV incident electrons. Noteworthy is the similarity in occupied states within ~8 eV of E_F for the three superconductors, i.e. a low density of states near the Fermi level, E_F, a central feature at ~3.4 eV, and shoulders at 2.1 and 5.4 eV. While all show modest emission at E_F, the Bi-containing superconductor exhibits the best defined cutoff because of the additional Bi-O states at E_F.[25] For $La_{1.85}Sr_{0.15}CuO_4$, two features are clearly seen at ~9 and ~12 eV.[13] For $YBa_2Cu_3O_{7-x}$, the ~9 eV feature is again visible but the 12 eV feature is obscured by overlap with the stronger Ba 5p emission (it can be seen in Cu 3p-3d resonance photoemission at hv ≅ 75 eV[7,11]). For $Bi_2Ca_{1+x}Sr_{2-x}Cu_2O_{8+y}$, the broad Bi 6s emission centered at ~11 eV overwhelms the 12 eV structure, which should be visible in resonance photoemission. This 12 eV feature is the most prominent of the Cu d^8 satellite multiplets,[26] created from a d^9 initial state and shifted out of the valence band region by Coulomb correlation energy for two holes on a Cu site. It has been discussed in detail and is a consequence of Cu-O hybridization that produces formally divalent Cu. The 9 eV peak has been observed in seven different rare earth-copper oxides (of 2-1-4 and 1-2-3 type),[13] and its origin is generally attributed to two holes on an oxygen site. It is likely to be present for the Bi superconductor as well.

Superimposed on each valence band are the calculated ground state density of states, DOS, from Refs. 25, 28, and 29. Comparison shows that the predicted state density at E_F is substantially larger than experimentally observed. In each case, the centroid of the calculated DOS is shifted ~1.5 eV relative to experiment. Again, this shift is a consequence of the Cu 3d-3d correlation and the fact that photoemission does not measure the ground state, as discussed by many authors.[5,27,30,31]

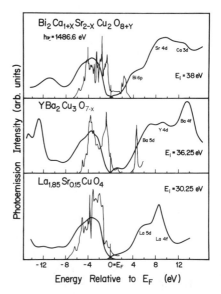

FIG. 1. Experimental and theoretical occupied and empty electronic states for the 2-1-4, 1-2-3, and 2-1-2-2 ceramic superconductors (calculations from Refs. 25, 28 and 29, top to bottom). The 1.5 eV shift of the occupied states reflects correlation effects. Emission at E_F is low, but not zero, and empty state features are labeled according to their orbital origin. XPS and IPES normalization was selected to give equal steps at E_F.

Inverse photoemission spectra for the empty states also show weak emission at E_F, with the most distinct edges in the 2-1-2-2's and 2-1-4's. For $La_{1.85}Sr_{0.15}CuO_4$, the structure above E_F can be assigned to the empty La 4f (at 8.7 eV) and La 5d (at ~5.3 eV) states, with likely overlap of the broader 5d bands with the more localized 4f levels.[16] For $YBa_2Cu_3O_{7-x}$, the empty Ba 5d and Y 4d bands extend from ~2 to ~12 eV, and the large feature at 13.5 eV is related to Ba 4f level states.[17] For $Bi_2Ca_{1+x}Sr_{2-x}Cu_2O_{8+y}$, the broad features at 3.5, 9.6, and ~12.5 eV are associated with Bi 6p, Ca 3d, and Sr 4d empty bands, respectively. (See Ref. 32 for details.)

Synchrotron radiation resonance photoemission has been used to advantage to highlight the occupied Ba, Cu, and lanthanide 4f features, as noted above. Inverse resonance photoemission can also be used to enhance empty states by scanning the incident electron energy through thresholds needed to create a hole in an appropriate shallow core level. In Fig. 2 we show photon distribution curves for $Bi_2Ca_{1+x}Sr_{2-x}Cu_2O_{8+y}$ (dashed) and $YBa_2Cu_3O_{7-x}$ (solid) for E_i =14 through 20 eV. Both show states at the Fermi level. For $Bi_2Ca_{1+x}Sr_{2-x}Cu_2O_{8+y}$, there is constant emission above the Fermi step until ~2 eV. For $YBa_2Cu_3O_{7-x}$, there is a broad peak centered at ~1.4 eV. For both superconductors, there is enhanced emission within ~2 eV of E_F for incident electron energies near 18 eV. This enhancement reflects electron excitation of O 2s electrons into 2p states, followed by radiative decay of the core hole, namely $O(2s^1 2p^6) \rightarrow O(2s^2 2p^5) + h\nu$. Enhancement results when this channel couples with continuum inverse photoemission decay channels. From XPS, the O 2s core level binding energy is ~20 eV and we associate the difference in energy, 18 vs. 20 eV, with intermediate O 2p screening states lowered in energy by a Coulomb interaction with the O 2s core hole. This enhancement reflects the presence of O 2p holes; it is not observed for 1-2-3 samples of low oxygen content.

The O 1s and Cu $2p_{3/2}$ spectra for these 2-1-4, 1-2-3, and 2-1-2-2 samples are shown in Fig. 3. The Cu $2p_{3/2}$ lineshape indicates a nominal 2+ valence state, analogous to that discussed by Sawatzky and coworkers[33] for a number of divalent Cu halides. The broad doublet at high binding energy is derived from 8 multiplets which reflect the interaction of a $\underline{2p}_{3/2}$ core hole (denoted by underlining) with the $3d^9$ configuration. This broad satellite is centered at 942 eV for each superconductor, consistent with the argument that the $\underline{2p}_{3/2}3d^9$ final state energies should be independent of the details of the Cu-ligand bonding. The main line, centered near 933 eV, is derived from ligand screening of the Cu 2p core hole, namely $\underline{2p}_{3/2}d^{10}\underline{L}$, where L notes a suitably symmetrized oxygen p orbital. In part, the width of the main line Cu core level features is due to mixing of

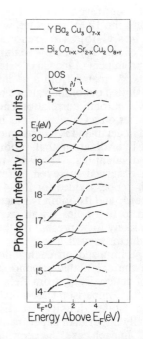

FIG. 2. Photon distribution curves for 1-2-3 and 2-1-2-2 superconductors for incident electron energies that straddle the O 2s-2p resonance at 18 eV. The results show enhancement of empty states within ~2 eV of E_F indicative of O 2p holes.

the d^9 and $d^{10}\underline{L}$ ground state configurations since they are close in energy. Additional broadening reflects the inequivalent sites for Cu in the crystal lattice, corresponding to Cu^{II} and Cu^{III} where the superscripts reflect coordination changes but not formal d^8 or d^9 valence. (There is no evidence for formal d^8 valence, consistent with the picture in which oxygen is "oxidized" at high oxygen content.) Experimentally, the main lines for $La_{1.85}Sr_{0.15}CuO_4$ and $YBa_2Cu_3O_{7-x}$ are of nearly equal width; for $Bi_2Ca_{1+x}Sr_{2-x}Cu_2O_{8+y}$ it is ~0.5 eV wider and there is a distinct asymmetry at both lower and higher energy. From Fig. 3, it can also be seen that the $YBa_2Cu_3O_{7-x}$ main line is shifted 0.2 eV to lower binding energy relative to that in $La_{1.85}Sr_{0.15}CuO_4$ while that in $Bi_2Ca_{1+x}Sr_{2-x}Cu_2O_{8+y}$ is shifted 0.3 eV to higher binding energy. This can be accounted for by slight differences in ligand screening, possibly due influences of neighboring atoms to the ligand oxygens or by oxygen defects.

At the right of Fig. 3, we show the O 1s spectra for these 2-1-4, 1-2-3, and 2-1-2-2 superconductors. In all cases, the spectra exhibit a wide, asymmetric main peak at ~528.5 eV, indicating the presence of more than a single chemical environment. These can be identified more readily after lineshape analysis, as given by the dashed lines. These fits were aided by examination of single crystals of CuO, which provided the full width at half maximum fitting parameter. Moreover, such analysis has been done for single crystals of other 2-1-4's and 1-2-3's where La or Y has been replaced by a different rare earth ion.[13]

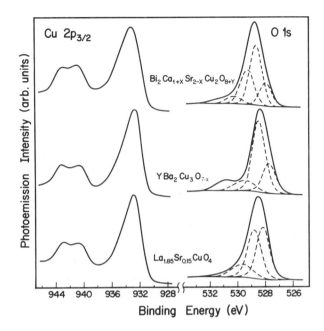

FIG. 3. Cu $2p_{3/2}$ emission for 1-2-3, 2-1-4, and 2-1-2-2 superconductors indicating nominal Cu 2+ valence. (Bi 4s emission distorts the Cu satellite lineshape in the 2-1-2-2's.) O 1s lineshape decomposition reveals the inequivalent oxygen sites, as discussed in the text.

For the 2-1-4's, the O 1s emission is made up of two main peaks split by ~0.7 eV and having approximately equal emission intensity. The shallower peak is derived from oxygen atoms in the Cu-O planes and can be written (O $\underline{1s}$ $2p^6$) $3d^9$ where the 1s hole is screened by the $2p^6$ orbitals and $3d^9$ denotes one of the two possible ground state configurations. The deeper of the two large peaks arises from oxygen atoms bound to La (and Sr) in the more ionic arrangement of the off-planes of the unit cell. We have attributed the small peak at ~1.5 eV higher binding energy to screening of the 1s hole by the second configuration, namely a (O $\underline{1s}$ $2p^6$) $3d^{10}\underline{L}$.[13] Fujimori et al.[34] have observed a similar feature and have attributed it to an energy loss satellite.

For the 1-2-3's, the overall O 1s emission is narrower than in the 2-1-4's. It is made up of a dominant component at ~529 eV derived from Cu-O planes (accounting for 4 of 7 oxygen atoms in a formula unit) and a shallower component attributed to Cu-O chains (accounting for the remainder oxygen atoms). Significantly, the relative intensity of the Cu-O chains feature varies with oxygen content, increasing as the oxygen content approaches 7. As for $La_{1.85}Sr_{0.15}CuO_4$, we identify the 530 eV feature with (O $\underline{1s}$ $2p^6$) $3d^{10}\underline{L}$ screening or the energy loss mechanism.

For the Bi 2-1-2-2 superconductors, the O 1s emission is broader, consistent with the bonding of oxygen in Cu-O planes, in Bi-O planes, and with Sr or Ca. Based on previous assignments and using equivalent linewidths for fitting, we associate the shallowest component with oxygen atoms of Sr-O or Ca-O structures. The next higher binding energy component reveals oxygen atoms of the Cu-O planes (and oxygen atoms bound to Bi atoms in off-plane sites) while oxygen atoms in the Bi-O planes make up the third component. The small, highest binding energy component shown is most likely due to the (O $\underline{1s}$ $2p^6$) $3d^{10}\underline{L}$ feature or the energy loss satellite, although this feature is not as intense as in the other spectra.[34]

Of great importance in understanding the surface properties of these high temperature superconductors is information related to their reactivity during interface formation. To provide this insight, we have investigated a number of reactive metal/superconductor interfaces, including Fe, Ti, and Al with $La_{1.85}Sr_{0.15}CuO_4$; Fe, Ti, Al, Cu, Pd, In, La, Bi, Ge, and Si with $YBa_2Cu_3O_{7-x}$; and Al, Cu, and Bi with $Bi_2Ca_{1+x}Sr_{2-x}Cu_2O_{8+y}$. Other studies have emphasized Ag and Au and nonreacting passivating overlayers, including CaF_2 and several oxides. Here, we review some of the critical issues raised in that work.[18-20,39] We note especially that when the adatoms are reactive with respect to oxide formation, they cause disruption of the superconductor surface and form oxide overlayers. Oxygen loss is observed by a dramatic change in the Cu $2p_{3/2}$ emission with reduction of the 2+ satellite and sharpening of the main line as the formal valence changes for Cu^{2+} to Cu^{1+}. In the following, Cu reduction is used to gauge for overlayer reactivity.

In Fig. 4 we show photoemission results for $Ti/YBa_2Cu_3O_{7-x}$ as an example of reactive interface behavior. With the deposition of 0.5 Å of Ti, one can clearly see the Ti $2p_{3/2,1/2}$ doublet shifted ~ 5 eV to higher binding energy relative to Ti metal. This doublet moves further to higher binding energy until a coverage of 4 Å. The lineshape and binding energy position of the Ti 2p core level at early deposition reflects the growth of a Ti-O reaction product, and reaches completion at a Ti coverage of 4 Å. The fully reacted product is TiO_2-like in bonding configuration.[37] A new shoulder appears at 4 Å coverage and it grows and shifts to lower binding energy with increasing coverage. This reflects an evolving Ti-metal component as diffusion of oxygen from the substrate becomes kinetically limited and reaction with Ti atoms is restricted. By 48 Å, the Ti 2p spectrum is that of Ti metal, consistent with the convergence of the surface region to nearly pure Ti.

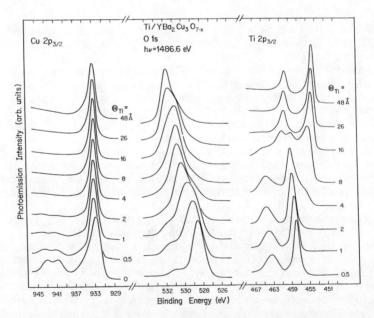

FIG. 4. Ti 2p, O 1s, and Cu 2p core level spectra, normalized to constant height, as a function of Ti adatom deposition on $YBa_2Cu_3O_{7-x}$. Substrate disruption and conversion from 2+ to 1+ is revealed by the loss of the Cu satellite. The oxygen and titanium results show the formation of Ti-O bonding configurations at low coverage, kinetic limitations to O outdiffusion, and the formation of Ti metal with oxygen in solution at high coverage. The result is a metastable, heterogeneous, multilayer system.

The O 1s spectrum in the central panel of Fig. 4 reveal changes which parallel the Ti 2p core level evolution. At low Ti deposition, a new O 1s component appears on the high binding energy side of the main substrate, and it grows and shifts until ~4 Å coverage as Ti-O reaction proceeds. Above ~4 Å, another O 1s feature appears and is attributed to the formation of a metal-rich oxide (oxygen in solution in increasingly-pure Ti).

At the left of Fig. 4 we show the Cu $2p_{3/2}$ core level emission. The most striking result is that very little Ti leads to the complete loss of Cu^{2+} satellite emission within the XPS probe depth (30-50 Å at these energies with our geometry). The deposition of 0.5 Å results in main line sharpening and satellite emission reduction by ~50%; 1 Å reduces the satellite to 25% of its initial intensity. These results go significantly beyond what could have been concluded based only on the fact that oxygen was lost and that Ti-O formed at the surface. They cannot be explained by simple out-diffusion of oxygen from an unmodified $YBa_2Cu_3O_x$ structure. Instead, they reveal a surprisingly long range structural destruction and the fragile nature of planar Cu^{2+} bonding configuration. (Results for Ba also show new bonding configurations.) Analogous effects for Ti overlayers deposited onto clean $La_{1.85}Sr_{0.15}CuO_4$, where there are no Cu-O chains,[38] confirm that oxygen is also leached from the Cu-O planes and structural changes result.

Similar surface reactivity has been observed for metal overlayers deposited on $YBa_2Cu_3O_{7-x}$ and $La_{1.85}Sr_{0.15}CuO_4$. However, comparisons of 1-2-3 and 2-1-2-2 interface reactivity show differences that can be attributed to the Bi-O planes. To highlight these differences, we reproduce in Fig. 5 the Cu $2p_{3/2}$ core level emission of $YBa_2Cu_3O_{7-x}$ and $Bi_2Ca_{1+x}Sr_{2-x}Cu_2O_{8+y}$ as a function of copper deposition. For equivalent depositions, the 2-1-2-2's exhibits a smaller loss of satellite emission (decreased to 37% of its original value by 1 Å deposition for $YBa_2Cu_3O_{7-x}$, compared to 67% for $Bi_2Ca_{1+x}Sr_{2-x}Cu_2O_{8+y}$). In the bottom panel of Fig. 5 we show Bi 4f core level changes for Cu overlayers on both polycrystalline (right) and single crystal 2-1-2-2 samples (left). The striking result is that a new Bi structure appears at lower binding energy as Cu deposition proceeds. The binding energy position of this new feature corresponds to Bi atoms released from the superconductor, and we attribute it to Bi segregation on the Cu overlayer.[37]

Finally, in Fig. 6 we show Cu $2p_{3/2}$ and O 1s spectra for a number of nonreactive overlayers on $YBa_2Cu_3O_{7-x}$. ($YBa_2Cu_3O_{7-x}$ was chosen since it is more reactive than the Bi compounds and overlayers that can passivate the surfaces of 1-2-3 materials should be suitable for the 2-1-2-2 material.[40]) Above the clean surface spectra are results obtained

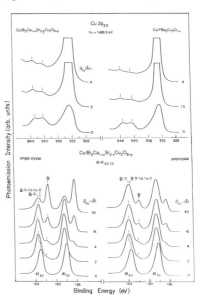

FIG. 5. Top: Cu $2p_{3/2}$ satellite emission for 2-1-2-2 and 1-2-3 superconductors as a function of Cu deposition showing the more extended conversion from Cu 2+ to 1+ nominal valence. Bottom: Bi 4f emission for single crystal and polycrystal 2-1-2-2 showing Cu-induced substrate disruption and the appearance of Bi atoms in the evolving overlayer.

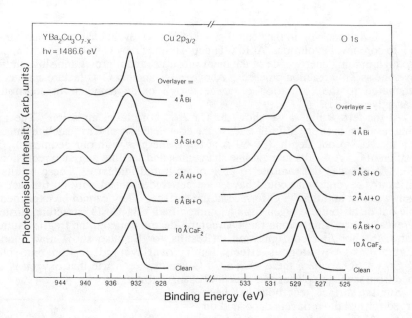

FIG. 6. Cu $2p_{3/2}$ and O 1s emission from $YBa_2Cu_3O_{7-x}$ and after deposition of nonreactive CaF_2. Deposition of Bi, Al, and Si in activated oxygen does not disrupt the superconductor. The complex O emission is related to adatom oxidation. Comparison with Bi atom deposition shows the effectiveness of oxygen activation in suppressing substrate modification.

following deposition of 10 Å of CaF_2, a large bandgap insulating dielectric. They show no change in the superconductor as compared to the clean surface, i.e. no reactions with oxygen were noted and no loss of Cu satellite emission was observed. Moreover, CaF_2 appears to cover the surface uniformly, as judged by the exponential attenuation of the substrate core level emission.

The results shown in Fig. 6 for Bi, Al, and Si deposition were obtained by evaporating these materials in an activated oxygen environment, as described in the experimental section. The data show that the Cu satellite structure persists and that there are new O 1s bonding configurations related to the oxides of these adatoms. From the stability of the Cu satellite, we conclude that the surface oxides of Bi-O, Al-O, and Si-O form without leaching oxygen from the superconductor substrate. Instead, O_2 activation provides a source of highly reactive oxygen which bonds with the evaporant to grow the oxide overlayer. This is supported by the results for 4 Å Bi deposition where there was sufficient oxygen withdrawal from the substrate to quench most of the Cu satellite. These results demonstrate an effective method for depositing a non-disruptive dielectric layer on the surfaces of high temperature superconductors.

In this paper, we have briefly reviewed results for clean surfaces and for overlayers formed in different ways with representative adatoms. Space limitations have prevented detailed comparisons with the now extensive literature (see Refs. 13 and 14 for exhaustive citations). From the studies described here, it should be clear that the surfaces of the ceramic Cu-O's are very fragile. Detailed understanding of these surfaces from the perspective of surface science is still in its infancy, restrained by the relative inaccessibility of single crystals and the inherent complexity of the ceramic structures.

ACKNOWLEDGMENTS

This work has been supported by the Office of Naval Research. We particularly acknowledge collaborative studies with D.L. Nelson (ONR); D.W. Capone II and K.C. Goretta (Argonne National Laboratory); Z. Fisk and D. Peterson (Los Alamos National Laboratory); N.D. Spencer (W.R. Grace); C. Gallo (3M and SuperconiX); and Y. Gao (Minnesota).

REFERENCES

1. *Thin Film Processing and Characterization of High Temperature Superconductors*, AIP Conference Proceedings No. **165**, ed. J.M.E. Harper, R.C. Colton, and L.C. Feldman (New York, NY) 1988. See references therein for a review of the field as of ~10/87.
2. N. Nucker, J. Fink, J.C. Fuggle, P.J. Durham, and W.M. Temmerman, Phys. Rev. B **37**, 5158 (1988). J.C. Fuggle, P.J.W. Weijs, R. Schoorl, G.A. Sawatzky, J. Fink, N. Nucker, P.J. Durham, and W.M. Temmerman, Phys. Rev. B **37**, 123 (1988).
3. D.D. Sarma, O. Strebel, C.T. Simmons, U. Neukirch, G. Kaindl, R. Hoppe, and H.P. Muller, Phys. Rev. B **37**, 9784 (1988).
4. D. van der Marel, J. van Elp, G.A. Sawatzky, and D. Heitmann, Phys. Rev. B **37**, 5136 (1988).
5. A. Fujimori, Phys. Rev. B (submitted); A. Fujimori, E. Takayama-Muromachi, Y. Uchida, and B. Okai, Phys. Rev. B **35**, 8814 (1987).
6. T. Takahashi, H. Matsuyama, H. Katayama-Yoshida, Y. Okabe, S. Hosoya, K. Seki, H. Fujimoto, M. Sato, and H. Inokuchi, Nature xxx (1988)
7. R.L. Kurtz, R.L. Stockbauer, D. Mueller, A. Shih, L.E. Toth, M. Osofksy, and S.A. Wolf, Phys. Rev. B **35**, 8818 (1987).
8. P. Steiner, J. Albers, V. Klinsinger, I. Sander, B. Siegward, S. Hüfner and C. Politis, Z. Phys. B, **67** 19 (1987); ibid 497.
9. R. S. List, A. Arko and coworkers, private communication.
10. P. Steiner, S. Hüfner, V. Klinsinger, I. Sander, B. Siegward, H. Schmitt, R. Schultz, S. Junk, G. Schwitzgebel, A. Gold, C. Politis, H.P. Muller, R. Hoppe, S. Kemmler-Sack and C. Kunz, Z. Phys. B, **69**, 449 (1988).
11. M. Onellion, Y. Chang, D.W. Niles, R. Joynt, G. Margaritondo, N.G. Stoffel, and J.M. Tarascon, Phys. Rev. B **36**, 819 (1987).
12. M. Onellion, M. Tang, Y. Chang, G. Margaritondo, J.M. Tarascon, D.A. Morris, W.A. Bonner, and N.G. Stoffel, Phys. Rev. B **38**, 881 (1988).
13. J.H. Weaver, H.M. Meyer III, T.J. Wagener, D.M. Hill, Y. Gao, D. Peterson, Z. Fisk, and A.J. Arko, Phys. Rev. B **38**, 4668 (1988).
14. H.M. Meyer III, D.M. Hill, T.J. Wagner, Y. Gao, J.H. Weaver, D.W. Capone II, and K.C. Goretta, Phys. Rev. B **38**, xxx (1988).
15. H.M. Meyer III, D.M. Hill, J.H. Weaver, D.L. Nelson, and C.F. Gallo, Phys. Rev. B **38**, xxx (1988).
16. T.J. Wagener, Y. Gao, J.H. Weaver, A. J. Arko, B. Flandermeyer, and D.W. Capone II, Phys. Rev. B **36**, 3899 (1987).
17. Y. Gao, T.J. Wagener, J.H. Weaver, A. J. Arko, B. Flandermeyer, and D.W. Capone II, Phys. Rev. B **36**, 3971 (1987).
18. H.M. Meyer III, D.M. Hill, J.H. Weaver, C.F. Gallo, and K.C. Goretta, J. Appl. Phys. (1988). Single crystal $YBa_2Cu_3O_{7-x}$ and $Bi_2Ca_{1+x}Sr_{2-x}Cu_2O_{8+y}$ surfaces and Ag overlayers.

19. H.M. Meyer III, D.M. Hill, J.H. Weaver, D.L. Nelson, C.F. Gallo, and K.C. Goretta, Appl. Phys. Lett. (1988). Bi adatoms on $YBa_2Cu_3O_{7-x}$ and $Bi_2Ca_{1+x}Sr_{2-x}Cu_2O_{8+y}$.
20. D.M. Hill, H.M. Meyer III, J.H. Weaver, and D.L. Nelson, Appl. Phys. Lett. (1988). Passivation of high T_c superconductor surfaces with CaF_2, and Bi, Al, and Si oxides.
21. J.D. Jorgensen, H. B. Schuttler, D. G. Hinks, D. W. Capone, K. Zhang, M. B. Brodsky, and D. J. Scalapino, Phys. Rev. Lett., **58** 1024 (1987).
22. J.L. Routbort, S.J. Rothman, L.J. Nowicki, and K.C. Goretta, Mat. Sci. Forum **34-36**, 315 (1988).
23. S. A. Chambers, D. M. Hill, F. Xu, and J. H. Weaver, Phys. Rev. B, **35** 634 (1987).
24. Y. Gao, M. Grioni, B. Smandek, J.H. Weaver, and T. Tyrie, J. Phys. E **21**, 489 (1988).
25. For a recent review of band structure calculations, see B.M. Klein *et al.* and A.J. Freeman *et al.* in Chapters 3 and 6 of *Physical Chemistry of High Temperature Superconductors*, ed. T.F. George and D.L. Nelson (ACS, Washington, D.C., 1988), ACS Symposium Series **377**. See also S. Massidda, J. Yu, and A.J. Freeman, Physica C **152**, 251 (1988).
26. M.R. Thuler, R.L. Benbow, and Z. Hurych, Phys. Rev. B **26**, 669 (1982).
27. Z. Shen, J.W. Allen, J.J. Yeh, J.S. Kang, W. Ellis, W. Spicer, I. Lindau, M.B. Maple, Y.D. Dalichaouch, M.S. Torikachvili, J.Z. Sun, and T.H. Geballe, Phys. Rev. B **36**, 8414 (1987).
28. L.F. Mattheiss, Phys. Rev. Lett. **58**, 1028 (1987).
29. L.F. Mattheiss and D. R. Hamann, Solid State Commun. **63**, 395 (1987).
30. A. Fujimori and J.H. Weaver, Phys. Rev. B **31**, 6345 (1985).
31. M.R. Norman, D.D. Koelling, and A.J. Freeman, Phys. Rev. B **32**, 7748 (1985).
32. T.J. Wagener, Y. Hu, M. Jost, J.H. Weaver, N.D. Spencer, and K.C. Goretta, Phys. Rev. B. Resonance inverse photoemission of 2-1-2-2's and 1-2-3's and unoccupied oxygen states.
33. G. van der Laan, C. Westra, C. Haas, and G. A. Sawatzky, Phys. Rev. B **23**, 4369 (1981).
34. A. Fujimori, S. Takekawa, E. Takayama-Murmachi, Y, Uchida, A. Ono, T. Takahashi, Y. Okabe, and H. Katayama-Yoshida, Phys. Rev. B (submitted). Photoemission study of $Bi_2Ca_{1+x}Sr_{2-x}Cu_2O_{8+y}$.
35. H.M. Meyer III, D.M. Hill, J.H. Weaver, D.L. Nelson, Chapter 21 in *Physical Chemistry of High Temperature Superconductors*, edited by T.F. George and D.L. Nelson (ACS, Washington, D.C., 1988), ACS Symposium Series **377**.
36. P.E. Larson, J. Electron. Spectros. and Related Phenom. **4**, 213 (1974).
37. G. Rocker and W. Gopel, Surface Science **181**, 530 (1987).
38. H.M. Meyer III, D.M. Hill, S.G. Anderson, J.H. Weaver, and D.W. Capone II, Appl. Phys. Lett. **51**, 1750 (1987).
39. D.M. Hill, H.M. Meyer III, J.H. Weaver, C.F. Gallo, and K.C. Goretta, Phys. Rev. B (in press). Cu adatom interactions with single- and polycrystalline $Bi_2Ca_{1+x}Sr_{2-x}Cu_2O_{8+y}$ and $YBa_2Cu_3O_{7-x}$.
40. R.S. Williams and S. Chaudhury, Chapter 22 in *Physical Chemistry of High Temperature Superconductors*, edited by T.F. George and D.L. Nelson (ACS, Washington, D.C., 1988), ACS Symposium Series **377**.

THEORY

Tight-binding Description of High-temperature Superconductors 410

Utilization of a Highly-correlated CuO_n Cluster Model to Interpret Electron Spectroscopic Data for the High-temperature Superconductors 418

Novel Applications of High Temperature Superconductivity: Neurocomputing   426

TIGHT-BINDING DESCRIPTION OF HIGH-TEMPERATURE SUPERCONDUCTORS

Brent A. Richert and Roland E. Allen
Center for Theoretical Physics, Department of Physics
Texas A&M University, College Station, Texas 77843

ABSTRACT

A single tight-binding model, with fully transferable parameters, provides a good description of the electronic structures of all currently-known high-temperature superconductors, including the copper oxides $La_{2-x}Sr_xCuO_4$, $YBa_2Cu_3O_7$, $Bi_2CaSr_2Cu_2O_8$, $Tl_2Ba_2CuO_6$, $Tl_2CaBa_2Cu_2O_8$, and $Tl_2Ca_2Ba_2Cu_3O_{10}$ and the bismuth oxide $Ba_{1-x}K_xBiO_3$ (within the one-electron approximation). The energy bands, local densities of states, and atomic valences have been calculated for each material.

INTRODUCTION

The discovery of high-temperature superconductivity[1,2] was followed by structural[3,4] and electronic[5,6] characterizations of $La_{2-x}Sr_xCuO_4$ and $YBa_2Cu_3O_7$. Both materials have CuO_2 planes separated by metal-oxide or ion layers, with $YBa_2Cu_3O_7$ also containing CuO chains.[4] The electronic properties near the Fermi energy E_F are thought to be dominated by antibonding $Cu(d)$–$O(p)$ bands.[5,6]

Recent discoveries have produced several new classes of high-T_c materials with properties similar to but different from those of the original materials. Following a report of superconductivity with $T_c \approx 22$ K in the Bi-Sr-Cu-O system,[7] a series of copper-oxide materials containing bismuth was found to have T_c's up to 114 K.[8] The superconducting phase with $T_c = 85$ K was identified as $Bi_2CaSr_2Cu_2O_8$.[9] X-ray diffraction analyses of powder[10] and single crystal[11] samples revealed a layered structure: there are two CuO_2 planes, separated by a Ca ion, together with SrO and double BiO layers, giving the metal stacking sequence Bi-Bi-Sr-Cu-Ca-Cu-Sr. The crystal substructure is approximately body-centered tetragonal, but with a long-range modulation along the orthorhombic b direction.[10] Substitution of Pb into this phase was found to increase the onset temperature to 107 K.[11] A second phase of composition $Bi_2Sr_2CuO_6$ was found to have a T_c of only 6 K.[12] This bct structure has only a single CuO_2 plane per unit cell,[12] as in $La_{2-x}Sr_xCuO_4$.

Another set of copper-oxide materials containing Tl has yielded the highest T_c's.[13–15] The phase $Tl_2Ba_2CuO_6$ is structurally very similar to $Bi_2Sr_2CuO_6$, but has a much higher T_c of 83 K.[13] The structure of the phase $Tl_2CaBa_2Cu_2O_8$ (with $T_c = 112$ K) is similar to that of $Bi_2CaSr_2Cu_2O_8$, but with no evidence of a long-range modulation.[14] The phase $Tl_2Ca_2Ba_2Cu_3O_{10}$ has $T_c = 125$ K, and has a triple layer of CuO_2 planes separated by Ca ions.[15] There are no one-dimensional CuO chains in any of the Bi or Tl materials, distinguishing them from $YBa_2Cu_3O_7$.[4]

Most remarkably, a copper-free bismuth-oxide compound has recently been found to be a true high-temperature superconductor.[16–18] The compound $Ba_{1-x}K_xBiO_3$ has $T_c \approx 30$ K for $x \approx 0.4$.[17,18] As in the compound $BaPb_{0.75}Bi_{0.25}O_3$ with a T_c of 13 K,[19] the conduction bands of this high-T_c material are formed from antibonding $Bi(s)$–$O(p)$ states, with no d character near E_F.[20] It is interesting to note that the undoped parent compound $BaBiO_3$ is diamagnetic, distinctly different from the antiferromagnetic parent compounds La_2CuO_4 and $YBa_2Cu_3O_6$.[21,22] Finally, $Ba_{1-x}K_xBiO_3$ has a cubic rather than anisotropic and layered structure.

All these materials have somewhat similar electronic properties near the Fermi energy, even though their structures and magnetic behavior are quite different. We present here the electronic energy bands and atomic valences for each material, with the densities of states for the representative cases $Bi_2CaSr_2Cu_2O_8$ and $Tl_2Ca_2Ba_2Cu_3O_{10}$. The complete results will be published in a later article.[23]

METHOD

We previously calculated the electronic energy bands, the local densities of states, and the atomic valences of $La_{1.85}Sr_{0.15}CuO_4$ and $YBa_2Cu_3O_7$ in the one-electron approximation, using a semiempirical tight-binding model with p and s orbitals included for the oxygen atoms and d and s orbitals for all the metal atoms.[24] We now apply this same model to the Bi- and Tl-containing copper oxides and the copper-free material $Ba_{1-x}K_xBiO_3$. In addition to the parameters previously fitted[24] to the energy bands of La_2CuO_4,[5] we need the s, p, and d energies for Bi and Tl, and the $sp\sigma$, $pp\sigma$, and $pp\pi$ interatomic matrix elements.[25] These parameters were fitted to the previous band structure calculations[20] for $BaBiO_3$ and $BaPbO_3$, with the Tl energy parameters extrapolated from those of Bi and Pb. The resulting atomic parameters are listed in Table I; the interatomic parameters are $\eta_{ss\sigma} = -1.1$, $\eta_{sp\sigma} = 1.4$, $\eta_{pp\sigma} = 1.5$, $\eta_{pp\pi} = -0.6$, $\eta_{sd\sigma} = -1.6$, $\eta_{pd\sigma} = -2.5$, and $\eta_{pd\pi} = 1.4$.

The local density of states for both spins is calculated from

$$\rho(E) = -\frac{2}{\pi} Tr\, Im\, G_0(E), \qquad (1)$$

where Tr indicates a trace over those orbitals associated with a given site. The Green's function is given by

$$G_0(E) = \sum_{\vec{k},n} w_{\vec{k}} \frac{\psi(\vec{k},n)\, \psi^\dagger(\vec{k},n)}{E - E(\vec{k},n) + i\delta}, \qquad (2)$$

where $E(\vec{k},n)$ and $\psi(\vec{k},n)$ are the electronic energy and wave function (in a tight-binding representation) for the nth band and one of the N sample wave vectors $\vec{k}$ (with weight $w_{\vec{k}}$) within the irreducible part of the Brillouin zone. In the present calculations, $N = 24$ for the primitive cell of the bct structures, and $N = 20$ for the cubic structures. A finite value $\delta = 0.2$ eV was used to smooth the results. The Fermi energy is calculated by integrating the total density of states up to the number of valence electrons for each material. The valence of each atom, defined to be the number of electrons lost by the atom when it is bonded in the solid, is calculated from

$$\Delta n_i = n_i - 2 \sum_{\vec{k},\alpha, E_n \leq E_F} w_{\vec{k}}\, |\psi_i^\alpha(\vec{k},n)|^2, \qquad (3)$$

where n_i is the number of valence electrons on the free atom i, $\psi_i^\alpha(\vec{k},n)$ is the component of the eigenfunction corresponding to the valence orbital α on this atom, and the summation is over the N sample wave vectors $\vec{k}$.

RESULTS AND DISCUSSION

We first consider the material $Bi_2CaSr_2Cu_2O_8$, using the crystal structure and atomic positions of Ref. 10. The band structure for this material has been calculated by several groups, all with similar results.[26-30] In order to fit these results with our tight-binding model, we have included the second-neighbor Bi-Bi interactions.[26] We regard these Bi-Bi interactions as justified by the large covalent radius of Bi and the relatively small lattice parameter a, even though only nearest-neighbor interactions were needed for the earlier materials.[24] We ignore the long-range structural modulation,[10,11] which should have only a minor effect on the electronic properties.

The energy bands of $Bi_2CaSr_2Cu_2O_8$ along the symmetry lines of the Brillouin zone for the bct crystal structure[31] are shown in Fig. 1, with the zero of energy shifted to E_F. We note that the bands are almost dispersionless along ΓZ, perpendicular to the copper-oxygen planes, and are dominated by antibonding bands of $Cu(d)-O(p)$ crossing the Fermi energy along ΓX. Two Bi p bands disperse down to the Fermi energy, with one band crossing below E_F to about -0.7 eV near the Brillouin zone boundary at the symmetry point D in the [100] direction. As in the other calculations,[26-30] these p bands form occupied electron pockets near the Brillouin zone boundary, donating holes to the system.

The atomic valences Δn calculated from (3) are presented in Table II. In our notation, O(1) is the oxygen site within the CuO_2 planes, O(2) is in the SrO region separating the Cu and

TABLE I. Atomic tight-binding parameters for high-temperature superconductors. The parameters for Tl are extrapolated from those of Pb and Bi. All unfitted parameters are taken from a standard table (Ref. 25).

	ϵ_s (eV)	ϵ_p (eV)	ϵ_d (eV)	r_d (Å)
Tl	-14.8	-8.3	-23.0	1.0
Pb[a]	-18.0	-9.4	-29.0	1.0
Bi[b]	-21.2	-10.5	-35.0	1.0
K	-4.2	...	-3.2	1.2
Ca	-5.4	...	-3.2	1.2
Sr	-5.0	...	-6.8	1.6
Y	-5.5	...	-6.8	1.6
Ba	-4.5	...	-6.6	1.6
La	-4.9	...	-6.6	1.6
Cu[c]	-12.0	...	-14.0	0.95
O	-29.0	-14.0	...	...

[a] Parameters fitted to BaPbO$_3$ (Ref. 20).
[b] Parameters fitted to BaBiO$_3$ (Ref. 20).
[c] Parameters fitted to La$_2$CuO$_4$ (Ref. 24).

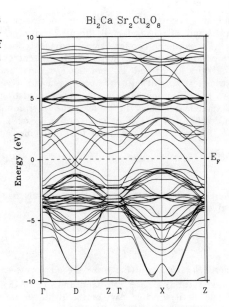

FIG. 1. Energy bands for Bi$_2$CaSr$_2$Cu$_2$O$_8$. The zero of energy has been shifted to the Fermi energy E_F. Two Cu(d)–O(p) antibonding bands cross E_F along the ΓX symmetry line. Two Bi p bands disperse to the Fermi energy, with one band dipping below E_F near the Brillouin zone boundary at the symmetry point D in the [100] direction, donating holes to the CuO$_2$ planes.

TABLE II. Valences Δn for high-temperature superconductors.

	Bi,Tl	Ca	Sr,Ba	Cu[a]	Cu(2)	O(1)	O(2)	O(3)	O(4)
Bi$_2$CaSr$_2$Cu$_2$O$_8$	1.80	1.51	1.09	0.84	...	-1.09	-1.14	-1.18	...
Tl$_2$Ba$_2$CuO$_6$	1.05	...	1.36	0.89	...	-1.10	-1.04	-0.72	...
Tl$_2$CaBa$_2$Cu$_2$O$_8$	1.05	1.45	1.37	0.86	...	-1.13	-1.03	-0.71	...
Tl$_2$Ca$_2$Ba$_2$Cu$_3$O$_{10}$	1.06	1.45	1.38	0.54	1.02	-1.27	-1.09	-1.08	-0.65
Ba$_{0.6}$K$_{0.4}$BiO$_3$	1.80	...	1.62	...	...	-1.01	-1.00	...	...

[a] Central Cu(1) site for Tl$_2$Ca$_2$Ba$_2$Cu$_3$O$_{10}$.

Bi layers, and O(3) is in the BiO layer. The valences for Cu and O(1) are quite similar to those found previously for the copper-oxide plane regions in $La_{1.85}Sr_{0.15}CuO_4$ and $YBa_2Cu_3O_7$.[24]

The local densities of states calculated from (1) for Bi, Ca, Sr, and Cu are shown in Fig. 2, and those for oxygen in Fig. 3. The Bi d bands lie approximately 24 eV below the Fermi energy, while the Bi p bands, centered slightly above E_F, interact strongly to produce some metallic contribution to $\rho(E_F)$. Both the Ca and Sr d states are unoccupied and quite ionic. The Cu and O(1) sites show strong interactions, related to the $pd\sigma$ bonds, resulting in the antibonding bands protruding 1.6 eV above E_F at X in Fig. 1. The O(2) and O(3) sites show primarily p character just below the Fermi energy (with s bands far below E_F), but some p character is distributed above the Fermi energy because of mixing with the s and p orbitals of neighboring Bi. These oxygen sites do not absorb all the holes donated by the occupied Bi p bands, so excess hole carriers are donated to the copper-oxygen planes.

The electronic energy bands for $Tl_2Ba_2CuO_6$ are given in Fig. 4. The single antibonding $Cu(d)$-$O(p)$ band crossing E_F arises from the single CuO_2 plane in the bct unit cell.[13] The Tl-Tl in-plane interactions are included, as in the Bi calculation,[26] causing the Tl p bands to disperse by about 3 eV from Γ to D along the [100] direction in Fig. 4. These bands do not dip below the Fermi energy as do the Bi p bands in $Bi_2CaSr_2Cu_2O_8$.[26-30] However, the $Tl(s)$-$O(p)$ hybrid states do cross below E_F to about –0.5 eV at Γ, forming occupied electron pockets. The valences for $Tl_2Ba_2CuO_6$ are listed in Table II. Our notation for the oxygen atoms places O(1) in the CuO_2 plane, O(2) in the BaO layer, and O(3) in the TlO layer.

The energy bands for $Tl_2CaBa_2Cu_2O_8$, with the crystal structure taken from Ref. 14, are given in Fig. 5. The two antibonding $Cu(d)$-$O(p)$ bands protruding above E_F correspond to the two adjacent CuO_2 planes in this phase. The $Tl(s)$-$O(p)$ bands disperse below the Fermi energy to –0.3 eV at the symmetry points Γ and Z. The valences for $Tl_2CaBa_2Cu_2O_8$ are given in Table II. Note that the valence of Cu is similar to that in $Tl_2Ba_2CuO_6$.

The electronic energy bands of the phase $Tl_2Ca_2Ba_2Cu_3O_{10}$, with the bct crystal structure of Ref. 15, are shown in Fig. 6. This 125 K superconductor has three antibonding $Cu(d)$-$O(p)$ states which cross E_F, corresponding to the triple layer of CuO_2 planes stacked with Ca ions. The $Tl(s)$-$O(p)$ hybrid states cross below E_F to about –0.6 eV at Γ and Z, again forming occupied electron pockets.

Figure 7 shows the local densities of states for Tl, Ca, Ba, and Cu, with those for oxygen given in Fig. 8. The notation for the oxygen atoms places O(1) in the central CuO_2 plane, O(2) in the outer CuO_2 planes, O(3) in the BaO layer, and O(4) in the TlO layer. The Tl d bands lie far below E_F, while the Tl p bands, centered about 3 eV above E_F, interact somewhat weakly with the neighboring O(4) oxygens. Both the Ca and the Ba d bands are above the Fermi energy, and these atoms are quite ionic. The copper and oxygen plane sites provide the metallic character at the Fermi energy, with a small contribution from $Tl(s)$. In the present model, the total density of states at E_F is 5.3 states/eV cell.

The atomic valences for $Tl_2Ca_2Ba_2Cu_3O_{10}$ in Table II show that Cu(2), in the outer CuO_2 planes adjacent to the BaO layers, has a higher valence than those of the other copper-oxide materials. However, the central copper site Cu(1) shows a decreased number of holes, with a valence of only 0.54.

These tight-binding results are quite similar to the bands of other one-electron calculations for the Tl materials.[32-34] The hole conduction bands also resemble the previous results for the planar CuO_2 regions of $La_{1.85}Sr_{0.15}CuO_4$,[5,24] $YBa_2Cu_3O_7$,[6,24] and $Bi_2CaSr_2Cu_2O_8$.[26-30]

We have also calculated the energy bands of $Ba_{1-x}K_xBiO_3$, with the doping treated in the virtual crystal approximation. We use the cubic structure of Ref. 17, with $a_0 = 4.293$ Å. Our energy bands in Fig. 9 are in good agreement with the recent calculation[35] for $Ba_{0.5}K_{0.5}BiO_3$. (We omit the very core-like Ba p states.) The valences for the Bi-based superconductor are listed in Table II. Our calculation indicates that doping $Ba_{1-x}K_xBiO_3$ at $x = 0.4$ gives a carrier density of about 0.26 holes per BiO_2 unit with respect to semiconducting $BaBiO_3$.[20]

In each of the copper-oxide high-temperature superconductors, we find that the dominant hole carrier states – the antibonding $Cu(d)$-$O(p)$ states – are quite similar in structure and carrier density. However, the bismuth-oxide material $Ba_{1-x}K_xBiO_3$ has $Bi(s)$-$O(p)$ character at the Fermi energy. Mechanisms for high-temperature superconductivity which rely on magnetic properties of Cu d electrons are not consistent with this fact. On the other hand, a mechanism

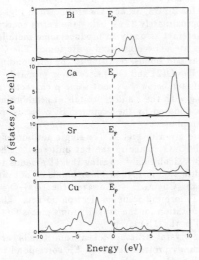

FIG. 2. Local densities of states for the metal atoms in $Bi_2CaSr_2Cu_2O_8$. Bi interacts with the neighboring O(2) and O(3) atoms, but Ca and Sr are ionic.

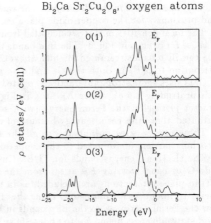

FIG. 3. Local densities of states for the oxygen atoms in $Bi_2CaSr_2Cu_2O_8$. The Cu–O(1) in-plane $pd\sigma$ interactions give metallic character at the Fermi energy, as in the other high-T_c copper-oxide superconductors.

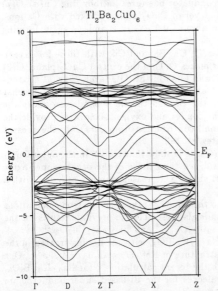

FIG. 4. Energy bands for $Tl_2Ba_2CuO_6$. The Cu(d)–O(p) antibonding band, with hole conduction states, crosses E_F along the ΓX symmetry line. A Tl(s)–O(p) band dips below E_F, making both the CuO_2 planes and TlO layers metallic.

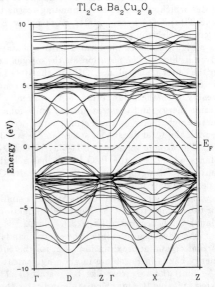

FIG. 5. Energy bands for $Tl_2CaBa_2Cu_2O_8$. Two Cu(d)–O(p) antibonding bands from the two adjacent CuO_2 planes protrude above E_F.

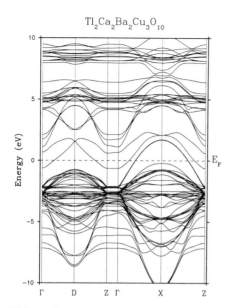

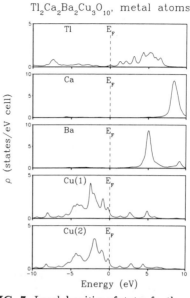

FIG. 6. Energy bands for $Tl_2Ca_2Ba_2Cu_3O_{10}$. The Cu–Ca–Cu–Ca–Cu layer structure gives three Cu(d)-O(p) hole conduction bands in this material.

FIG. 7. Local densities of states for the metal atoms in $Tl_2Ca_2Ba_2Cu_3O_{10}$. Tl interacts with the neighboring O(4) atoms, while Ca and Ba remain relatively ionic.

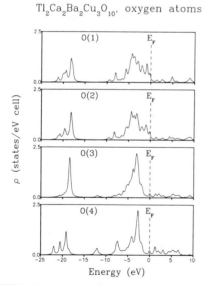

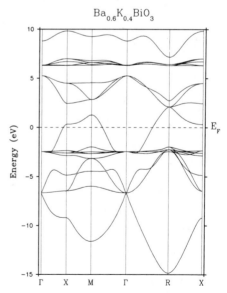

FIG. 8. Local densities of states for the oxygen atoms in $Tl_2Ca_2Ba_2Cu_3O_{10}$.

FIG. 9. Energy bands for $Ba_{0.6}K_{0.4}BiO_3$. The Bi(s)–O(p) antibonding bands provide the conduction states for this $T_c \approx 30$ K superconductor.

involving LaO, BaO, or SrO $O(p)$-metal(d) excitons provides a consistent interpretation of high-temperature superconductivity in both the copper-oxide and bismuth-oxide materials.[36]

CONCLUSIONS

A single tight-binding model, with fully transferable parameters, can be used to describe the electronic structures of all currently-known high-temperature superconductors. The calculated electronic structures of the copper-oxide materials are in good agreement with other one-electron calculations,[26-30,32-34] and are similar to those of $La_{2-x}Sr_xCuO_4$ and $YBa_2Cu_3O_7$. The Bi p bands are found to cross below E_F in the Bi–Ca–Sr–Cu–O system, while Tl s bands form occupied electron pockets in the Tl–Ca–Ba–Cu–O system, doping the copper-oxygen planes with hole carriers.

The calculations for copper-free $Ba_{1-x}K_xBiO_3$ demonstrate that this material has valence bands of antibonding $Bi(s)$–$O(p)$ states, as in $BaPb_{1-y}Bi_yO_3$. The change from Cu d to Bi s conduction states in these materials indicates that magnetic properties do not provide a viable interpretation of the observed high-temperature superconductivity.

Recent photoemission studies by Meyer et al.[37] and Onellion et al.[38] of $Bi_2CaSr_2Cu_2O_8$ confirm the basic occupied electronic states of the earlier one-electron calculations.[26-30] Also, angle-resolved resonant photoemission studies by Takahashi et al.[39] of $Bi_2CaSr_2Cu_2O_8$ indicate an energy band dispersion of 0.2 to 0.5 eV near E_F along the ΓX symmetry line.

ACKNOWLEDGMENTS

This work was supported by the Office of Naval Research (No. N00014-82-K-0447). Additional support was provided by the Robert A. Welch Foundation. B. A. R. was partially supported by the U.S. Air Force Institute of Technology.

REFERENCES

1. J. G. Bednorz and K. A. Müller, Z. Phys. B **64**, 189 (1986).
2. M. K. Wu, J. R. Ashburn, C. J. Torng, P. H. Hor, R. L. Meng, L. Gao, Z. J. Huang, Y. Q. Wang, and C. W. Chu, Phys. Rev. Lett. **58**, 908 (1987).
3. R. M. Fleming, B. Batlogg, R. J. Cava, and E. A. Reitman, Phys. Rev. B **35**, 7191 (1987); R. J. Cava, A. Santoro, D. W. Johnson, Jr., and W. W. Rhodes, Phys. Rev. B **35**, 6716 (1987).
4. M. A. Beno, L. Soderholm, D. W. Capone, II, D. G. Hinks, J. D. Jorgensen, J. D. Grace, I. K. Schuller, C. U. Segre, and K. Zhang, Appl. Phys. Lett. **51**, 57 (1987); T. Siegrist, S. Sunshine, D. W. Murphy, R. J. Cava, and S. M. Zahurak, Phys. Rev. B **35**, 7137 (1987); J. E. Greedan, A. H. O'Reilly, and C. V. Stager, Phys. Rev. B **35**, 8770 (1987); J. J. Capponi, C. Chaillout, A. W. Hewat, P. Lejay, M. Marezio, N. Nguyen, B. Raveau, J. L Soubeyroux, J. L. Tholence, and R. Tournier, Europhys. Lett. **3**, 1301 (1987).
5. L. F. Mattheiss, Phys. Rev. Lett. **58**, 1028 (1987); J. Yu, A. J. Freeman, and J. -H. Xu., Phys. Rev. Lett. **58**, 1035 (1987); T. Oguchi, Jpn. J. Appl. Phys. **26**, L417 (1987); K. Takegahara, H. Harima, and A. Yanase, Jpn. J. Appl. Phys. **26**, L352 (1987).
6. L. F. Mattheiss and D. R. Hamann, Solid State Commun. **63**, 395 (1987); T. Fujiwara and Y. Hatsugai, Jpn. J. Appl. Phys. **26**, L716 (1987); S. Massidda, J. Yu, A. J. Freeman, and D. D. Koelling, Phys. Lett. A **122**, 198 (1987); F. Herman, R. V. Kasowski, and W. Y. Hsu, Phys. Rev. B **36**, 6904 (1987).
7. C. Michel, M. Hervieu, M. M. Borel, A. Grandin, F. Deslandes, J. Provost, and B. Raveau, Z. Phys. B **68**, 421 (1987).
8. H. Maeda, Y. Tanaka, M. Fukutomi, and T. Asano, Jpn. J. Appl. Phys. **27**, L209 (1988); C. W. Chu, J. Bechtold, L. Gao, P. H. Hor, Z. J. Huang, R. L. Meng, Y. Y. Sun, Y. Q. Wang, and Y. Y. Xue, Phys. Rev. Lett. **60**, 941 (1988).
9. R. M. Hazen, C. T. Prewitt, R. J. Angel, N. L. Ross, L. W. Finger, C. G. Hadidiacos, D. R. Veblen, P. J. Heaney, P. H. Hor, R. L. Meng, Y. Y. Sun, Y. Q. Wang, Y. Y. Xue, Z. J. Huang, L. Gao, J. Bechtold, and C. W. Chu, Phys. Rev. Lett. **60**, 1174 (1988); M. A. Subramanian, C. C. Torardi, J. C. Calabrese, J. Gopalakrishnan, K. J. Morrissey, T. R. Askew, R. B. Flippen, U. Chowdhry, and A. W. Sleight, Science **239**, 1015 (1988).

10. J. M. Tarascon, Y. Le Page, P. Barboux, B. G. Bagley, L. H. Greene, W. R. McKinnon, G. W. Hull, M. Giroud, and D. M. Hwang, Phys. Rev. B **37**, 9382 (1988).
11. S. A. Sunshine, T. Siegrist, L. F. Schneemeyer, D. W. Murphy, R. J. Cava, B. Batlogg, R. B. van Dover, R. M. Fleming, S. H. Glarum, S. Nakahara, R. Farrow, J. J. Krajewski, S. M. Zahurak, J. V. Waszczak, J. H. Marshall, P. Marsh, L. W. Rupp, Jr., and W. F. Peck, Phys. Rev. B **38**, 893 (1988).
12. J. B. Torrance, Y. Tokura, S. J. LaPlaca, T. C. Huang, R. J. Savoy, and A. I. Nazzal, Solid State Commun. **66**, 703 (1988).
13. Z. Z. Sheng and A. M. Hermann, Nature **332**, 55 (1988); C. C. Torardi, M. A. Subramanian, J. C. Calabrese, J. Gopalakrishnan, E. M. McCarron, K. J. Morrissey, T. R. Askew, R. B. Flippen, U. Chowdhry, and A. W. Sleight, Phys. Rev. B **38**, 225 (1988).
14. Z. Z. Sheng, A. M. Hermann, A. El Ali, C. Almasan, J. Estrada, T. Datta, and R. J. Matson, Phys. Rev. Lett. **60**, 937 (1988); M. A. Subramanian, J. C. Calabrese, C. C. Torardi, J. Gopalakrishnan, T. R. Askew, R. B. Flippen, K. J. Morrissey, U. Chowdhry, and A. W. Sleight, Nature **332**, 420 (1988).
15. Z. Z. Sheng and A. M. Hermann, Nature **332**, 138 (1988); S. S. P. Parkin, V. Y. Lee, E. M. Engler, A. I. Nazzal, T. C. Huang, G. Gorman, R. Savoy, and R. Beyers, Phys. Rev. Lett. **60**, 2539 (1988); C. C. Torardi, M. A. Subramanian, J. C. Calabrese, J. Gopalakrishnan, K. J. Morrissey, T. R. Askew, R. B. Flippen, U. Chowdhry, and A. W. Sleight, Science **240**, 631 (1988).
16. L. F. Mattheiss, E. M. Gyorgy, and D. W. Johnson, Jr., Phys. Rev. B **37**, 3745 (1988).
17. R. J. Cava, B. Batlogg, J. J. Krajewski, R. Farrow, L. W. Rupp, Jr., A. E. White, K. Short, W. F. Peck, and T. Kometani, Nature **332**, 814 (1988).
18. D. G. Hinks, B. Dabrowski, J. D. Jorgensen, A. W. Mitchell, D. R. Richards, Shiyou Pei, and Donglu Shi, Nature **333**, 836 (1988).
19. A. W. Sleight, J. L. Gillson, and P. E. Bierstedt, Solid State Commun. **17**, 27 (1975).
20. L. F. Mattheiss and D. R. Hamann, Phys. Rev. B **28**, 4227 (1983).
21. G. Shirane, Y. Endoh, R. J. Birgeneau, M. A. Kastner, Y. Hidaka, M. Oda, M. Suzuki, and T. Murakami, Phys. Rev. Lett. **59**, 1613 (1987).
22. J. M. Tranquada, D. E. Cox, W. Kunnmann, H. Moudden, G. Shirane, M. Suenaga, P. Zolliker, D. Vaknin, S. K. Sinha, M. S. Alvarez, A. J. Jacobson, and D. C. Johnston, Phys. Rev. Lett. **60**, 156 (1988).
23. B. A. Richert and R. E. Allen, to be published.
24. B. A. Richert and R. E. Allen, Phys. Rev. B **37**, 7869 (1988).
25. W. A. Harrison, *Electronic Structure and the Properties of Solids* (Freeman, San Francisco, 1980).
26. M. S. Hybertsen and L. F. Mattheiss, Phys. Rev. Lett. **60**, 1661 (1988).
27. H. Krakauer and W. E. Pickett, Phys. Rev. Lett. **60**, 1665 (1988).
28. F. Herman, R. V. Kasowski, and W. Y. Hsu, Phys. Rev. B **38**, 204 (1988).
29. S. Massidda, J. Yu, and A. J. Freeman, Physica C (in press).
30. P. A. Sterne and C. S. Wang, to be published.
31. A. W. Luehrmann, Adv. Phys. **17**, 1 (1968).
32. J. Yu, S. Massidda, and A. J. Freeman, Physica C (in press).
33. R. V. Kasowski, W. Y. Hsu, and F. Herman, to be published.
34. D. R. Hamann and L. F. Mattheiss, Phys. Rev. B **38**, 5138 (1988).
35. L. F. Mattheiss and D. R. Hamann, Phys. Rev. Lett. **60**, 2681 (1988).
36. B. A. Richert and R. E. Allen, in *Proceedings of the 18th International Conference on Low-Temperature Physics, Part III*, edited by Y. Nagaoka [Jpn. J. Appl. Phys. **26**, Suppl. 26-3, 2047 (1987)]; in *Proceedings of the Drexel International Conference on High-Temperature Superconductors*, edited by S. M. Bose and S. D. Tyagi (World Scientific, Singapore, 1988), p. 147 [Rev. Solid State Sci. **1**, 295 (1987)]; and to be published.
37. H. M. Meyer III, D. M. Hill, J. H. Weaver, D. L. Nelson, and C. F. Gallo, to be published.
38. M. Onellion, M. Tang, Y. Chang, G. Margaritondo, J. M. Tarascon, P. A. Morris, W. A. Bonner, and N. G. Stoffel, Phys. Rev. B **38**, 881 (1988).
39. T. Takahashi, H. Matsuyama, H. Katayama-Yoshida, Y. Okabe, S. Hosoya, K. Seki, H. Fujimoto, M. Sato, and H. Inokuchi, Nature **334**, 691 (1988).

UTILIZATION OF A HIGHLY-CORRELATED CuO_n CLUSTER MODEL TO INTERPRET ELECTRON SPECTROSCOPIC DATA FOR THE HIGH-TEMPERATURE SUPERCONDUCTORS

David E. Ramaker*
Department of Chemistry, George Washington University
Washington, DC 20052, USA

ABSTRACT

We have consistently interpreted electron spectroscopic data for the high temperature superconductors utilizing a highly-correlated CuO_n cluster model, and an extended Hubbard Hamiltonian which includes the inter-site Cu-O U_{pd} and O-O U_{pp} parameters. The data indicate much larger U_p and U_{pp} values than found in other typical highly conductive metals. Previously unassigned features in the data are now assigned within the model.

INTRODUCTION

In this work we summarize results of an interpretation of electron spectroscopic data for the high temperature superconductors. The data interpreted include the valence band (VB), Cu 2p, and O 1s photoelectron data (UPS and XPS), the Cu $L_{23}VV$, Cu $L_{23}M_{23}V$, and O KVV Auger data, and the O K and Cu L_{23} x-ray absorption and emission (XANES and XES) data. Published data for polycrystalline and single crystal samples of $La_{2-x}Ba_xCuO_4$ and $YBa_2Cu_3O_{7-x}$ (herein referred to as the La and 123 HTSC's) are considered along with that for CuO and Cu_2O.

The basic electronic structure of the HTSC's can be described with the Anderson Hamiltonian utilized by Sawatzky and coworkers[1,2]. It includes the transfer or hopping integral t, the Cu and O orbital energies ε_d and ε_p, the core polarization energy Q_d, and the intra-site Coulomb repulsion energies U_d and U_p (the latter sometimes are assumed to be zero). This model is most useful when the U's are large relative to the band widths[1], i.e. when correlation effects dominate hybridization effects. A $CuO_n^{(2n-2)-}$ cluster model, which is also reasonably valid when U >> t, simplifies the model further[1]. We utilize an extended Hubbard model by adding the inter-site repulsion energies U_{dp} and U_{pp}^o (i.e. between neighboring Cu-O and O-O atoms). The addition of these interactions is important for understanding many of the features in the data.

RESULTS

Our results for the Hubbard parameters are summarized at the top of Table 1. Other estimates of these Hubbard U and ε parameters have been reported previously for the HTSC's[2-4]. These were obtained empirically from the Cu 2p XPS and the VB UPS data utilizing the Anderson model. Our optimal extended Hubbard

*Supported in part by the Office of Naval Research.

TABLE 1   Summary of hole states revealed in the spectroscopic data, and estimated energies using the following optimal values for the Hubbard parameters in eV[a]:

$\delta_1 = 2$ $\varepsilon_d = 2$ $U_p = 12, 13$ $U_d = 9.5, 10.2$
$\delta_2 = 0.5, 0.8$ $\varepsilon_p = 2, 3$ $U_{pp}^o = 4.5, 4$ $U_{dp} = 1$
$\Gamma = 2$ $U_{pp}^p = 0$ $U_{cp}^o = 2$ $Q_d = 9$
$\alpha = 1, 0.5$ $\beta = 2$ $\Delta = 0, 1$ $K = 4$

State[b]		Energy expression	Calc. E. eV[c,d]	Exp. E. eV[c]	Remark
	G.S. and IPES, v				
Ψ_a)	d	$\varepsilon_d - \delta_1 \mp \Gamma$	0 ∓ 2	–	heavily
Ψ_b)	p	$\varepsilon_p + \delta_1 \mp \Gamma$	4 ∓ 2	–	mixed
	UPS and XES, v²				
1)[e]	pp^p	$\varepsilon_p + \Delta - \delta_2 + \alpha$	2.5	2.5	heavily
2)[e]	dp	$\varepsilon_p + U_{dp} + \delta_2 + \alpha$	4.5	4.2	mixed
3)	pp$^o{}_a$	$\varepsilon_p + \Delta + U_{pp}^o - \Gamma + \alpha$	5.5	5.	
4)	pp$^o{}_b$	$\varepsilon_p + \Delta + U_{pp}^o + \Gamma + \alpha$	9.5	9.5	mystery peak
5)	d²	$\varepsilon_d + U_d + \alpha$	12.5	12.5	Cu sat.
6)	p²	$\varepsilon_p + \Delta + U_p + \alpha$	15	16	
	Cu 2p XPS, cv				
d →	cp	$\varepsilon_c + \Delta + \alpha$	$\varepsilon_c + 1$	E_{2p}	main
	cd	$\varepsilon_c + Q_d + \alpha$	$\varepsilon_c + 10$	$E_{2p} + 9.2$	sat.
	Cu 2p XPS for NaCuO₂, pp^p → cv²				
pp^p→	cpp^p	$\varepsilon_c + \delta_2 + \beta$	$\varepsilon_c + 2.5$	$\varepsilon_c + 2.2$	main
	cpp$^o{}_a$	$\varepsilon_c + U_{pp}^o - \Gamma + \delta_2 + \beta$	$\varepsilon_c + 4.5$	$\varepsilon_c + 5$	?
	cpp$^o{}_b$	$\varepsilon_c + U_{pp}^o + \Gamma + \delta_2 + \beta$	$\varepsilon_c + 8.5$	$\varepsilon_c + 9$	?
	cdp	$\varepsilon_c - \Delta + Q_d + U_{dp} + \delta_2 + \beta$	$\varepsilon_c + 11.5$	$\varepsilon_c + 11$	sat.
	cp²	$\varepsilon_c + U_p + \delta_2 + \beta$	$\varepsilon_c + 15.5$	$\varepsilon_c + 14$	sat.?
	O 1s XPS, cv				
d →	cd	$\varepsilon_c + \alpha$	$\varepsilon_c + 1$	E_{1s}	main
	cp^p	$\varepsilon_c + \Delta + \alpha$	$\varepsilon_c + 1$	E_{1s}	main
	cp^o	$\varepsilon_c + \Delta + U_{cp}^o + \alpha$	$\varepsilon_c + 3$	$E_{1s} + 2$?	tail
	cp	$\varepsilon_c + \Delta + Q_p + \alpha$	?	?	not obs
pp^p→	cdp^p	$\varepsilon_c - \Delta + U_{dp} + \delta_2 + \beta$	$\varepsilon_c + 3.5$	$E_{1s} + 2$?	tail
	Cu L₃VV AES, v³				
	dpp^p	$2\varepsilon_p + 2U_{dp} + \alpha$	7	7	2 cent.
	dpp^o	$2\varepsilon_p + U_{pp}^o + 2U_{dp} + \alpha$	11.5	–	no mix
	d²p	$\varepsilon_d + \varepsilon_p + U_d + 2U_{dp} - \delta_2 + \alpha$	16	15.5	main
	dp²	$2\varepsilon_p + U_p + 2U_{dp} + \delta_2 + \alpha$	19.5	18-25	sat.

TABLE 1 (cont.)

State[b]	Energy expression	Calc. E. eV[c,d]	Exp. E. eV[c]	Remark
Cu $L_3M_{23}V$ AES, cv^2				
cdp	$\varepsilon_c + \varepsilon_p + Q_d + U_{dp} \mp K + \alpha$	$\varepsilon_c +9$	$E_{3p}+10$	main, ^1L
		$\varepsilon_c +17$	$E_{3p}+18$	main, ^3L
cp^2	$\varepsilon_c + \varepsilon_p + \Delta + U_p + \alpha$	$\varepsilon_c +15$	-	not
cd^2	$\varepsilon_c \infty \varepsilon_d + U_d + 2Q_d + \alpha$	$\varepsilon_c +30.5$	-	obs.
Cu L_{23} EELS, c				
d → c	$\varepsilon_c - \varepsilon_d + \delta_1$	$E_{2p}-1$	$E_{2p}-1.4$	edge
cpCB	$\varepsilon_c + \Delta - CB + \alpha$	$E_{2p}-CB$	$E_{2p}+1.2$	upper
pp^p → cp	$\varepsilon_c - \varepsilon_p + \delta_2 + \beta$	$E_{2p}-0.5$	E_{2p}	middle
O K EELS, c				
d → c	$\varepsilon_c - \varepsilon_d + \delta_1$	$E_{1s}-1$	E_{1s}	edge
cdCB	$\varepsilon_c - CB + \alpha$	$E_{2s}-CB$	$E_{1s}+1.7$	upper
pp^p → cd	$\varepsilon_c - \Delta - \varepsilon_p + \delta_2 + \beta$	$E_{1s}-0.5$	-	not obs

[a]Parameters for 123 indicated first, those for CuO second.
[b]The dominant character in the hybridized states is given.
[c]The Calc. E and Exp. E columns indicate the results for 123, except for the "Cu 2p XPS, pp^p → cv^2" section, which is for $NaCuO_2$.
[d]The calculated E is defined relative to the ground v^1 (d) state energy = $\varepsilon_d - \alpha$, or to the v^2 (pp^p) ground state energy = $2\varepsilon_p - \delta_2 - \beta$. The v^1(d) energy defines the Fermi level relative to the vacuum level at zero.
[e]The dominant character switches as described in the text, and thus the sign in front of δ_2 is the opposite for CuO.

parameters in Table 1 were obtained by considering this same data plus XANES, Auger and XES data. Although we are in general agreement with the reported magnitudes for most of the parameters, our U_d value is larger by about 2-3 eV so that it is consistent with the AES data. In Table 1, we indicate the location of two valence holes by d (Cu 3d) or p (O 2p). In the case of two holes on the oxygens, we distinguish two holes on the same O (p^2), on ortho neighboring O atoms (pp^o), or on para O atoms (pp^p) of the cluster. Furthermore, neighboring pp^o holes can dimerize[5], so we distinguish between two holes in bonded (pp^o_b) and antibonded (pp^o_a) O pairs.

The magnitudes of the U parameters are critical to the mechanism for the superconductivity. As a consequence, much effort has also gone into theoretically calculating these parameters, but wide disagreement still exists over the magnitudes. Theoretical values for U_d in the range 6.5-10 eV, U_p (actually $U_p-U_{pp^o}$) in the range 7-14 eV, and U_{dp} in the range 0.6-1.6 eV have been reported[6], with the smaller results favored based on the quality of the calculations. No results for U_{pp^o} have been reported. Our empirical results indicate that U_p = 12 and U_{pp^o} = 4.5 eV for 123. These are much larger than previously thought for these metallic systems, although U_p-U_{pp^o} is in agreement with the best theoretical results above.

An upper estimate of the two-center pp^o hole-hole repulsion, U_{pp^o}, can be obtained from the Klopman approximation[7],

$$U_{ij} = e^2/(r_{ij}^2 + (2e^2/\{U_i + U_j\})^2)^{1/2}, \qquad (1)$$

where r_{ij} is the interatomic distance and U_i and U_j are the corresponding intra-atomic repulsion energies. Equation 1 gives a value for U_{pp^o} around 4.8 eV assuming r_{o-o} is 2.7 A°. The experimental energies of 9.5 and 5.0 eV for the $pp^o{}_b$ and $pp^o{}_a$ features in 123 suggests that the pp^o final state energy is 7.2 eV. This gives an empirical estimate for U_{pp^o} of 4.2 eV, very close to the Klopman theoretical result, which does not include the effects of interatomic screening.

The above result shows that metallic screening of two holes, which are spatially separated on neighboring O atoms, is not very significant. This is in contrast to two Cu-O holes, where Table 1 indicates the optimal U_{dp} = 1 eV, while eq. 1 estimates U_{dp} at 6.1 eV assuming r_{Cu-O} is 1.9 A°. This large reduction in U_{dp} may result from charge transfer into the Cu 4sp levels to screen the Cu-O holes. Although metallic screening, which results from virtual electron-hole (e-p) pair excitations at the Fermi level, is not expected to be large in an insulator such as CuO, screening effects are expected to be much larger in metals, such as the HTSC's. The above results show that U_{dp} is significantly reduced in both, and U_{pp^o} remains large in both. The lack of a significant change in the U's between CuO and the HTSC's indicates that the DOS at the Fermi level in the HTSC's must be very small.

Table 1 also correlates the calculated energies of the excited states with features in the experimental data. The $CuO_n{}^{(2n-2)}$-cluster has one hole shared between the Cu 3d and O 2p shells in the ground state, which we term the v (valence) states. The spectroscopic final states reflect multi-hole states, e.g. v^2, cv (c = core) etc. The v states, as reflected by the theoretical DOS[6], have the Cu-O bonding (Ψ_b) and antibonding (Ψ_a) orbitals centered at 4 and 0 eV, respectively, and the nonbonding Cu and O orbitals at 2 eV. The O features each have a width 2Γ = 4 eV due to the O-O bonding and antibonding character and the Cu-O dispersion. We also define the Cu-O hybridization shift $\delta_1 = [(\Delta^2+4t^2)^{1/2} - \Delta]/2$, which is utilized in Table 1 to give the energies. Thus, the ground state of an average CuO_n cluster is located at 1 eV having the energy $\varepsilon_d - \delta_1 + \Gamma/2 = \varepsilon_d - \alpha$, which we use as a reference energy for the excited states. In CuO, the hybridization shift Γ is smaller, and we shall see below that $\Delta = \varepsilon_p - \varepsilon_d$ has increased to 1 eV.

Those clusters containing additional charge carrier holes (these exists in doped La, and 123 when x > 0.5) actually have two holes per CuO_n cluster. The average v^2 ground state, which is dominated by the pp^p configuration, i.e. the charge carrier holes spend most of their time on the O atoms, so we indicate this ground state by the notation pp^p. We use $2\varepsilon_p - \delta_2 - \beta$ as the energy of the pp^p ground state relative to the vacuum level, where β = 2 eV is the energy shift between the principal pp^p UPS final state at 2.5 eV and the lowest ground pp^p states around 0.5 eV from the Fermi level.

The correlation between the calculated energies and experimental features, utilizing the indicated optimal Hubbard parameters is very good. Details of this work are published elsewhere[9]. Figs. 1 and 2 show examples of the UPS and Cu AES data for 123 and CuO$_2$, which reveal some of the features itemized in Table 1; the remaining data are published elsewhere.

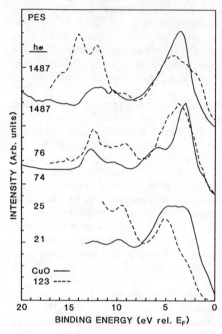

Figure 1. Comparison of UPS spectra for CuO and 123 taken with the indicated photon energies in eV. Data for CuO from refs. 10 (hν = 1487), 11 (hν = 74) and 12 (hν = 21). Data for 123 from ref. 13 (hν = 25 and 74) and 14 (hν = 1487).

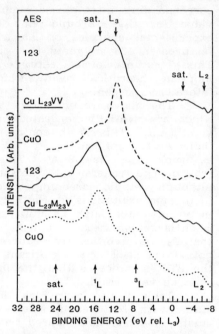

Figure 2. Comparison of Auger data for the materials indicated. L$_{23}$VV data for CuO and 123 from ref. 15. L$_{23}$M$_{23}$V data for CuO from ref. 16 and for 123 from Ref. 9. The L$_{23}$VV data is on a 2-hole binding energy scale = $E_{L3} - E_{k,,}$, and the L$_{23}$M$_{23}$V on a 1-hole scale = $E_{L3} - E_k - E_{M3}$, where E_{L3} = 933.4 and E_{M3} = 77.3 eV[11,12].

SUMMARY AND CONCLUSIONS

This work has allowed us to assign some previously unassigned features in the data, and greatly increased our understanding of the dynamical electronic processes which produce these features. We itemize our conclusions as follows:

1) A switch in the character of state 1 (see Table 1) from more dp to pp^p and vice versa for state 2 between CuO and 123 arises because Δ decreases from 1 eV to 0 eV. The smaller Δ in 123, due to a smaller ε_p polarization energy, is consistent with the Cu 2p XPS and XES data (the latter showing this effect dramatically)[17,18]. Since state 1 is more of pp^p character in the SC's, the "charge carrier holes" (present in the La after Sr doping and in the 123 when 7-x is greater than 6.5) spend more time on the oxygens in 123 than in CuO.

2) The pp^o$_b$ state is believed to be responsible for the "mystery" peak found at 9.5 eV in the UPS. Figure 1 indicates that such a feature also appears for CuO[11,12]. This feature does not appear for Cu$_2$O, as expected since UPS reflects the one-hole DOS in Cu$_2$O. Thus this feature is not unique to the SC's; it naturally appears for those systems with two-hole photoemission final states.

3) Although Cu$_2$O, CuO, and NaCuO$_2$ have a formal Cu valence of +1, +2, and +3, in the current picture they reflect the cv^n DOS, with n=0, 1, and 2. Furthermore, we consistently predict the "chemical shifts" in the primary Cu 2p XPS peaks. Whereas, Cu$_2$O exhibits just a primary core hole c state at energy ε_c, CuO has its primary cd feature energy shifted by $\Delta+\alpha$ relative to ε_c, and the primary cpp^p feature for NaCuO$_2$ by $\delta_2+\beta$ (Table 1), which is consistent with the experimental data[17]. The width of the primary feature is seen to correlate with the intensity of the satellite, and is not due to the O p band width as suggested by others[18].

4) The increased "satellite" feature at 19 eV in the Cu L$_{23}$VV Auger line shape for the HTSC's compared with CuO[15,16] (see Fig. 2) arises because of increased final-state configuration mixing between the d^{2}p and dp^2 states. Its intensity is increased in 123 relative to CuO because the energy separation (before hybridization) between d^{2}p and dp^2 has decreased from 3.8 eV in CuO to 2.5 eV in 123. We have indicated this mixing in Table 1 by adding the hybridization shifts δ_2 to the energy expressions for these two states.

5) We find that the initial-core shakeup (ICSU) process, which is known to be responsible for the satellite features in the Cu 2p XPS data[1], does not produce satellites in the AES or XES data, because the ICSU states generally "relax" to the primary states of the same symmetry before the core level decay. Such a relaxation is expected when the ICSU excitation energy is larger than the core level width[19]. Previously, vanderLaan et al.[1], for the Cu halides, suggested that the intensity of these ICSU states in the XPS should be quantitatively reflected in the intensity of the Auger satellites found in the L$_{23}$VV lineshapes. The data do not indicate this however. We previously[15] indicated that a fraction of these ICSU states probably resulted in Auger satellites for the HTSC's, and that this fraction becomes larger as the the covalency of the HTSC material increases. This work indicates rather that the ICSU states relax before the core level decay to states of the same symmetry, provided they have a ICSU excitation energy that is much greater

than the core level width. We believe this to be a general result, at least in the Cu^{+2} materials.

6) The EELS and XANES data[20,21] reflect the contributions from three possible transitions; the dominant d → c contribution nearest the Fermi level, the pp^p → cv (v = d or p) contribution resulting from the carrier hole states, as well as the cvCB contribution well above the Fermi level[22]. Here CB represents an electron present in the higher Cu 4sp or O 3p "conduction band". The latter two contributions are not always resolved, and sometimes have been confused in the literature[20-22].

7) All of the temperature effects seen in the spectroscopic data[23-26] can be attributed to a single phenomenon, namely a decrease in ε_p due to increased metallic screening, or long range polarization. This is consistent with the decrease in the primary cp peak energy in the Cu 2p XPS, while the cd satellite remains unshifted. The larger energy separation between the cd and cp states decreases the mixing which causes the satellite to decrease in intensity and the main peak to get narrower. Although the primary cd peak does not shift in the O 1s XPS, a slight shift to lower energy is seen in the cp^p and cp^o contributions at lower temperature, as expected with a decrease in ε_p. The UPS spectra show a skewing toward the Fermi level at lower temperature, as expected with a decrease in ε_p. Finally the growth of the satellite intensity in the Cu L$_{23}$VV Auger lineshape is consistent with a decrease in e$_p$. The increased metallic like screening or polarization which appears to occur at lower temperature, reducing ε_p, probably involves the grain boundaries, since the more recent data for the single crystal samples do not change with temperature[27].

In summary, an interpretation of the data utilizing a highly correlated CuO$_n$ cluster model shows that a single set of Hubbard parameters predicts all of the state energies. Changes in the data between CuO and the HTSC's arises primarily from a reduction in ε_p; this reduction continues with decreasing temperature in the HTSC's due to increased metallic screening. Compared with CuO, the HTSC's show an increased covalent interaction between the Cu-O bonds. The large size of U$_{pp}$^{o}, and the temperature dependence, reveal that metallic screening is incomplete, and hence that the DOS at the Fermi level in the HTSC's is relatively small.

REFERENCES

1. G. vanderLaan et al., Phys. Rev. **24**, 4369 (1981); G.A. Sawatzky et al., Phys. Rev. Lett. **53**, 2339 (1985); J. Zaanen et al., Phys. Rev. **B33**, 8060 (1986).
2. Z. Shen et al., Phys. Rev. **B36**, 8414 (1987).
3. A. Fujimori et al., Phys. Rev. **B35**, 8814 (1987).
4. J.C. Fuggle et al., Phys. Rev. **B37**, 1123 (1988).
5. R.A. de Groot, H. Gutfreund, and M. Weger, Sol. State Commun. **63**, 451 (1987); W. Folkerts et al., J. Phys. C: Solid State Phys. **20**, 4135 (1987); A. Manthiram, X.X. Tang, and J.B. Goodenough, Phys. Rev. **B37**, 3734 (1988).

6. C.F. Chen et al., unpublished; A.K. McMahan, R.M. Martin, and S. Satpathy, unpublished; M. Schluter, M.S. Hybertsen, and N. E. Christensen, Proc. Intn. Conf. High T_c Superconductors and Materials and Mechanisms of Superconductivity, J. Muller and J.L. Olsen, Eds., (Interlaken, Switzerland, 1988).
7. G. Klopman, J. Am. Chem. Soc. 86, 4550 (1964).
8. J. Redinger et al., Phys. Lett. 124, 463 and 469 (1987).
9. D.E. Ramaker, N.H. Turner, and F.L. Hutson, submitted.
10. A. Rosencwaig and G.K. Wertheim, J. Elect. Spectrosc. Related Phenom. 1, 493 (1972/73).
11. M.R. Thuler, R.L. Benbow, and Z. Hurych, Phys. Rev. B26, 669 (1982).
12. C. Benndorf et al., J. Electron. Spectrosc. Related Phenom. 19, 77 (1980).
13. N.G. Stoffel et al., Phys. Rev. B37, 7952 (1988); B38, July (1988).
14. D.C. Miller et al., in Thin Film Processing and Characterization of High Temperature Superconductors, J.M. Harper, J.H. Colton, and L.C. Feldman, Eds., AVS Series No. 3 (AIP: New York, 1988) p 336.
15. D.E. Ramaker et al., Phys. Rev. 36, 5672 (1987).
16. P.E. Larson, J. Electron Spectrosc. Related Phenom. 4, 213 (1974).
17. P. Steiner et al., Z. Phys. B- Condensed Matter 67, 497 (1987).
18. D.D. Sarma, Phys. Rev. B37, 7948 (1988).
19. J.W. Gadzuk and M. Sunjic, Phys. Rev. B12, 524 (1975).
20. D.D. Sarma et al., Phys. Rev. B37, 9784 (1988).
21. N. Nucker et al., Z. Phys. B: Cond. Matter 67, 9 (1987); Phys. Rev. 37, 5158 (1988).
22. A. Bianconi et al., Solid State Commun. 63, 1009 (1987); Intn. J. Modern Phys. 131, 853 (1987).
23. N.S. Kohiki and T. Hamada, Phys. Rev. B36, 2290 (1987).
24. B. Dauth et al., Z. Phys. B- Condensed Matter 68, 407 (1987).
25. D.H. Kim et al., Phys. Rev. B37, 9745 (1988).
26. A. Balzarotti et al., Phys. Rev. B36, 8285 (1987).
27. J. Weaver and P. Steiner, private communication.

NOVEL APPLICATIONS OF HIGH TEMPERATURE SUPERCONDUCTIVITY: NEUROCOMPUTING

R. Singh, J.R. Cruz, F. Radpour, and M.J. Semnani
School of Electrical Engineering and Computer Science
The University of Oklahoma, Norman, OK, 73019

ABSTRACT

The newly discovered phenomena of high temperature superconductivity can be used for certain revolutionary applications. One such application involves neurocomputing. Based on a digital architecture, high performance, neural systems can be developed. The heart of the proposed signal processor is an ultra fast multiplexed multiplier accumulator (MMA). In addition to conventional known devices, a new three terminal hybrid superconductor/semiconductor resonant tunnel transistor is the key building block of the MMA. From the materials and processing point of view, the ideal fabrication scheme involves the deposition of epitaxial dielectrics, epitaxial superconductors and epitaxial semiconductors on Si wafers. Metal organic chemical vapor deposition based on rapid isothermal processing is ideally suited for the deposition of superconductor and semiconductor layers.

I. INTRODUCTION

The recent discovery of high temperature superconductivity has received world wide attention within academia, government and industry. This is primarily due to the possibility of 77K and higher temperature applications of superconductivity. The nature of these applications varies from evolutionary to revolutionary. One such novel application of superconducting electronics involves neurocomputers, which are essentially non–programmable adaptive information processing systems. Our main interest is in implementing proper neural networks for digital signal processing. The ultra high speed feature of superconducting electronics can be exploited for the realization of neurocomputing systems. In this paper, we present our preliminary results on the framework of neural networks based on high temperature superconducting electronics.

The next section deals with the question of evolutionary and revolutionary applications. The following section addresses the issue of digital versus analog neural networks. Proposed digital architecture is briefly described in section IV. In section V, we have presented the essential features of proposed circuits and devices. Performance evaluation of the proposed superconducting systems in described in section VI. Materials and processing related issues are discussed in section VII. Finally, the paper is concluded in section VIII.

© 1989 American Institute of Physics

II. REVOLUTIONARY VERSUS EVOLUTIONARY APPLICATIONS

The discovery of high temperature superconductivity has given new life to this 77 year old quiet field. If certain potential applications can be put into practice, the discovery of high temperature superconductivity may bring about the next industrial revolution. In order to formulate a possible strategy for new product development, we have used a most efficient method of advancing a technology: analyze the state of the art, and proposed new schemes which build on any strengths and/or remove any limitations that are revealed by the analysis. This approach is especially necessary in a field such as electronics where the immense body of research and practice is simply too great to ignore.

In order to illustrate our point of view, we have selected a particular application of high temperature superconducting electronics namely advanced signal processors. As shown in Fig. 1, the proposed superconducting signal processor is essentially an evolutionary modification of the current technology. Due to marginal improvement over the performance of the existing product, this kind of approach has low chances of market penetration. Fig. 2 shows the revolutionary approach where in addition to the processing issues, device and circuit details, the question of appropriate architecture is brought into the picture at an early stage of product development. In this approach, we have higher probability of market penetration, since the performance of the product will be far superior compared to other similar available in the market. In order to implement, a neural network, we have to decide whether to use a digital or an analog approach. This is discussed in the following section.

III. DIGITAL VERSUS ANALOG APPROACH

Neurocomputing deals with non-programmed adaptive information processing systems (neural network). A neural network consists of a collection of computational units (neurons) that models some of functionality of the human nervous systems and attempts to capture some of its computational strength[1]. Although, most of the published literature on neurcomputing deals with analog implementations, recent work[2] shows that future implementations will move towards digital based designs. Both digital and analog neural networks have their own strength and weakness[1]. However, based on the existing knowledge of digital electronics, coupled with the ultra high speed features of superconducting electronics, we believe that digital neural networks can be successfully used to perform signal processing tasks beyond the reach of conventional sequential machines.

IV. DIGITAL ARCHITECTURE

As stated above, in this work we have selected a digital approach for implementing neural networks. The generic structure of each computing cell is shown in Fig. 3. Our proposed architecture, is similar to the structure reported in Ref. 1. Regardless of the selection of updating scheme, the key mathematical expression in defining both the dynamic and static characteristics of the neural system is the weighted summation S such that:

$$S = \sum_{j=0}^{n-1} T_{ij} V_j \qquad (1)$$

In the above expression, n is the number of neurons in the neural network, V_j is the state of neuron j, and T_{ij} is the synaptic weight of the neuron i related to input V_j.

As shown in Fig. 4, two approaches can be used to calculate the sum of products shown in equation (1). In the first approach shown in Fig. 4(A), a large number of bit-serial pipelined multipliers are used. In the second approach, an ultra high speed multiplier accumulator in a multiplexed mode is used. Due to the absence of interchip communications, as well as speed advantages we have selected the second approach, in our work. The multiplexed multiplier accumulator is the heart of the advanced signal processor. The other parts of the hardware such as supporting memory, etc. can be implemented in a conventional manner.

V. PROPOSED CIRCUIT AND DEVICES

The organization of a computing cell with the multiplexed multiplier accumulator (MMA) is shown in Fig. 3. The local microprocessor provides the necessary addressing for random access memory (RAM) in this organization. At the end of each solution cycle the microprocessor reads the contents of the output, and initiates the next solution cycle. The key element of the multiplier accumulator is an asynchronous multiplier array which carries out parallel multiplication and may be realized by combinational logic functions such as full and half-adders.

In the design of MMA we have used a new hybrid resonant tunneling transistor (RTT). The resonant tunneling transistor structure proposed previously[3,4], and shown in Fig 4(a) and 4(b), has several of the characteristics desirable for future superconducting electronics. Except for the use of a gate potential rather than biasing to modulate the resonant tunneling current, the structure and operation of this device are virtually identical to that of the simplest resonant tunneling (RT) device[5], the resonant tunneling diode. Numerical simulation of the RTD yield a propagation delay or switching time of about 100 fs[6]. Because of the similarity of the proposed RTT to the RTD, it should achieve a comparable switching time.

The superconductor/semiconductor contacts are inherently indirect in this device. A tunnel barrier separates the semiconductor layer from the superconductor interconnects. This can greatly ease processing and fabrication considerations. The semiconductor layer shown in Fig. 4(a) can be replaced by the superconductor, provided the new material has a low density of states. ($N < 1 \times 10^{18}/cm^3$). The proposed device because of its vertical structure, has no device area beyond the contacts. Thus, it would gain the maximum possible packing density benefit. Unlike Josephson Junctions, the proposed transistor will have adequate power gain.

VI. PERFORMANCE EVALUATION

Our proposed superconducting MMA will occupy much less area than a conventional MOSFET based MMA. This is due to the fact that the hybrid resonant tunneling transistor, because of its vertical structure; has no device area beyond its contacts. In order to estimate the speed advantage of our proposed neural network based digital signal processor, we have used an 8 x 8 bit parallel multiplier. The calculated propagation delay per gate of hybrid RTT is 10^{-13} sec[6]. Neglecting delay due to interconnections, which is justified due to the superconducting interconnections, the delay time of 3.8ps is estimated. For an 8 x 8 CMOS multiplier operating at 77K the delay time of 8ns is observed[7]. Thus compared to advanced CMOS technology, a speed advantage of roughly 2000 is estimated.

VII. MATERIALS AND PROCESSING ISSUES

Realization of neural networks for digital signal processing requires three basic building blocks, namely, (1) semiconductors, (b) superconductors, and (c) dielectrics. These materials should have the features of selectively deposition, epitaxial growth, submicron patterning and be processed at low temperatures. Additionally, atomic layer epitaxial deposition can provide improved interfaces as well as new advanced materials based on concept such as super strained lattices.

In our opinion, semiconductors, epitaxial dielectrics (e.g. II- A fluorides and their mixtures) and epitaxial superconductors should be deposited by metal-organic chemical vapor deposition method. If the appropriate organometallic precursors are not available (e.g. thermodynamical reasons) then the ultra high vacuum processing should be used. As a low thermal budget processing technique, rapid isothermal processing (RIP)[8] based on incoherent sources of light should be used. The processing systems should be modified to take advantage of in–situ rapid isothermal processing. Thus, RIP MOCVD can emerge as a major processing technique for the deposition of advanced semiconductor and superconductor materials. Advanced ion beam lithography should be used for selective material removal, doping and if needed for the deposition of the material. Any damage to the surface will be annealed by an in–situ rapid isothermal processor incorporated in the ion beam system.

VIII. CONCLUSION

In this paper, we have presented a novel application of high temperature superconducting electronics namely an advanced digital signal processor based on neural networks. The heart of the proposed signal processor is an ultra fast multiplexed multiplier accumulator (MMA). In addition to conventional known devices, a new three terminal hybrid superconductor/semiconductor tunnel transistor is the building block of MMA. Proposed digital neural networks can successfully be used to perform signal processing tasks beyond the reach of conventional sequential machines. From the materials point of view, deposition of epitaxial materials (semiconductor, dielectric, and superconductor) on Si substrate is an ideal choice. Rapid isothermal processing based metal organic chemical vapor deposition (MOCVD) system can emerge as the most suitable technique for the deposition of superconductors and semiconductors. Advanced ion beam lithography should be explored for selective material removal, doping and deposition of material.

ACKNOWLEDGEMENT

The authors greatfully acknowledge several useful discussions with Dr. Fernand Bedard of National Security Agency. Part of this work is supported by Defense Advanced Research Project Agency (DARPA) under contract No. MDA 972-88-K-0004.

REFERENCES

1. A.F. Murray and A.V.W. Smith, IEEE Journal of Solid State Circuits, 235, 688 (1988).
2. A. Wilson, The Electronic System Design Magazine, 18(7), 29 (1988).
3. R. Singh and B.A. Biegel in Thin Film Processing and Characterization of HIgh Temperature Superconductors, edited by J.M.E. Harper, R.J. Coulton, and Fieldman, AVS Proc. Vol. 165, p. 211 (1988).
4. B.A. Biegel, R. Singh, and F. Radpour, in High-T_c Superconductivity: Thin Films and Devices, edited by R.B. Van Dover, C.C. Chi, SPIE, Proc. Vol. 948, p. 3 (1988).
5. F. Capasso, K. Mohammad, and A.Y. Cho, IEEE J. Quantum Electron., QE 22, 1853 (1986).
6. B.A. Biegel, M.S. Thesis, University of Oklahoma, 1988.
7. S. Hanamura, IEEE Trans. on Electron Devices, 34, 94 (1987).
8. R. Singh, J. App. Phys., 63, R59 (1988).

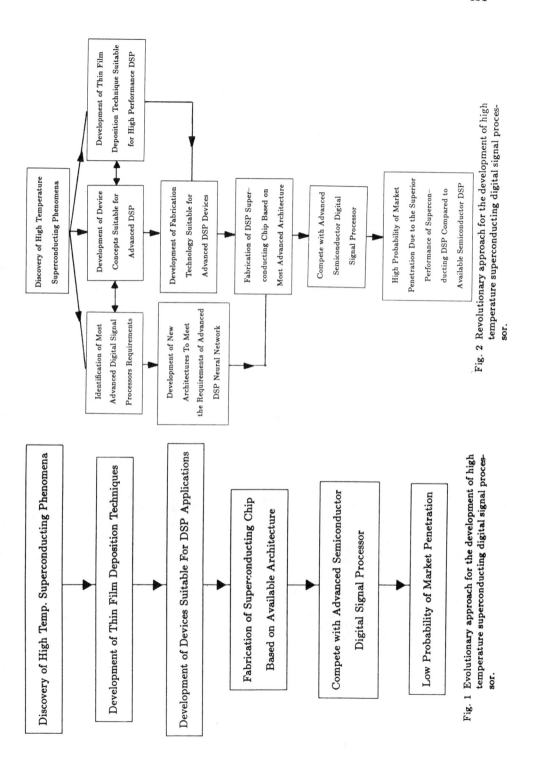

Fig. 1 Evolutionary approach for the development of high temperature superconducting digital signal processor.

Fig. 2 Revolutionary approach for the development of high temperature superconducting digital signal processor.

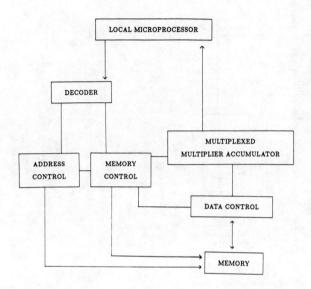

Fig. 3 Generic structure of a computing cell of a neural network.

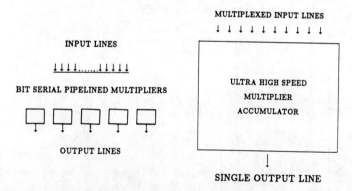

Fig. 4 Two different approaches for the design of the multiplier case (A) represents bit-serial pipelined multiplier and (B) represents multiplier accumulaor in a multiplexed mode.

Author Index

A

Aarnink, W., 208
Allen, J. W., 330
Allen, Roland E., 410
Ameens, M. S., 61
Apfelstedt, I., 208
Archuleta, T., 45
Arendt, P. N., 45
Arie, Y., 2
Arko, A. J. 283
Asano, T., 330
Aspnes, D. E., 318
Auciello, O., 61
Azamar, J. A., 53

B

Bacon, D. D., 107,122
Barr, T. L., 216
Bartlett, R. J., 283
Beasley, M. R., 163
Berghuis, P., 172
Bhasin, Kul B., 147
Bonner, W. A., 248, 252, 257
Brundle, C. R., 216
Budhani, R. C., 155
Byrne, A. S., 352

C

Campuzano, J. C., 283
Caracciolo, R., 232
Caudano, R., 240
Chandrashekhar, G. V., 289
Chang, Chin-An, 384
Chang, Y., 248, 252, 257
Chen, C. H., 107, 360
Chen, D. H., 74, 368
Chen, J., 368
Chen, L. M., 216
Cheong, S.-W., 283
Chung, D. W., 8
Clarke, Peter J., 16
Clemens, B. M., 82
Cline, James P., 53
Conard, T., 240
Conradson, S. D., 283
Crow, J. E., 368
Cruz, J. R., 426
Cuomo, J. J., 115

D

Dam, B., 172
Davis, Matthew F., 140
Del Castillo, L., 53
Deline, V., 8
Dessau, D. S., 391
Dew, S. K., 99
Dhere, N. G., 26
Dhere, R. G., 26
Di Marzio, D., 74, 360
Dimos, D., 194
Doyle, J. P., 115

E

Elliot, N. E., 45
Eom, C. B., 312

F

Farrell, C. E., 384
Felder, R. J., 122
Fisanick, G., 180
Fisk, Z., 283
Flipse, C. F. J., 172
Foote, M. C., 376

G

Gardner, Bill, 16
Geballe, T. H., 312
Ginley, D. S., 262
Goldman, A. M., 33
Graettinger, T. M., 61
Gray, K. E., 269
Griessen, R. P., 172
Guo, S. Q., 172

H

Halbritter, J., 208
Hammond, R. H., 163
Heald, S. M., 74, 360
Hedge, M. S., 232
Hill, D. M., 399

Himpsel, F. J., 289
Holloway, P. H., 65
Hong, M., 107, 122
Hu, Y., 399
Huang, T. C., 8
Huber, D. L., 248, 252
Hunt, B. D., 376
Hunter, A. T., 82

I

Ignatiev, Alex, 297
Inam, A., 232

J

Jack, W. D., 82
Josefowicz, J. Y., 82
Joynt, Robert, 248, 252

K

Kampwirth, R. T.,
Kang, J. H., 269
Kapitulnik, A., 312, 391
Karimi, R., 8
Kazmerski, L. L., 269
Kelly, M. K., 318
Kerber, G., 82
Kim, D. H., 33
Kim, Y. H., 33
Kimura, H., 82
Kingon, A. I., 61
Klumb, A., 216
Kortan, A. R., 107
Kreider, Kenneth G., 53
Kubo, K., 130
Kurtz, Richard L., 276
Kvaas, R. E., 82
Kwo, J., 107, 122

L

Lee, V. Y., 8
Lee. W. Y., 8
Lensink, J. G., 172
Leskela, M., 65
Lichtenberg, Christopher L., 140
Lin, C. L., 368

Lindau, I., 312, 330, 391
Lindberg, P. A. P., 312, 330, 391
Liou, S. H., 107
List, R. S., 283
Liu, Y. L. 222, 216
Lu, C., 163

M

Maldonado, L., 53
Manini, P., 90
Mannhart, J., 194
Margaritondo, G., 248, 252, 257
Matey, J. R., 2
Mathes, H.-J., 208
Matijasevic, V., 163
McLean, A. B., 289
Meyer, H. M., 399
Miller, A., 122
Missert, N., 163
Mitzi, D. B., 312, 391
Miura, T., 130
Mizushima, K., 130
Moodenbaugh, A. R., 360, 368
Mooij, J. E., 163
Moreland, J., 26
Morosin, B., 262
Morris, P. A., 248, 252, 257
Mueller, C. H., 65
Mueller, D., 276
Muenchausen, R. E., 45
Mulhern, P. J., 99
Murray, P. Terence, 297
Myroszynyk, J. M., 82

N

Nassau, K., 122
Nastasi, M., 45
Nelson, A. J., 269
Nichols, D., 368
Nigro, A., 90

O

O'Rourke, J. A., 283
Olson, C. G., 283
Onellion, M., 248, 252, 257
Osborne, N. R., 99
Osofsky, M., 276

P

Parsons, R. R., 99
Paulikas, A. P., 283
Pena, J. L., 53
Peterson, D. E., 283
Pi, T.-W., 283
Pierce, C. B., 283
Pireaux, J. J., 240

Q

Qui, S. L., 74, 360, 368

R

Radpour, F., 426
Rahman, M. M., 352
Ramaker, David E., 418
Rensch, D. B., 82
Richert, Brent A., 410
Robey, Steven, W., 276
Rogers, Jr., J. W., 262
Rohrer, Norman, J., 147
Rojas, A., 53
Romano, P., 90
Rosenberg, R. A., 304
Rosenthal, P., 163
Rou, S. H., 61
Roy, R. A., 115
Ruckman, M. W., 155, 360, 368

S

Sabatini, R. L., 74, 155
Sagoi, M., 130
Salem, J., 8
Savoy, R., 8
Schirber, J. E., 262, 283
Schmidt, Howard K., 297
Schultz, J. Albert, 297
Semnani, M. J., 426
Shafer, M. W., 289
Shapiro, Alexander, 53
Shen, Z.-X., 312, 330, 391
Shih, A., 276
Shinn, N. D., 262, 283
Singh, A. K., 276
Singh, R., 426
Soblel, C., 61
Soukiassian, P., 312
Spargo, J., 82

Spicer, W. E., 312, 330, 391
Stickle, W. F., 352
Stockbauer, Roger L., 276
Stoffel, N. G., 248, 252, 257
Stollman, G. M., 172
Strongin, Myron, 358
Sullivan, B. T., 99
Swartzlander, A., 269

T

Taleb-Ibrahimi, A., 289
Tang, Ming, 240
Tarascon, J. M., 248, 252, 257
Terashima, Y., 130
Thompson, J. D., 283
Toth, L., 276
Truman, J. K., 65
Tsai, J. A., 384
Tsuei, C. C., 194

V

Vaglio, R., 90
Valco, George J., 147
Vasquez, R. P., 376
Veal, B. W., 283
Venketesan, T., 232
Venturini, E. L., 262
Vohs, J. M., 240

W

Wachtman, Jr., J. B., 232
Wagener, T. J., 399
Walk, P., 208
Warner, Joseph D., 147
Weaver, J. H., 399
Wehner, G. K., 33
Wells, B. O., 312, 391
Wen, C.-R., 304
Wiener-Avnir, E., 82
Wiesmann, H., 74, 155
Wilson, J. A., 82
Wolfe, J. C., 140
Wosik, Jaroslaw, 140

X

Xu, Youwen, 368

Y

Yang, A.-B., 283
Yang, C. Y., 352, 360
Yee, D. S., 115
Yeh, J.-J., 122

Yin, M. P., 216
Yoshida, J., 130

Z

Zanoni, R., 248, 252, 257

AIP Conference Proceedings

		L.C. Number	ISBN
No. 1	Feedback and Dynamic Control of Plasmas – 1970	70-141596	0-88318-100-2
No. 2	Particles and Fields – 1971 (Rochester)	71-184662	0-88318-101-0
No. 3	Thermal Expansion – 1971 (Corning)	72-76970	0-88318-102-9
No. 4	Superconductivity in d- and f-Band Metals (Rochester, 1971)	74-18879	0-88318-103-7
No. 5	Magnetism and Magnetic Materials – 1971 (2 parts) (Chicago)	59-2468	0-88318-104-5
No. 6	Particle Physics (Irvine, 1971)	72-81239	0-88318-105-3
No. 7	Exploring the History of Nuclear Physics – 1972	72-81883	0-88318-106-1
No. 8	Experimental Meson Spectroscopy –1972	72-88226	0-88318-107-X
No. 9	Cyclotrons – 1972 (Vancouver)	72-92798	0-88318-108-8
No. 10	Magnetism and Magnetic Materials – 1972	72-623469	0-88318-109-6
No. 11	Transport Phenomena – 1973 (Brown University Conference)	73-80682	0-88318-110-X
No. 12	Experiments on High Energy Particle Collisions – 1973 (Vanderbilt Conference)	73-81705	0-88318-111-8
No. 13	π-π Scattering – 1973 (Tallahassee Conference)	73-81704	0-88318-112-6
No. 14	Particles and Fields – 1973 (APS/DPF Berkeley)	73-91923	0-88318-113-4
No. 15	High Energy Collisions – 1973 (Stony Brook)	73-92324	0-88318-114-2
No. 16	Causality and Physical Theories (Wayne State University, 1973)	73-93420	0-88318-115-0
No. 17	Thermal Expansion – 1973 (Lake of the Ozarks)	73-94415	0-88318-116-9
No. 18	Magnetism and Magnetic Materials – 1973 (2 parts) (Boston)	59-2468	0-88318-117-7
No. 19	Physics and the Energy Problem – 1974 (APS Chicago)	73-94416	0-88318-118-5
No. 20	Tetrahedrally Bonded Amorphous Semiconductors (Yorktown Heights, 1974)	74-80145	0-88318-119-3
No. 21	Experimental Meson Spectroscopy – 1974 (Boston)	74-82628	0-88318-120-7
No. 22	Neutrinos – 1974 (Philadelphia)	74-82413	0-88318-121-5
No. 23	Particles and Fields – 1974 (APS/DPF Williamsburg)	74-27575	0-88318-122-3
No. 24	Magnetism and Magnetic Materials – 1974 (20th Annual Conference, San Francisco)	75-2647	0-88318-123-1
No. 25	Efficient Use of Energy (The APS Studies on the Technical Aspects of the More Efficient Use of Energy)	75-18227	0-88318-124-X

No. 26	High-Energy Physics and Nuclear Structure – 1975 (Santa Fe and Los Alamos)	75-26411	0-88318-125-8
No. 27	Topics in Statistical Mechanics and Biophysics: A Memorial to Julius L. Jackson (Wayne State University, 1975)	75-36309	0-88318-126-6
No. 28	Physics and Our World: A Symposium in Honor of Victor F. Weisskopf (M.I.T., 1974)	76-7207	0-88318-127-4
No. 29	Magnetism and Magnetic Materials – 1975 (21st Annual Conference, Philadelphia)	76-10931	0-88318-128-2
No. 30	Particle Searches and Discoveries – 1976 (Vanderbilt Conference)	76-19949	0-88318-129-0
No. 31	Structure and Excitations of Amorphous Solids (Williamsburg, VA, 1976)	76-22279	0-88318-130-4
No. 32	Materials Technology – 1976 (APS New York Meeting)	76-27967	0-88318-131-2
No. 33	Meson-Nuclear Physics – 1976 (Carnegie-Mellon Conference)	76-26811	0-88318-132-0
No. 34	Magnetism and Magnetic Materials – 1976 (Joint MMM-Intermag Conference, Pittsburgh)	76-47106	0-88318-133-9
No. 35	High Energy Physics with Polarized Beams and Targets (Argonne, 1976)	76-50181	0-88318-134-7
No. 36	Momentum Wave Functions – 1976 (Indiana University)	77-82145	0-88318-135-5
No. 37	Weak Interaction Physics – 1977 (Indiana University)	77-83344	0-88318-136-3
No. 38	Workshop on New Directions in Mossbauer Spectroscopy (Argonne, 1977)	77-90635	0-88318-137-1
No. 39	Physics Careers, Employment and Education (Penn State, 1977)	77-94053	0-88318-138-X
No. 40	Electrical Transport and Optical Properties of Inhomogeneous Media (Ohio State University, 1977)	78-54319	0-88318-139-8
No. 41	Nucleon-Nucleon Interactions – 1977 (Vancouver)	78-54249	0-88318-140-1
No. 42	Higher Energy Polarized Proton Beams (Ann Arbor, 1977)	78-55682	0-88318-141-X
No. 43	Particles and Fields – 1977 (APS/DPF, Argonne)	78-55683	0-88318-142-8
No. 44	Future Trends in Superconductive Electronics (Charlottesville, 1978)	77-9240	0-88318-143-6
No. 45	New Results in High Energy Physics – 1978 (Vanderbilt Conference)	78-67196	0-88318-144-4
No. 46	Topics in Nonlinear Dynamics (La Jolla Institute)	78-57870	0-88318-145-2
No. 47	Clustering Aspects of Nuclear Structure and Nuclear Reactions (Winnepeg, 1978)	78-64942	0-88318-146-0
No. 48	Current Trends in the Theory of Fields (Tallahassee, 1978)	78-72948	0-88318-147-9

No. 49	Cosmic Rays and Particle Physics – 1978 (Bartol Conference)	79-50489	0-88318-148-7
No. 50	Laser-Solid Interactions and Laser Processing – 1978 (Boston)	79-51564	0-88318-149-5
No. 51	High Energy Physics with Polarized Beams and Polarized Targets (Argonne, 1978)	79-64565	0-88318-150-9
No. 52	Long-Distance Neutrino Detection – 1978 (C.L. Cowan Memorial Symposium)	79-52078	0-88318-151-7
No. 53	Modulated Structures – 1979 (Kailua Kona, Hawaii)	79-53846	0-88318-152-5
No. 54	Meson-Nuclear Physics – 1979 (Houston)	79-53978	0-88318-153-3
No. 55	Quantum Chromodynamics (La Jolla, 1978)	79-54969	0-88318-154-1
No. 56	Particle Acceleration Mechanisms in Astrophysics (La Jolla, 1979)	79-55844	0-88318-155-X
No. 57	Nonlinear Dynamics and the Beam-Beam Interaction (Brookhaven, 1979)	79-57341	0-88318-156-8
No. 58	Inhomogeneous Superconductors – 1979 (Berkeley Springs, W.V.)	79-57620	0-88318-157-6
No. 59	Particles and Fields – 1979 (APS/DPF Montreal)	80-66631	0-88318-158-4
No. 60	History of the ZGS (Argonne, 1979)	80-67694	0-88318-159-2
No. 61	Aspects of the Kinetics and Dynamics of Surface Reactions (La Jolla Institute, 1979)	80-68004	0-88318-160-6
No. 62	High Energy e^+e^- Interactions (Vanderbilt, 1980)	80-53377	0-88318-161-4
No. 63	Supernovae Spectra (La Jolla, 1980)	80-70019	0-88318-162-2
No. 64	Laboratory EXAFS Facilities – 1980 (Univ. of Washington)	80-70579	0-88318-163-0
No. 65	Optics in Four Dimensions – 1980 (ICO, Ensenada)	80-70771	0-88318-164-9
No. 66	Physics in the Automotive Industry – 1980 (APS/AAPT Topical Conference)	80-70987	0-88318-165-7
No. 67	Experimental Meson Spectroscopy – 1980 (Sixth International Conference, Brookhaven)	80-71123	0-88318-166-5
No. 68	High Energy Physics – 1980 (XX International Conference, Madison)	81-65032	0-88318-167-3
No. 69	Polarization Phenomena in Nuclear Physics – 1980 (Fifth International Symposium, Santa Fe)	81-65107	0-88318-168-1
No. 70	Chemistry and Physics of Coal Utilization – 1980 (APS, Morgantown)	81-65106	0-88318-169-X
No. 71	Group Theory and its Applications in Physics – 1980 (Latin American School of Physics, Mexico City)	81-66132	0-88318-170-3
No. 72	Weak Interactions as a Probe of Unification (Virginia Polytechnic Institute – 1980)	81-67184	0-88318-171-1
No. 73	Tetrahedrally Bonded Amorphous Semiconductors (Carefree, Arizona, 1981)	81-67419	0-88318-172-X

No. 74	Perturbative Quantum Chromodynamics (Tallahassee, 1981)	81-70372	0-88318-173-8
No. 75	Low Energy X-Ray Diagnostics – 1981 (Monterey)	81-69841	0-88318-174-6
No. 76	Nonlinear Properties of Internal Waves (La Jolla Institute, 1981)	81-71062	0-88318-175-4
No. 77	Gamma Ray Transients and Related Astrophysical Phenomena (La Jolla Institute, 1981)	81-71543	0-88318-176-2
No. 78	Shock Waves in Condensed Matter – 1981 (Menlo Park)	82-70014	0-88318-177-0
No. 79	Pion Production and Absorption in Nuclei – 1981 (Indiana University Cyclotron Facility)	82-70678	0-88318-178-9
No. 80	Polarized Proton Ion Sources (Ann Arbor, 1981)	82-71025	0-88318-179-7
No. 81	Particles and Fields –1981: Testing the Standard Model (APS/DPF, Santa Cruz)	82-71156	0-88318-180-0
No. 82	Interpretation of Climate and Photochemical Models, Ozone and Temperature Measurements (La Jolla Institute, 1981)	82-71345	0-88318-181-9
No. 83	The Galactic Center (Cal. Inst. of Tech., 1982)	82-71635	0-88318-182-7
No. 84	Physics in the Steel Industry (APS/AISI, Lehigh University, 1981)	82-72033	0-88318-183-5
No. 85	Proton-Antiproton Collider Physics –1981 (Madison, Wisconsin)	82-72141	0-88318-184-3
No. 86	Momentum Wave Functions – 1982 (Adelaide, Australia)	82-72375	0-88318-185-1
No. 87	Physics of High Energy Particle Accelerators (Fermilab Summer School, 1981)	82-72421	0-88318-186-X
No. 88	Mathematical Methods in Hydrodynamics and Integrability in Dynamical Systems (La Jolla Institute, 1981)	82-72462	0-88318-187-8
No. 89	Neutron Scattering – 1981 (Argonne National Laboratory)	82-73094	0-88318-188-6
No. 90	Laser Techniques for Extreme Ultraviolt Spectroscopy (Boulder, 1982)	82-73205	0-88318-189-4
No. 91	Laser Acceleration of Particles (Los Alamos, 1982)	82-73361	0-88318-190-8
No. 92	The State of Particle Accelerators and High Energy Physics (Fermilab, 1981)	82-73861	0-88318-191-6
No. 93	Novel Results in Particle Physics (Vanderbilt, 1982)	82-73954	0-88318-192-4
No. 94	X-Ray and Atomic Inner-Shell Physics – 1982 (International Conference, U. of Oregon)	82-74075	0-88318-193-2
No. 95	High Energy Spin Physics – 1982 (Brookhaven National Laboratory)	83-70154	0-88318-194-0
No. 96	Science Underground (Los Alamos, 1982)	83-70377	0-88318-195-9

No. 97	The Interaction Between Medium Energy Nucleons in Nuclei – 1982 (Indiana University)	83-70649	0-88318-196-7
No. 98	Particles and Fields – 1982 (APS/DPF University of Maryland)	83-70807	0-88318-197-5
No. 99	Neutrino Mass and Gauge Structure of Weak Interactions (Telemark, 1982)	83-71072	0-88318-198-3
No. 100	Excimer Lasers – 1983 (OSA, Lake Tahoe, Nevada)	83-71437	0-88318-199-1
No. 101	Positron-Electron Pairs in Astrophysics (Goddard Space Flight Center, 1983)	83-71926	0-88318-200-9
No. 102	Intense Medium Energy Sources of Strangeness (UC-Sant Cruz, 1983)	83-72261	0-88318-201-7
No. 103	Quantum Fluids and Solids – 1983 (Sanibel Island, Florida)	83-72440	0-88318-202-5
No. 104	Physics, Technology and the Nuclear Arms Race (APS Baltimore –1983)	83-72533	0-88318-203-3
No. 105	Physics of High Energy Particle Accelerators (SLAC Summer School, 1982)	83-72986	0-88318-304-8
No. 106	Predictability of Fluid Motions (La Jolla Institute, 1983)	83-73641	0-88318-305-6
No. 107	Physics and Chemistry of Porous Media (Schlumberger-Doll Research, 1983)	83-73640	0-88318-306-4
No. 108	The Time Projection Chamber (TRIUMF, Vancouver, 1983)	83-83445	0-88318-307-2
No. 109	Random Walks and Their Applications in the Physical and Biological Sciences (NBS/La Jolla Institute, 1982)	84-70208	0-88318-308-0
No. 110	Hadron Substructure in Nuclear Physics (Indiana University, 1983)	84-70165	0-88318-309-9
No. 111	Production and Neutralization of Negative Ions and Beams (3rd Int'l Symposium, Brookhaven, 1983)	84-70379	0-88318-310-2
No. 112	Particles and Fields – 1983 (APS/DPF, Blacksburg, VA)	84-70378	0-88318-311-0
No. 113	Experimental Meson Spectroscopy – 1983 (Seventh International Conference, Brookhaven)	84-70910	0-88318-312-9
No. 114	Low Energy Tests of Conservation Laws in Particle Physics (Blacksburg, VA, 1983)	84-71157	0-88318-313-7
No. 115	High Energy Transients in Astrophysics (Santa Cruz, CA, 1983)	84-71205	0-88318-314-5
No. 116	Problems in Unification and Supergravity (La Jolla Institute, 1983)	84-71246	0-88318-315-3
No. 117	Polarized Proton Ion Sources (TRIUMF, Vancouver, 1983)	84-71235	0-88318-316-1

No. 118	Free Electron Generation of Extreme Ultraviolet Coherent Radiation (Brookhaven/OSA, 1983)	84-71539	0-88318-317-X
No. 119	Laser Techniques in the Extreme Ultraviolet (OSA, Boulder, Colorado, 1984)	84-72128	0-88318-318-8
No. 120	Optical Effects in Amorphous Semiconductors (Snowbird, Utah, 1984)	84-72419	0-88318-319-6
No. 121	High Energy e^+e^- Interactions (Vanderbilt, 1984)	84-72632	0-88318-320-X
No. 122	The Physics of VLSI (Xerox, Palo Alto, 1984)	84-72729	0-88318-321-8
No. 123	Intersections Between Particle and Nuclear Physics (Steamboat Springs, 1984)	84-72790	0-88318-322-6
No. 124	Neutron-Nucleus Collisions – A Probe of Nuclear Structure (Burr Oak State Park - 1984)	84-73216	0-88318-323-4
No. 125	Capture Gamma-Ray Spectroscopy and Related Topics – 1984 (Internat. Symposium, Knoxville)	84-73303	0-88318-324-2
No. 126	Solar Neutrinos and Neutrino Astronomy (Homestake, 1984)	84-63143	0-88318-325-0
No. 127	Physics of High Energy Particle Accelerators (BNL/SUNY Summer School, 1983)	85-70057	0-88318-326-9
No. 128	Nuclear Physics with Stored, Cooled Beams (McCormick's Creek State Park, Indiana, 1984)	85-71167	0-88318-327-7
No. 129	Radiofrequency Plasma Heating (Sixth Topical Conference, Callaway Gardens, GA, 1985)	85-48027	0-88318-328-5
No. 130	Laser Acceleration of Particles (Malibu, California, 1985)	85-48028	0-88318-329-3
No. 131	Workshop on Polarized ^{3}He Beams and Targets (Princeton, New Jersey, 1984)	85-48026	0-88318-330-7
No. 132	Hadron Spectroscopy–1985 (International Conference, Univ. of Maryland)	85-72537	0-88318-331-5
No. 133	Hadronic Probes and Nuclear Interactions (Arizona State University, 1985)	85-72638	0-88318-332-3
No. 134	The State of High Energy Physics (BNL/SUNY Summer School, 1983)	85-73170	0-88318-333-1
No. 135	Energy Sources: Conservation and Renewables (APS, Washington, DC, 1985)	85-73019	0-88318-334-X
No. 136	Atomic Theory Workshop on Relativistic and QED Effects in Heavy Atoms	85-73790	0-88318-335-8
No. 137	Polymer-Flow Interaction (La Jolla Institute, 1985)	85-73915	0-88318-336-6
No. 138	Frontiers in Electronic Materials and Processing (Houston, TX, 1985)	86-70108	0-88318-337-4
No. 139	High-Current, High-Brightness, and High-Duty Factor Ion Injectors (La Jolla Institute, 1985)	86-70245	0-88318-338-2

No. 140	Boron-Rich Solids (Albuquerque, NM, 1985)	86-70246	0-88318-339-0
No. 141	Gamma-Ray Bursts (Stanford, CA, 1984)	86-70761	0-88318-340-4
No. 142	Nuclear Structure at High Spin, Excitation, and Momentum Transfer (Indiana University, 1985)	86-70837	0-88318-341-2
No. 143	Mexican School of Particles and Fields (Oaxtepec, México, 1984)	86-81187	0-88318-342-0
No. 144	Magnetospheric Phenomena in Astrophysics (Los Alamos, 1984)	86-71149	0-88318-343-9
No. 145	Polarized Beams at SSC & Polarized Antiprotons (Ann Arbor, MI & Bodega Bay, CA, 1985)	86-71343	0-88318-344-7
No. 146	Advances in Laser Science–I (Dallas, TX, 1985)	86-71536	0-88318-345-5
No. 147	Short Wavelength Coherent Radiation: Generation and Applications (Monterey, CA, 1986)	86-71674	0-88318-346-3
No. 148	Space Colonization: Technology and The Liberal Arts (Geneva, NY, 1985)	86-71675	0-88318-347-1
No. 149	Physics and Chemistry of Protective Coatings (Universal City, CA, 1985)	86-72019	0-88318-348-X
No. 150	Intersections Between Particle and Nuclear Physics (Lake Louise, Canada, 1986)	86-72018	0-88318-349-8
No. 151	Neural Networks for Computing (Snowbird, UT, 1986)	86-72481	0-88318-351-X
No. 152	Heavy Ion Inertial Fusion (Washington, DC, 1986)	86-73185	0-88318-352-8
No. 153	Physics of Particle Accelerators (SLAC Summer School, 1985) (Fermilab Summer School, 1984)	87-70103	0-88318-353-6
No. 154	Physics and Chemistry of Porous Media—II (Ridge Field, CT, 1986)	83-73640	0-88318-354-4
No. 155	The Galactic Center: Proceedings of the Symposium Honoring C. H. Townes (Berkeley, CA, 1986)	86-73186	0-88318-355-2
No. 156	Advanced Accelerator Concepts (Madison, WI, 1986)	87-70635	0-88318-358-0
No. 157	Stability of Amorphous Silicon Alloy Materials and Devices (Palo Alto, CA, 1987)	87-70990	0-88318-359-9
No. 158	Production and Neutralization of Negative Ions and Beams (Brookhaven, NY, 1986)	87-71695	0-88318-358-7

No. 159	Applications of Radio-Frequency Power to Plasma: Seventh Topical Conference (Kissimmee, FL, 1987)	87-71812	0-88318-359-5
No. 160	Advances in Laser Science–II (Seattle, WA, 1986)	87-71962	0-88318-360-9
No. 161	Electron Scattering in Nuclear and Particle Science: In Commemoration of the 35th Anniversary of the Lyman-Hanson-Scott Experiment (Urbana, IL, 1986)	87-72403	0-88318-361-7
No. 162	Few-Body Systems and Multiparticle Dynamics (Crystal City, VA, 1987)	87-72594	0-88318-362-5
No. 163	Pion–Nucleus Physics: Future Directions and New Facilities at LAMPF (Los Alamos, NM, 1987)	87-72961	0-88318-363-3
No. 164	Nuclei Far from Stability: Fifth International Conference (Rosseau Lake, ON, 1987)	87-73214	0-88318-364-1
No. 165	Thin Film Processing and Characterization of High-Temperature Superconductors	87-73420	0-88318-365-X
No. 166	Photovoltaic Safety (Denver, CO, 1988)	88-42854	0-88318-366-8
No. 167	Deposition and Growth: Limits for Microelectronics (Anaheim, CA, 1987)	88-71432	0-88318-367-6
No. 168	Atomic Processes in Plasmas (Santa Fe, NM, 1987)	88-71273	0-88318-368-4
No. 169	Modern Physics in America: A Michelson-Morley Centennial Symposium (Cleveland, OH, 1987)	88-71348	0-88318-369-2
No. 170	Nuclear Spectroscopy of Astrophysical Sources (Washington, D.C., 1987)	88-71625	0-88318-370-6
No. 171	Vacuum Design of Advanced and Compact Synchrotron Light Sources (Upton, NY, 1988)	88-71824	0-88318-371-4
No. 172	Advances in Laser Science–III: Proceedings of the International Laser Science Conference (Atlantic City, NJ, 1987)	88-71879	0-88318-372-2
No. 173	Cooperative Networks in Physics Education (Oaxtepec, Mexico 1987)	88-72091	0-88318-373-0
No. 174	Radio Wave Scattering in the Interstellar Medium (San Diego, CA 1988)	88-72092	0-88318-374-9
No. 175	Non-neutral Plasma Physics (Washington, DC 1988)	88-72275	0-88318-375-7

No. 176	Intersections Between Particle Land Nuclear Physics (Third International Conference) (Rockport, ME 1988)	88-62535	0-88318-376-5
No. 177	Linear Accelerator and Beam Optics Codes (La Jolla, CA 1988)	88-46074	0-88318-377-3
No. 178	Nuclear Arms Technologies in the 1990s (Washington, DC 1988)	88-83262	0-88318-378-1
No. 179	The Michelson Era in American Science: 1870–1930 (Cleveland, OH 1987)	88-83369	0-88318-379-X
No. 180	Frontiers in Science: International Symposium (Urbana, IL, 1987)	88-83526	0-88318-380-3
No. 181	Muon-Catalyzed Fusion (Sanibel Island, FL, 1988)	88-83636	0-88318-381-1